Nonlinear Dy

Nonlinear Dynamics

A Two-way Trip from Physics to Math

H G Solari

Departamento de Física,
Universidad de Buenos Aires, Argentina

M A Natiello

Department of Mathematics,
Royal Institute of Technology, Sweden

G B Mindlin

Departamento de Física,
Universidad de Buenos Aires, Argentina

Institute of Physics Publishing
Bristol and Philadelphia

British Library Cataloguing-in-Publication Data

A catalogue record for this book is available from the British Library.

ISBN 0 7503 0379 4 hbk
ISBN 0 7503 0380 8 pbk

Library of Congress Cataloging-in-Publication Data
Solari, Hernán G.
    Nonlinear dynamics : a two-way trip from physics to math / Hernán
G. Solari, Mario A. Natiello, Gabriel B. Mindlin.
        p.   cm.
    "December 14, 1995"
    Includes bibliographical references and index.
    ISBN 0-7503-0379-4 (hardbound : alk. paper). - - ISBN 0-7503-0380-8
(pbk. : alk. paper)
    1. Nonlinear theories. 2. Dynamics. 3. Mathematical physics.
I. Natiello, Mario A. II. Mindlin, Gabriel B. III. Title.
QC20.7.N6S65 1996
003′ .85 - - dc20
                                                                96-15831
                                                                CIP

Published by Institute of Physics Publishing, wholly owned by The Institute of Physics, London

Institute of Physics Publishing, Techno House, Redcliffe Way, Bristol BS1 6NX, UK

US Editorial Office: Institute of Physics Publishing, The Public Ledger Building, Suite 1035, 150 South Independence Mall West, Philadelphia, PA 19106, USA

Typeset in TEX using the IOP Bookmaker Macros
Printed in the UK by J W Arrowsmith Ltd, Bristol BS3 2NT

There are more things in heaven and earth, Horatio
Than are dreamt of in your philosophy...
W Shakespeare, Hamlet, Act 1, Scene 5.

To:
Bárbara, Patrizia and Silvia
Florencia and Julia
and **Cristal**

Contents

Acknowledgments

... si acaso fuera a quedar de mis deudas un haber...

José Larralde, Herencia pa' un hijo gaucho

The manuscript has been produced by the authors using L^AT_EX and *Canvas* (for most of the figures). It has travelled a distance comparable to the size of the orbit of the Moon around the Earth in the form of binary pulses. We thank all the PTTs and relay operators involved for providing an almost daily connection across the world (and for allowing us to test the limits of our patience).

Grants from Fundación Antorchas, Argentina, from Universidad de Buenos Aires, Argentina and from Naturvetenskapliga Forskningsrådet, Sweden have partially supported this project. Travel grants from the Departamento de Física, Universidad de Buenos Aires, Argentina, from Uppsala University, Sweden and from the International Centre for Theoretical Physics, Trieste, Italy for supporting about half a dozen face-to-face discussion sessions during these three years are also acknowledged. H G S and G B M are members of the Carrera del Investigador Científico, CONICET, Argentina. We thank the librarians at Beurling and NTC libraries, Uppsala University for the kindness and patience with which they answered our wildest requests.

We are indebted to many people, some of them known to us and many of them not known, since every one of our ideas has been possible only because of the ideas of other people. We have certainly have received a lot more than we will ever be able to give.

We would like to thank those friends that have helped us in one way or another in producing this book, in particular (in alphabetic order): Neal Abraham, Celia Dibar de Ure, Billy Dussel, Sascha Firle, Robert Gilmore, Rafael Gonzalez, Osvaldo Goscinski, Lotten Hägg, Hector Mancini, Carlos Perez, Silvina Ponce Dawson, Marcos Saraceno, Jorge Tredicce and Martin Zimmermann.

H G Solari, M A Natiello, G B Mindlin

Preface

About the book

The aim of this book is to render available to readers the tools that nonlinear dynamics provides for the exploration of new problems in all fields of physics. In research we deal with open problems: problems for which, at the beginning, we have no solutions but at most a set of hunches, feelings and guesses based on our previous experience with other problems. From there, we work our path to the solutions (though we do not always succeed). In finding our way we take what seems to us the *most natural* approach just as we would bush-walk in the forest avoiding as much as possible the difficult paths in our hike towards an interesting place.

We will follow in the presentation the same procedure that we follow when doing research, i.e., we begin with the problem and find one path to the solution, we work inductively proposing new paths, checking them and redrawing our route according to the experience we gain in successive efforts.

We will avoid the temptation of selecting the problems according to the tools we possess. Rather, we prefer to construct the tools along with the problems. We sustain the idea that our tools (theories) and our problems evolve hand in hand. The best pages of physics have been written in this way. Consider for example the pairs calculus–mechanics (Newton) or Hilbert spaces–quantum theory (Von Neumann) and, as we will see, nonlinear dynamics–topology (Poincaré).

This is a book to be read with paper and pencil at hand. Our intention is to furnish readers with enough knowledge to be able to do research in nonlinear dynamics after having read the book (or better, *while* they are reading the book). We prefer to convey the key ideas within their mathematical framework rather than doing lengthy demonstrations. Therefore, the calculations around the results presented in the book are usually only sketched or left more or less as guided exercises.

At the end of the day, we would like the reader to finish this book with the feeling (or certainty) that, given enough time, she/he would have come up with the same answers to the problems as those we have shown (well ... perhaps just better answers). After all *the answers are dictated by the problems*, the two of them evolve in interaction, and our task is to read them from nature.

About nonlinear dynamics

Nonlinear dynamics is a subject at least as old in physics as Newton's mechanics. The dynamics of the planetary system, a primary concern for Newton, Poincaré and many others, turns out to be nonlinear in general, but fortunately the simplest examples can be solved exactly (two-body systems are completely integrable). In the context of Hamiltonian mechanics a completely integrable system is 'almost' a linear system in an appropriate set of coordinates (those given by the Hamilton–Jacobi theory for integrable systems [arno89]).

In contrast, three-body problems are also nonlinear, but in general very complex and non-integrable. It was while studying the three-body problem that Poincaré gave a new and important impulse to nonlinear dynamics at the beginning of the 20th century. However, the new physics of the atom (later the nucleus, then quarks, ...) caught the attention of physicists. There was apparently no use for nonlinear dynamics in quantum mechanics since the latter rests on Hilbert spaces (linear spaces after all). The excess of zeal with quantum mechanics caused the (almost complete) disappearance of nonlinear dynamics from physics and especially from textbooks.

The emergence of computers as new tools for theoretical physics in the late 1950s and early 1960s favoured a comeback of nonlinear dynamics. Numerical simulations made accessible to the intuition of physicists and non-physicists the richness of nonlinear models. The graphical output added an artistic touch.

Commensurate with its earlier neglect, the impact of the phenomenology of nonlinear dynamics rocked the physics community in the early 1980s, to the point that people even talked of a 'new science' [glei87].

Today, we have a calmer perspective. We recognize that many situations can only be described with nonlinear interactions. There is a growing consciousness that the tools, methods and phenomenology of nonlinear dynamics will be increasingly necessary for the study of most subjects in physics and natural science in general.

The present text is far from being a complete guide to nonlinear dynamics but it covers the basic ideas for general systems. Some topics, though important for historical and even practical reasons for physicists, like one-dimensional maps and Hamiltonian mechanics, have not been emphasized, *on purpose*. We are certain that if we were to stress these special (singular) cases we would induce the wrong generalization, just as our generation was induced to think that Hamilton–Jacobi theory applied to all systems.

The discussion is presented in most cases having in mind low-dimensional systems, i.e., systems where the spatial aspects behave coherently. In general, spatio-temporal dynamics is known to a lesser degree than low-dimensional dynamics and the authors' knowledge of the subject is correspondingly more limited.

Chapter 1

Nonlinear dynamics in nature

'man never regards what he possesses as so much his own, as what he does; and the labourer who tends a garden is perhaps in a truer sense its owner, than the listless voluptuary who enjoys its fruits... In view of this consideration, it seems as if all peasants and craftsmen might be elevated into artists; that is, men who love their labour for its own sake, improve it by their own plastic genius and inventive skill, and thereby cultivate their intellect, ennoble their character, and exalt and refine their pleasures. And so humanity would be ennobled by the very things which now, though beautiful in themselves, so often serve to degrade it... But, still freedom is undoubtedly the indispensable condition, without which even the pursuits most congenial to individual human nature can never succeed in producing such salutary influences. Whatever does not spring from a man's free choice, or is only the result of instruction and guidance, does not enter into his very being, but remains alien to his true nature; he does not perform it with truly human energies, but merely with mechanical exactness... we admire what he does, but despise what he is.'

Wilhelm Von Humboldt [chom87]

A sound theory starts usually with experiment. The cornerstone of natural sciences is precisely its validation procedure, namely the fact that relevant assertions can be put to test by way of suitable experiments and eventually be rejected (if the experiment *proves* them wrong). Consequently, the aim of this chapter is to present nonlinear problems from the experimental point of view. By *nonlinear* we just mean systems that demand to be modelled by nonlinear differential equations or nonlinear discrete-time mappings.

We will review some experiments that illustrate the kind of phenomenon we expect to find in nonlinear systems. The choice of examples is unavoidably biased and incomplete. We combine everyday experiences with well known problems, all of them sharing the characteristic features of nonlinearity. We want to emphasize that for the physicist nonlinear dynamics is far more than a theoretical entertainment. Moreover, we hope to make clear that *almost everything in nature is nonlinear.*

1.1 Hiking among rabbits

A few years ago one of us went to a National Park near Ushuaia, Argentina on a 2 week hiking trip. The park is placed in the southernmost part of *Tierra del Fuego*, the mysterious island that enchanted sailors, poets and natural scientists [darw88]. At that time, it was striking to notice that there were plenty of rabbits, and that many of them looked rather ill.

In fact, rabbits are not 'original' to the place, but were introduced (from Europe) by sailors as a way of having fresh meat when the ports of Tierra del Fuego were a natural stop in the route from the Atlantic to the Pacific, before the opening of the Panama Canal. Rabbits reproduced and spread all around the region (which suggests that they encountered no natural predators).

Towards the end of the 19th century the region was substantially occupied by cattle-farmers. Rabbits entered into competition with the cattle (sheep) and the farmers searched for a way of getting rid of them.

The farmers learned that a similar situation had arisen in Australia, where the matter was 'solved' by inoculating rabbits with a specific virus, which provoked a dramatic decrease in their population. The same approach was tried in Tierra del Fuego despite the opposition of the Park authorities and rangers. The inoculation was not as massive as in Australia, since the Park became virtually a protection sanctuary and the farmers did not find collaboration for their project.

After a sharp drop in population, rabbits in Australia recovered their previous population levels, developing a virus-resistant breed. In Argentina, after an initial decrease in population, a fluctuating state has been reached. There is an apparently periodic alternation between years with high levels of mortality and sickness and years with low levels of mortality.

The tale has many morals. However, we will concentrate on some physically relevant consequences, although they may not be the most important. To begin with we shall assume that the comparison between the Australian and the Argentine 'experiments' is scientifically valid.

To assess which are the reasons behind oscillatory behaviour in population dynamics is an interesting topic of research in itself. For our purposes, one would expect that periodic changes in rabbit populations would have a 'natural' period such as that of the season cycle or the reproduction cycle.

We begin noticing that *seemingly similar systems (rabbit populations) starting from apparently different initial states (the ratio between infected and healthy rabbits) can lead to the occurrence of different final (asymptotic) states.* This result may not be surprising recalling better known problems such as the ideal pendulum or a two-body Kepler problem.

The dramatically new result is the existence of oscillating cycles which extend over several years. What is surprising is not the periodic character of the solution but rather the fact that the period is *not* the 'natural' one. A recurrence of many times a given period is called *subharmonic behaviour*. Such behaviour cannot be achieved within a linear model.

Population dynamics, epidemiology and mathematical biology in general count among the most important contributors to the 'comeback' experienced by nonlinear science in recent decades. For example, very important names associated with nonlinear dynamics such as S Smale [smal76] and R May [may 75] have worked on this subject. A discussion concerning the example of the rabbits can be found in [dwye90]. Needless to say, the example of the rabbits is far from being unique.

1.2 Turbulence

The 1883 paper by Osborne Reynolds [reyn83] established the law of similarity that now bears his name (recall the *Reynolds number*). The article was concerned with two issues, one practical and one philosophical: 'the law of resistance of the motion of water in pipes' and 'that the general character of the motion of fluids in contact with solid surfaces depends on the relation between a physical constant of the fluid and the product of the linear dimensions of the space occupied by the fluid and the velocity'. Discussing these issues, Reynolds gave us the first study of the transition from laminar motion to turbulent motion in a fluid (*direct* and *sinuous* motion, as they are called in the paper). The exposition hardly needs further comments.

Reynolds states that 'The internal motion of water assumes one or another of two broadly distinguishable forms — either the elements of the fluid follow one another along the lines of motion which lead in the most direct manner to their destination, or they eddy about in sinuous paths the most indirect possible'. These are the laminar and turbulent forms of motion.

Further, we read that 'Certain circumstances have been definitely associated with the particular laws of force. Resistance, [varying] as the square of the velocity, is associated with motion in tubes of more than capillary dimensions, and with the motion of bodies through the water at more than insensibly small velocities, while resistance [varying] as the velocity is associated with capillary tubes and small velocities'. This is illustrated by figure 1.1 taken from Prandtl and Tietjens [pran34].

There are other circumstances that distinguish turbulent motion from laminar motion. For example, while laminar motion does not mix different stream lines, turbulent motion does. Therefore, if we colour one of the streamlines (see figure 1.2 taken from [reyn83]), or heat it up as in [barn04], only this streamline will be altered while, by contrast, in turbulent motion the whole fluid downstream of the alteration will be changed. Reynolds' description of this experiment is an excellent piece of scientific literature.

Having established that the transition between laminar and turbulent flow

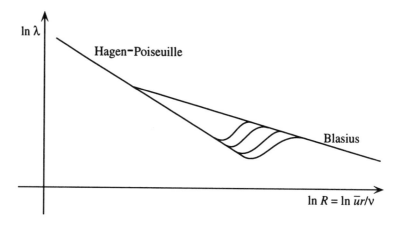

Figure 1.1. Resistance to motion in tubes against Reynolds number. Linear relation for laminar flow and quadratic relation for turbulent flows (from Prandtl and Tietjens, figure 17. See text).

was likely to depend on the dimensionless quantity $R = dUr/v$ (d = density, U = velocity, r = radius of the tube and v = viscosity. R is now called the *Reynolds number*), Reynolds studied the transition for different tubes at different temperatures and established the following facts:

(a) The transition depends on the initial disturbance of the water; the smaller the disturbance the larger the critical value of R.
(b) There is a lower R below which any disturbance will die out ($R \approx 1100$ is usually considered).
(c) The eddies appear with full strength all of a sudden instead of growing gradually with increasing R.
(d) The flow will behave *intermittently*, i.e.: 'The disturbance would suddenly come on through a certain length of the tube and pass away and then come again, giving the appearance of flashes, and these flashes would often commence successively at one point in the pipe'.

Reynolds tried to go further and wondered whether there was a critical size or kind of disturbance capable of triggering the transition. He leaned towards a positive answer to this question, although he was not completely convinced of this fact since he failed to devise a critical test for this matter.

Reynolds' paper is the first comprehensive study of the transition to turbulence. We can recognize in it several typical features of nonlinear systems, that underline the radical difference between turbulence and linear problems (or even certain 'well-organized' nonlinear problems), namely:

(i) Transitions from organized to unorganized motion.
(ii) Amplification of small disturbances (unstable regimes).

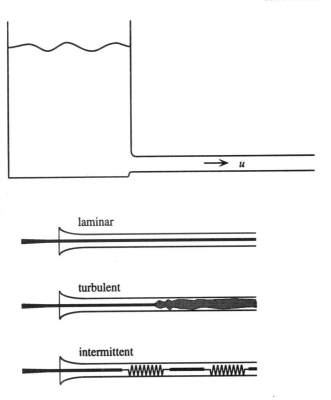

Figure 1.2. Experimental arrangement in Reynolds' experiment (figures 3, 4, 5 from Reynolds, see text).

(iii) Intermittence between laminar motion and turbulent motion in the regime of transition.

(iv) A minimum size of disturbance is required to produce the transition[†].

Although turbulence appeared suddenly in this experiment, Reynolds realized there was no reason for this to be always the case. He made a small experiment with two layers of water (one coloured) moving in opposite directions (see figure 1.3 taken from [reyn83]) and studied the behaviour of the surface of separation as a function of the velocity. He observed that for low velocities the surface became wavy. Increasing the velocity those waves developed peaks and finally for a larger velocity they broke. This was certainly a much smoother transition than the one from laminar motion to turbulent motion in tubes.

The transition to turbulence is far from being completely understood.

[†] This can be rephrased in modern terms as the size of the *basin of attraction* (set of initial conditions having the 'same' final (asymptotic) state) when there is *multistability* (coexistence of solutions).

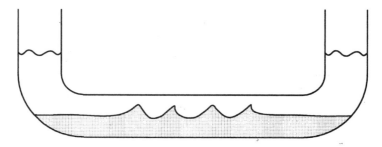

Figure 1.3. Two flows in counter motion (figure 6 of Reynolds, see text).

In the following chapters we shall see that the major advances in nonlinear dynamics have been made in a somewhat different direction. For the physicist, it is advisable to keep in sight the challenging problems that originated this development.

Turbulent flow is by no means the first nonlinear problem that physics encountered. Newton's treatment of the two-body Kepler problem had already been known for about 200 years. This problem, however, has a kind of regularity that renders the turbulent transition even more striking. For instance, different elliptic orbits are quite similar, all of them lying on a 3-torus when viewed from the reference frame of the centre of mass, the torus being defined by the z-component of the angular momentum, the total angular momentum and the energy of the interaction. The orbits traverse the 3-torus at constant angular speeds and, moreover, 'close-lying' orbits lie on close-lying tori and have close frequencies. Although the 'transition' between elliptic and hyperbolic orbits can be traced down to one point in energy space (i.e., negative or positive energy for fixed angular momentum), the dynamics is organized on both sides of the transition. In addition, to perform a somewhat reproducible experiment in which an elliptic orbit in a two-body gravitational system evolves into a hyperbolic orbit was unthinkable before artificial satellites. Unlike the two-body system, a three-body Kepler problem may present complex motion.

1.3 Bénard instability

The transition to turbulence in the case of a sheet of fluid heated from below has been the subject of a number of studies of both experimental and theoretical kinds (Henri Bénard [bena01], Lord Rayleigh [rayl16], R Krishnamurti [kris70a, kris70b], F H Busse [buss78]). In contrast to the experiment by Reynolds, the transition develops in a series of changes which can be studied on an individual basis, simplifying the problem thereon.

The experiment consists of a tank containing a thin layer of fluid. The bottom of the tank is made of a good heat conductor (say copper). The lateral walls are usually made also of a conductor, while the upper surface is either a free surface (in contact with the atmosphere) or a conductor. Although Bénard used both kinds of upper surface his observations rely mostly on the free surface case. The conducting surface is used in [kris70a, kris70b].

The tank is heated from below allowing the system to reach equilibrium after each temperature increment. The stability of the environment is carefully controlled during the course of the experiment (in the case of [kris70a, kris70b] the times involved range between hours and weeks).

The experiments by Bénard showed that convective currents appear. The surface of the fluid adopts a pattern composed by hexagonal cells, the fluid ascending through the centre of each hexagon and flowing downward following the (common) walls of the hexagons. The regime obtained is asymptotically stable, i.e., at the beginning hexagons coexist with squares, pentagons and other shapes of cells, but as time goes on the hexagons became increasingly dominant. Bénard also studied the typical length of the cells for different fluids and the deformation of the free surface that takes place.

The experiment was analysed by Lord Rayleigh [rayl16] in terms of dimensionless quantities (of the kind introduced by Reynolds, see above). The physical constants involved are gravity, g (the experiment requires heating from below), the density, r, the internal dissipation of the fluid, i.e., the kinematic viscosity, v, the thermal diffusivity, κ, the difference of temperature between surfaces, δT, and the velocity of the motion, U. The geometry of the problem enters only through the thickness of the cell, d.

In order to start the convection, the energy loss due to the viscous forces has to be compensated by the potential energy available when the gradient of density is directed opposite to gravity.

Convective motion starts when the transport of heat by convection becomes as efficient as thermal diffusion. The heat flow by diffusion, E_{dif}, is proportional to the temperature gradient and the thermal diffusivity

$$E_{dif} \approx \kappa \delta T / d$$

while the convection E_{conv} involves a volume of fluid moving with velocity U and carrying an amount of heat proportional to the temperature difference δT

$$E_{conv} \approx U \delta T.$$

Convection will therefore depend on E_{conv}/E_{dif} which is characterized by the dimensionless quantity

$$R = U d / \kappa$$

called the Rayleigh number.

What would be a typical value for U? Since the forces in action are the viscous force, $\approx U r_0 v / d^2$, and the buoyancy created by the inverse density gradient, $\approx (r_{top} - r_{bottom})g$, the typical velocity will be characterized by the balance of these two forces

$$U = d^2 g (r_{\text{top}} - r_{\text{bottom}})/(v r_0).$$

If the density changes linearly with T,

$$r = r_0 (1 - \gamma (T - T_1)),$$

Rayleigh's expression is then

$$R = \gamma (T_2 - T_1) g d^3 / (v \kappa).$$

The analysis of the Bénard experiment is made in terms of the Rayleigh number and the Prandtl number, $P = v/\kappa$, which enters only when nonlinear effects are important. The transition from the diffusive mode of conduction to the convective mode will be achieved for the same critical Rayleigh number for different fluids.

The experiment has been repeated a number of times obtaining different results in the pattern of convection. Even more important, by incrementing the temperature gradient new transitions have been found.

According to theoretical studies [buss78] there are three characteristic patterns for the initial convective state (once in equilibrium): bi-dimensional rolls (see figure 1.4), hexagons with upward flow in the centre (see figure 1.5) and hexagons with downward flow in the centre. Which of these patterns is realized depends on the asymmetry of the fluid layer with respect to the plane representing its mean depth. In the symmetric case, bi-dimensional rolls are obtained while in the asymmetric cases hexagons are the stable patterns. In the experiments by Krishnamurti the flow was confined between two plates of copper and the regime obtained was of the symmetric kind (i.e., rolls).

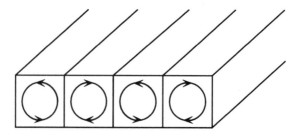

Figure 1.4. Bi-dimensional convection in Rayleigh–Bénard flow.

When the Rayleigh number is increased beyond the critical value, the size of the rolls starts to increase. Later, a second instability appears and a tri-dimensional stationary flow is achieved. The tri-dimensionality is given by a wavy deformation of the rolls. Together with this change in the flow pattern there is a change in the slope of the transport of heat. This effect is manifested

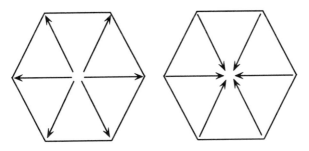

Figure 1.5. Hexagonal convection in Rayleigh–Bénard flow.

in a plot of heat flux times R (see figure 1.6 taken from [kris70a], figure 2). The critical value of the Rayleigh number for the transition, R_{II} in figure 1.6, is not given exactly since the transition exhibits hysteresis (i.e., approached from below it occurs for a higher value of R_{II} than when approached from above).

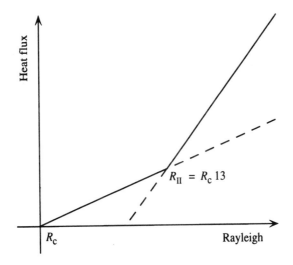

Figure 1.6. Heat flux versus R (taken from Krishnamurti, see text).

There are further developments in the route to turbulent convection. When R is increased even further, the rolls start to oscillate in a periodic fashion. There is a tilt (or noise) of the rolls superimposed with this oscillation. The two modes have always been observed coexisting (see figure 1.7). The transition displays hysteresis and presents again a change in the slope of the curve of heat flux versus R. The location in parameter space of all these different motions is represented in figure 1.8, taken from [kris70b], figure 9.

The work by A Libchaber and J Maurer [libc82] goes beyond this instability. The approach taken consists of selecting the oscillating mode constructing a cell

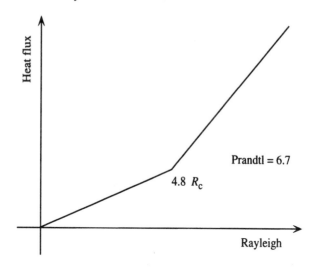

Figure 1.7.   Heat flux versus R in the oscillatory instability in Rayleigh–Bénard convection.

that can allocate only two rolls. With these conditions a full series of successive instabilities can be studied in close agreement with theoretical results concerning the development of instabilities in low-dimensional systems. We will see some of them in our study of the laser system.

The new feature that the Bénard experiment makes evident is that a number of different patterns, with different degrees of organization, can be achieved by varying one of the system parameters (e.g., the temperature difference between the lower and upper surface). Again, this wide richness of behaviours contrasts drastically with more conventional physics.

1.4   Dynamics of a modulated laser

In this experiment [tred86] an electro-optic modulator is inserted within the cavity of a single-mode CO_2 laser. The modulator is controlled with an external sinusoidal signal producing a periodic modulation of the cavity losses. The laser is thus driven by this external force, producing as output, at low modulation amplitude, a modulated intensity of the same periodicity as the external frequency, very much as would happen with a driven damped linear oscillator.

The varying parameter (independent variable) of the experiment is the amplitude of the modulation. All other parameters of the laser as well as the modulation frequency Ω are kept constant throughout the experiment. Typical times in the laser are of the order of microseconds, which are considerably smaller than the week-long characteristic times of the Bénard experiment.

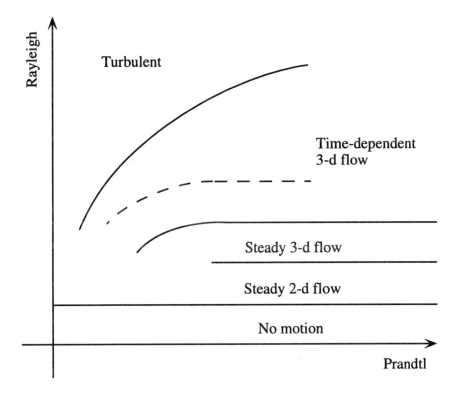

Figure 1.8. Bifurcation diagram for the Bénard experiment (taken from Krishnamurti, see text).

The study is mainly based on the observation of the output intensity, $I(t)$, expressed as a time-trace, and a stroboscopic sampling of I with the frequency of the modulation. Two other quantities that concentrate a great deal of the information present in the data are used as well, namely the Fourier transform (*spectrum*) of the intensity and the *phase space portrait* defined with the pair of variables $(I, \dot{I})$. Curves in the phase space portrait, i.e., time evolution of the two-dimensional array $(I(t), \dot{I}(t))$ are called *orbits*. It is the behaviour of these orbits which is the central subject of analysis.

For small modulation amplitudes one finds periodic orbits having the period of the modulation. This is reflected by peaks in the spectrum at Ω and its harmonics, as well as by a constant yield in the stroboscopic sampling. The orbit itself appears on the phase portrait as a single loop (see figure 1.9(a), taken from [tred86]).

When the amplitude of the modulation is increased beyond a certain threshold a qualitative change is observed. The orbit acquires frequency components of $\Omega/2$, i.e., *subharmonic* frequencies (remember the rabbits!). The

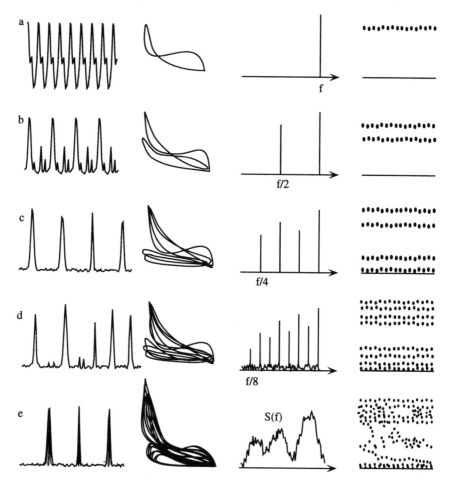

Figure 1.9. Period doubling sequence in the CO_2 laser: (a) period one, (b) period two, (c) period four, (d) period eight, (e) aperiodic (from Tredicce *et al*, see text).

stroboscopic sampling gives *two* different intensity values (identical at threshold), and the phase portrait displays a double loop (see figure 1.9(b)). Upon further increase of the modulation amplitude, another threshold is found: the spectrum acquires frequencies of $\Omega/4$, the stroboscopic sampling returns four different values alternating in a fixed pattern and the phase portrait shows now a four-loop orbit (see figure 1.9(c)).

This sequence of *subharmonic bifurcations* (or *period doublings*) continues with the occurrence of *period eight* ($\Omega/8$), represented by an eight-valued stroboscopic sampling and an eight-looped phase portrait (see figure 1.9(d)). Successive 2^{-n} subharmonic frequencies make their appearance when the modulation amplitude is increased. The process reaches a point beyond which the signal can no longer be recognized as periodic. This is called the *aperiodic attractor* (see figure 1.9(e)). We can meet this *accumulation point* of the sequence of bifurcations in a variety of models: bifurcation sequences are well known from various fields in nonlinear research as well as from popular essays on *chaos*.

Continuing with the sweeping of the modulation amplitude the aperiodic attractor suddenly disappears, giving way to a structure of period 3 (*frequency* = $\Omega/3$), with three alternating values in the stroboscopic sampling, while the trace of the intensity presents one peak every three periods. The peak is larger than those at previous amplitudes (see figure 1.10(a), taken from [tred86]). The transition between the aperiodic attractor and the period three presents *hysteresis*. Upon further increase of the modulation this period 3 pattern presents a similar scenario of period doubling (see figure 1.10(b)). Transitions from aperiodic behaviour to one-peak orbits of periodicity 4, 5, 6, 7 and 10 are reported (see figure 1.10(c), (d), (e), (f)).

The process of increasing the modulation amplitude is summarized in a diagram (*bifurcation diagram*) which shows the output(s) of the stroboscopic sample as the amplitude is increased (see figure 1.11, taken from [tred86]).

So far we have sketched only one of the possible asymptotic behaviours of the system. Depending on the initial dynamical conditions of the laser the asymptotics could be different since several structures can compete as final states for the system [tred86], i.e., the system is *multistable* in some regions of parameter space (which in this case consists of the frequency of the forcing, Ω, and the amplitude of the modulation). Again, the concept of *basin of attraction* (the set of initial conditions having the 'same' final (asymptotic) state) appears in a natural way.

1.5   Tearing of a plasma sheet

A *plasma focus* experiment consists of a machine designed to study the formation of a hot spot of plasma. The machine consists of a cylindrical chamber containing a low-pressure plasma between two cylindrical concentric electrodes. Since there is no reason to suppose that the fluid is distributed anisotropically, the system is assumed to have rotational symmetry.

By passing currents through the electrodes, two cylindrical sheets of plasma are created in the proximity of the electrodes. The sheets travel one towards the other dragging the neutral particles between them and creating a focus (this was the goal of the experiment) when they collide. This focus lasts for some oscillations of the plasma sheets but evolves from the cylindrical symmetry to a situation of lower symmetry represented by a *tearing* of the plasma (see figure 1.12, taken from [ande69]).

This result is somehow surprising since in linear theories (like electromagnetism) we are rather used to (a) unique solutions for a given problem:

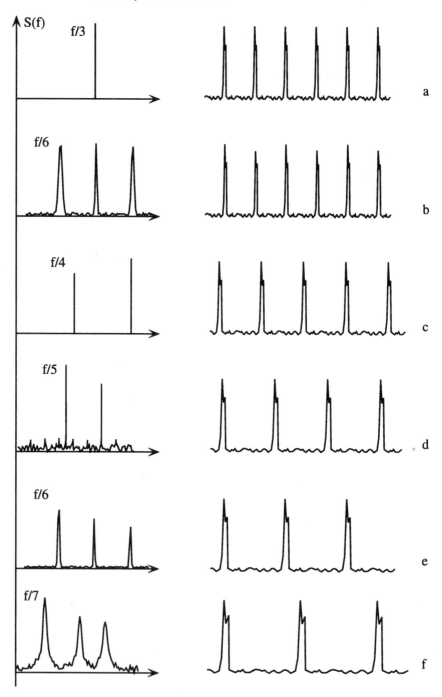

Figure 1.10. One-peak periodic orbits in the CO_2 laser (from Tredicce *et al*).

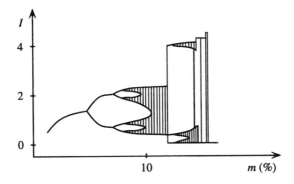

Figure 1.11. Experimental bifurcation diagram for the CO_2 laser (from Tredicce *et al*). The modulation amplitude (see text) is represented by the magnitude *m*.

Figure 1.12. Tearing of a plasma sheet (from Anderson and Kunkel, see text).

the solutions are governed by the conditions at the boundaries and the charge distribution, and (b) that *symmetries in the system are reflected in symmetries of the solutions*. In the present problem the symmetry is spontaneously broken, an effect ultimately due to the lack of stability of the symmetric solution. In any case, the symmetry breaking must be 'symmetrical' in the sense that the chances of obtaining a given (asymmetric) pattern or any other arbitrarily rotated (asymmetric) pattern have to be the same. In other words, if an asymmetric solution is present for a symmetric system, transforming the asymmetric solution (with the transformations associated to the symmetry) we have to find another equivalent solution.

This plasma experiment illustrates a nonlinear phenomenon associated with systems presenting symmetries, namely *spontaneous symmetry breaking*. Roughly speaking this concept deals with problems in which the system under study somehow 'unexpectedly' has *less* symmetry (is more unsymmetric) than the environment (a pure theoretician may say: 'when the solution has less symmetry than the equations'). A simple theoretical example could be a (classical) particle in a symmetric double-well potential, such as $V(x) = (x^2 - 1)^2$. The potential is reflection symmetric (interchanging x and $-x$ we obtain the same potential and thus the same global picture). For energies above the top of the well, the particle follows symmetric trajectories extending from negative to positive values of x, while for energies below the top of the well, it follows asymmetric trajectories confined to the left or to the right of the origin. If we add friction to the system we verify that the orbits acquire an increasing asymmetry no matter how close they start from the symmetric state $x = 0$. Something that we sometimes cannot (or do not bother to) account for is allowing the solution to have less symmetry than the equations (in this case a one-dimensional Newton equation for a conservative force with potential V).

Spontaneous symmetry breaking appears in a number of physical phenomena. This should not be confused with external or parametric symmetry breaking: for example, although in the Bénard experiment there is (ideally) translational symmetry, in the actual experiment the presence of the walls breaks that symmetry. The influence of the walls has been established experimentally [kris70a, kris70b].

1.6 Summary

What did we learn in this chapter? We intended to present a sample of the most exciting 'fingerprints' of nonlinear phenomena, taking examples from a variety of fields ranging from population biology to laser physics:

- Subharmonicity
- Amplification of small disturbances (unstable regimes)
- Transitions from simple organization to complex organization of the motion
- The existence of different basins of attraction leading to different final (asymptotic) states (multistability)
- The appearance of qualitative changes in the features of the solutions as parameters are changed, such as intermittence between simple and complex motion in the regime of transition
- Evidence for spontaneous symmetry breaking.

These fingerprints, which are absent or unusual in linear dynamics, are repeatedly encountered in nonlinear dynamical systems. They all have in common the fact that they render all long-term predictions about the time evolution of systems less accurate (often *in*accurate, for practical purposes). This intrinsic inability of making long-term predictions has come in a time in history when the confidence in science is (was?) perhaps exaggeratedly large, and it will hopefully contribute to help us understand where we really are.

One might also say that the examples presented here have somewhat different characteristics. For the case of turbulent flow, a comprehensive theory or analysis is yet to come. For the Bénard instability, experiments preceded the theory at the beginning and lately they have developed simultaneously. Finally, for the laser system many of the observed phenomena have been, if not predicted specifically for the case in study, pre-existent as an abstract theory. In some sense, the older the experiment, the harder it is to explain, and in a truer sense the larger the number of degrees of freedom involved in the description, the less we know.

The variety of examples illustrates the fact that nonlinear dynamics cuts across almost all branches of physics and natural sciences. Although it is not an 'established' branch of physics in the same way as nuclear physics or optics, it is a subject that the modern physicist cannot afford to ignore.

Chapter 2

Linear dynamics

2.1 Introduction

A central task for the physicist is to identify the 'relevant' dynamical variables of a natural system and to describe its time evolution. This identification process is a delicate matter that depends of a number of factors including the system, the observer and the models she/he has in mind. The *intrinsic* properties of the dynamics, i.e., those properties that do not depend on the observer, cannot depend on arbitrary choices, in particular the choice of a system of coordinates. However, it often happens that some variables are experimentally accessible while others are not, or some have a special importance beyond their dynamical meaning. In our presentation we favour the intrinsic point of view of the dynamics over the extrinsic point of view associated with the physical, biological, economical or any other meaning of the system under study. It shall be kept in mind that the dynamics will be described 'up to a change of coordinates'. In this context, for example, a circle can be easily transformed into an ellipse by rescaling the coordinates.

Although many 'real-life' systems can be better described by *partial* differential equations (consider, for example, the extent of application of Newton's equation as compared with that of the Navier–Stokes, Maxwell or Schrödinger equations in physics) a large number of problems admit an approximate description in terms of a (small-sized) set of *first-order ordinary differential equations*, i.e., a *dynamical system*. The degree of complication of such models is likely to increase with the accuracy of the description.

This book will deal mainly with *nonlinear dynamical systems*. To what extent and within which limits the systems presented in chapter 1 can be described by dynamical systems will be further discussed in chapter 3. Before that, we need a little preparation.

19

2.2 Why linear dynamics?

There are several reasons for giving a brief account of *linear dynamics* (from now on linear dynamical systems), before we start with *non*linear dynamics.

For the physicist, the natural example of a *linear* dynamical system would be the harmonic oscillator:

$$\begin{aligned} \dot{x} &= y \\ \dot{y} &= -\omega^2 x. \end{aligned} \tag{2.1}$$

The word *linear* arises from the fact that the right hand side is a linear function of the variables (i.e., it satisfies the following two conditions: (i) $f(ax) = af(x)$ for any real or complex number a and (ii) $f(z + w) = f(z) + f(w)$). As a consequence, a linear combination of two solutions, i.e., $\alpha x_1(t) + \beta x_2(t)$, is also a solution (this is known in physics as the *superposition principle*).

We may start by asking: Why are nonlinearities necessary? In other words, which elements of the natural phenomena—in particular of the cases presented in chapter 1—cannot be described with a (finite-dimensional) linear system of ordinary differential equations? To answer this question we need to have an idea of the scope of linear dynamics.

Secondly, some of the methods and ideals of nonlinear dynamics are taken directly from the linear case.

Thirdly, we will later see that even when the system is nonlinear we can learn about it by *linearizing* the equations (i.e., making a linear approximation) in the vicinity of certain particular solutions. Thus, some knowledge of elementary linear dynamics will be necessary.

In the coming sections we will review the case of general autonomous linear systems (*affine* systems to be precise) in the homogeneous and inhomogeneous cases, linear normal forms and changes of coordinates. We will examine the structure of the fixed point subspaces and the dynamics near these points (local dynamics). Finally we will discuss additively forced systems and parametrically forced systems; the latter already share some 'non-trivial' features with the nonlinear flows. All these concepts will become clear in the course of the discussion.

2.3 Linear flows

As mentioned above, we will include in our discussion a slightly more general class of dynamical systems than just linear, namely *affine systems*. We will be concerned with the set of solutions of such systems (intuitively, the words 'flow' and 'set of solutions of the system' are, in a loose sense, interchangeable). The most general form of an affine dynamical system reads

$$dx/dt = Ax + b, \tag{2.2}$$

with $x, b \in \mathbb{R}^n$, and $A \in \mathbb{R}^{n \times n}$, where n is the dimension of the system. A and b may possibly depend on time. Linear systems are a 'special case' of affine systems having $b = 0$.

Even if the system equation (2.2) is not *linear*, in the sense defined above, its solutions are essentially determined by its associated linear system (the *homogeneous* system, i.e., when $b = 0$). The most immediate example can be formulated as follows:

Exercise 2.1: Show that any affine dynamical system equation (2.2), where A is a constant matrix having non-zero determinant and b is a constant vector, can be rewritten as a linear system by a change of coordinates. For this problem, the denomination *inhomogeneous linear system* (see below) instead of affine system is usually found.

We remark that when we speak of nonlinear systems we are thinking of what could be called *intrinsically nonlinear* systems, or *non-affine* systems, i.e., problems described by a non-affine (and therefore nonlinear) system. The key point is that such systems cannot be reduced to linear systems by a change of coordinates, be it linear or not.

Rather than addressing the general problem of affine systems we will study some restricted cases which appear to be the most frequently found in practice and, consequently, the best known.

2.3.1 Autonomous flows

Autonomous flows are those where both A and b are constant in time. The *homogeneous* case is when $b = 0$, and the *in*homogeneous case, well... you guess.

2.3.1.1 Autonomous homogeneous flows

This case can be regarded as the simplest one. Since the vector b is identically zero and the matrix A is constant in time, equation (2.2) becomes

$$dx/dt = Ax. \qquad (2.3)$$

Equation (2.3) can be formally (and actually) solved in the form of an exponential

$$x(t) = e^{At}x(0) = \left(\sum_{n=0}^{\infty} A^n t^n / n! \right) x(0) \qquad (2.4)$$

where $x(0)$ is the initial condition.

Although from a computational point of view equation (2.4) appears to be a reasonable algorithm (it is actually quite inefficient), we can hardly say that we 'understand' the solutions of homogeneous linear equations for an arbitrary matrix A just because we have obtained this formal solution. The problem is that equation (2.4) does not translate into a new (and more intuitive or better known) representation of the problem. Nor can we (in general) read out directly the consequences of equation (2.4). What we usually call 'to understand' consists in fact of these two (unachieved) properties! (for a related discussion about understanding see [glas89]).

An alternative approach to the exponential form equation (2.4) is to introduce a linear change of coordinates in order to reduce equation (2.3) to its simplest expression. This amounts to reducing the matrix A to its *Jordan normal* (or *canonical*) *form* [gant59, most63, arno73, hirs78], i.e. to transform A with a linear change of coordinates into the form

$$
A = \begin{pmatrix} A_1 & 0 & 0 & 0 & 0 \\ \cdots & \cdots & \cdots & \cdots & \cdots \\ 0 & 0 & A_i & 0 & 0 \\ \cdots & \cdots & \cdots & \cdots & \cdots \\ 0 & 0 & 0 & 0 & A_m \end{pmatrix}, \tag{2.5}
$$

where each submatrix is of the form

$$
A_i = \begin{pmatrix} \lambda_i & \epsilon_i & 0 & 0 & 0 \\ \cdots & \cdots & \cdots & \cdots & \cdots \\ 0 & 0 & \lambda_i & \epsilon_i & 0 \\ \cdots & \cdots & \cdots & \cdots & \cdots \\ 0 & 0 & 0 & 0 & \lambda_i \end{pmatrix} \tag{2.6}
$$

and $\epsilon_i = 0$ or 1 and λ_i are complex numbers. The submatrices have order d_i. The order of the matrix is $n = \sum_{i=1}^{m} d_i$.

We will bypass the discussion of the special case where the matrix A is completely diagonalizable (i.e., the Jordan normal form is just a diagonal matrix). We just mention that the change of coordinates induced by the diagonalization transforms the problem into n uncoupled one-dimensional linear problems, where the time evolution in each transformed coordinate is determined by the associated eigenvalue.

Exercise 2.2: Find the solutions of the following linear problems:

(i) $\begin{cases} dx/dt = x + y \\ dy/dt = y \end{cases}$

(ii) $\begin{cases} dx/dt = -x + y \\ dy/dt = -y + x \end{cases}$

(iii) $\begin{cases} dx/dt = y \\ dy/dt = z \\ dz/dt = x. \end{cases}$

Let us explore the relations between the Jordan normal form and the (generalized) eigenvectors of a square matrix[†]. A proof of the normal form theorem can be found in most books on the theory of matrices, for example G D Mostov, J H Sampson and J–P Meyer [most63].

Consider the *spectrum* σ of a matrix A, i.e., the set of complex numbers λ such that $(A - \lambda I)$, regarded as a linear operator $T : \mathbb{R}^n \to \mathbb{R}^n$ is not one-to-one. This is the same as saying that the matrix $(A - \lambda I)$ does not have an inverse for those values of λ (I is the identity matrix).

Exercise 2.3:

- Show that for λ in the spectrum of A there exists a non-zero vector $x \in \mathbb{R}^n$ such that $Tx \equiv (A - \lambda I)x = 0$ (in other words, every eigenvalue has at least one associated eigenvector).
- Show that λ is in the spectrum of A iff the characteristic polynomial, $P(y)$, has a zero at $y = \lambda$, where $P(y) = \det (A - yI)$.

With each eigenvalue λ we can associate an index ν, defined as the smallest non-negative integer such that $(A - \lambda I)^\nu x = 0$ for every vector x for which $(A - \lambda I)^{\nu+1}x = 0$. We can also associate with each λ in the spectrum of A an eigenspace $\mathcal{M}^\nu(\lambda)$:

$$\mathcal{M}^\nu(\lambda) = \{x \in \mathbb{R}^n \ : \ (A - \lambda I)^\nu x = 0\}. \tag{2.7}$$

In general, we say that a subspace, S, is A *stable* when $Ax \in S$ for every $x \in S$.

Exercise 2.4:

- Show that the eigenspace $\mathcal{M}^\nu(\lambda)$ has A-stable subspaces of the form

$$\mathcal{M}^k = \{x \in \mathcal{M}^\nu \ : \ (A - \lambda I)^k x = 0\}$$

for all positive integers $0 \le k \le \nu$ and

$$\mathcal{M}^\nu \supseteq \ldots \supseteq \mathcal{M}^k \supseteq \mathcal{M}^{k-1} \supseteq \ldots \supseteq \mathcal{M}^1 \supseteq \mathcal{M}^0 = \{0\}.$$

Note that $\mathcal{M}^1$ contains the eigenvector(s) associated with λ.

[†] Let us suggest the reader follows this discussion formulating and checking each statement for

$$A = \begin{pmatrix} a & 1 & 0 & c \\ 0 & a & 1 & 0 \\ 0 & 0 & a & 0 \\ 0 & 0 & 0 & b \end{pmatrix}.$$

- Show that the eigenspace $\mathcal{M}^\nu(\lambda)$ has A-stable subspaces of the form

$$\mathcal{L}^k = \{x \in \mathcal{M}^\nu \;:\; x = (A - \lambda I)^k y, \; y \in \mathcal{M}^\nu \equiv \mathcal{L}^0\}$$

for all positive integers $0 \le k \le \nu$ and

$$\{0\} = \mathcal{L}^\nu \subseteq \ldots \subseteq \mathcal{L}^k \subseteq \mathcal{L}^{k-1} \subseteq \ldots \subseteq \mathcal{L}^0.$$

Observe that $\mathcal{L}^j(\lambda)$ contains all the vectors in $\mathcal{M}^\nu(\lambda)$ that have a pre-image by $(A - \lambda_i I)^j$ in $\mathcal{M}^\nu(\lambda)$.

Following a similar procedure, we can associate eigenspaces of $A^\dagger$, the adjoint of the matrix A, with each of its eigenvalues γ. We denote these eigenspaces by $\bar{\mathcal{M}}^\nu(\gamma)$. The role played by the eigenvalue λ of A will be played by λ^* in the case of $A^\dagger$.

Exercise 2.5:

- Show that $\sigma(A) = \sigma(A^\dagger)^*$.
- Show that the subspaces $\mathcal{M}^\nu(\lambda)$ and $\bar{\mathcal{M}}^\nu(\gamma)$ are orthogonal iff $\lambda \ne \gamma^*$. Hint: Use that $y^\dagger(A - \lambda I)^{\nu(\lambda)}x = 0$ for all $x \in \mathcal{M}^\nu(\lambda)$ and all y.

Our goal is to obtain an orthonormal set of basis vectors in order to perform the change of coordinates that will transform the matrix A to the form (2.6). The first step to obtain this basis set is to choose a basis for each (generalized) eigenspace $\mathcal{M}^{\nu(\lambda)}$ and for the corresponding eigenspaces $\bar{\mathcal{M}}^{\nu(\lambda)}$ of the adjoint problem. We call the basis vectors $\{\bar{v}_j^i, \; j = 1, \ldots, d_i ; \; i = 1, \ldots, m\}$ and $\{v_j^i, \; j = 1, \ldots, d_i ; \; i = 1, \ldots, m\}$, m is the number of different eigenvalues and d_i the dimension of the associated eigenspace ($\sum_{i=1}^m d_i = n$). The vectors $\{v_j^i, \; j = 1, \cdots, d_i\}$ span the eigenspace $\mathcal{M}^\nu(\lambda_i)$ while the vectors $\{\bar{v}_j^i, \; j = 1, \cdots, d_i\}$ span $\bar{\mathcal{M}}^\nu(\lambda_i)$.

In the new basis the matrix A becomes block diagonal, since $(\bar{v}^i)^\dagger A v^k = A_i \delta_{ik}$ (cf equation (2.6)).

Exercise 2.6: Prove the formula above.

Having reduced the matrix A to block diagonal form we can use the remaining freedom in the choice of our coordinates (basis set) to put each block in its Jordan canonical form. The proper choice of basis vectors in each eigenspace is partially illustrated in the following exercise (actually a part of the theory):

Exercise 2.7: Let $y \in \mathcal{L}^0(\lambda)$ ($= \mathcal{M}^\nu(\lambda)$) and $x \in \bar{\mathcal{M}}^\nu(\lambda^*)$ (we drop the λ from now on to lighten the notation).

- Show that if $x^\dagger y = 0$ for all $y \in \mathcal{L}^j$ then $x \in \bar{\mathcal{M}}^j$.
- Show that if $y \in \mathcal{L}^j$ then $x^\dagger y = 0$ for all $x \in \bar{\mathcal{M}}^j$.

The previous exercise shows that for each λ_i, the stable subspace $\mathcal{M}^\nu = \mathcal{L}^0$ can be decomposed using a chain of orthogonal pairs $(\mathcal{L}^k, \bar{\mathcal{M}}^k)$, $k = 1, \ldots, \nu(\lambda_i)$. Recall that $\bar{\mathcal{M}}^1$ is the subspace containing all the eigenvectors of $A^\dagger$ associated with λ^*. In other words, $\bar{\mathcal{M}}^1$ has those elements of $\mathcal{M}^\nu$ that *do not* have a pre-image by $(A - \lambda I)$ (i.e. those which are not in $\mathcal{L}^1$).

A basis for $\bar{\mathcal{M}}^1$ can be obtained in the 'traditional' way, i.e., by Gram–Schmidt orthonormalization of a maximal linearly independent set of vectors. This is equally valid for $\mathcal{M}^1$ in the original problem. Given a vector $x \in \mathcal{M}^\nu$ we can decompose it into its $\mathcal{L}^1$ and $\bar{\mathcal{M}}^1$ parts. Assuming for simplicity that $\bar{\mathcal{M}}^1$ is one-dimensional, we get

$$x = x_1 + y_1$$
$$x_1 = (\bar{v}_1^{i\dagger} x) v_1^i \in \bar{\mathcal{M}}^1$$
$$y_1 = x - (\bar{v}_1^{i\dagger} x) v_1^i \in \mathcal{L}^1.$$

Exercise 2.8:

- Show that every $x \in \mathcal{M}^\nu$ can be written as

$$x = x_1 + x_2 + \ldots + x_\nu$$

where x_i belongs to $\bar{\mathcal{M}}^i$. Hint: To get x_2 apply the procedure described above to the vector y_1.
- Show that $x \in \mathcal{M}^\nu$ can also be partitioned in terms of the pairs $(\bar{\mathcal{L}}^k, \mathcal{M}^k)$.

After having changed coordinates to the basis $\{\bar{v}_j^i, v_j^i\}$ in order to render A into Jordan canonical form, we abandon our algebraic detour and come back to the dynamics.

The solutions of equation (2.3) can now be written in the form of a sum of exponentials multiplied by polynomials in time. Within each Jordan block (of order d_i) we have

$$y(t) = e^{\lambda_i t} \left(\sum_{k=0}^{\nu-1} (A_i - \lambda_i I_i)^k \frac{t^k}{k!} \right) y(0) \tag{2.8}$$

where $A_i - \lambda_i I_i$ is a nilpotent matrix of order d_i having ones in the first upper diagonal and zeros elsewhere.

In the new coordinate system we immediately realize that the motions in the different eigenspaces are independent but the motion inside each subspace may not be further decomposed into independent components by means of linear transformations. A natural question that may arise is: what is the set of fixed points, i.e., the points x where the time-derivatives $\dot{x}$ are zero? We note that the set of such *fixed points* for autonomous homogeneous systems consists of *subspaces* of *phase space* (the dynamical space). It follows that a fixed point is either isolated and the only fixed point of the flow or it belongs to an *n*-dimensional subspace ($n \neq 0$).

2.3.1.2 *Autonomous inhomogeneous flows*

The problem of linear inhomogeneous flows with constant coefficients can be reduced to the case of homogeneous flows except in one case, where a particular solution of the inhomogeneous problem must be explicitly considered. We only need to consider the case where the vector b in equation (2.2) belongs to the (generalized) eigenspace associated with the zero eigenvalue of $A^\dagger$ (the space $\bar{\mathcal{M}}^1(0)$ in the previous exercises). All other inhomogeneities can be removed by shifting the origin with the constant vector $A^{-1}b$ (as proved in a previous exercise). Hereafter we will assume when needed that the fixed points are located at the origin.

Exercise 2.9: Show that if $A^\dagger b = 0$ (for a non-zero vector b) then there is no vector $x \in \mathbb{R}^n$ such that $Ax = b$.

When b has non-zero projection onto $\bar{\mathcal{M}}^1(0)$, we use a two-step procedure. First, change coordinates in order to separate the matrix A in two orthogonal blocks, one of which is associated with the eigenvalue zero (it is sufficient to put A into Jordan canonical form). Note that the remaining block (where A has no zero eigenvalues) can be treated as in the previous case. The resulting equation in the zero subspace can be written in the form

$$\mathrm{d}y/\mathrm{d}t = Ay + b_0. \tag{2.9}$$

Finally, solutions of equation (2.9) can be found as the sum of an arbitrary solution of the homogeneous equation and a particular solution of the inhomogeneous equation (2.9). It is always possible to find a particular solution with the additional property of being zero at $t = 0$. The reader can verify that such a particular solution, y_p, can be written as

$$y_\mathrm{p}(t) = \sum_{j=0}^{\nu(0)-1} A^j b_0 t^{j+1}/(j+1)! \tag{2.10}$$

($\nu(0)$ is defined as above), recalling that $A^{\nu(0)} = 0$.

Exercise 2.10: Find the change of coordinates that transforms

$$
\begin{cases} dx/dt = y + a \\ dy/dt = z + b \\ dz/dt = c \end{cases} \quad \text{into} \quad \begin{cases} dx'/dt = y' \\ dy'/dt = z' \\ dz'/dt = c. \end{cases}
$$

Show that the vectors $(0, 0, c)^\dagger$ are the only vectors in the zero eigenspace associated with

$$
\begin{pmatrix} 0 & 1 & 0 \\ 0 & 0 & 1 \\ 0 & 0 & 0 \end{pmatrix}^\dagger .
$$

Find a general solution for the equation.

2.3.1.3 Singular points in bi-dimensional flows

Apart from fully diagonalizable linear systems, which can be reduced to a set of independent one-dimensional linear equations, the simplest case that can be considered is that of flows living in a manifold of dimension two. We are concerned with equation (2.3) where A is a real 2×2 matrix, i.e.

$$
\begin{pmatrix} dx/dt \\ dy/dt \end{pmatrix} = \begin{pmatrix} a & b \\ c & d \end{pmatrix} \begin{pmatrix} x \\ y \end{pmatrix}. \tag{2.11}
$$

As we mentioned above, we call a *singularity*, or a *fixed point*, any point of phase-space where the time-derivatives are zero, i.e., those points that do not evolve in time into other points under the action of the flow or, more precisely, those points that evolve into themselves.

We would like to make a first attempt at classifying singularities using the local behaviour of the linear flows around them. We distinguish between *local* and *global* properties. Local refers here to a vicinity of the fixed point $(0, 0)$. However, x and/or y could be angular variables and the global topology be a bi-dimensional torus, $\mathbb{T}_{\Bbbk}$, or a cylinder, $\mathbb{R} \times \mathbb{S}^1$, rendering the global behaviour of the flow completely different.

We need only to study a few cases that we can sort according to the linear normal form associated with each one of them and the nature (real or complex) of the eigenvalues.

The general case corresponds to two different eigenvalues having non-zero real parts $\text{Re}(\lambda_j)$ $j = 1, 2$, which admits five different flow types:

(a, b) **Node.** We call the fixed point a node when both eigenvalues are real and both of them *simultaneously* greater (*unstable node, source*) or smaller (*stable node, sink*) than zero (figure 2.1(a) and (b), respectively).

$$
dx/dt = \lambda_1 x \tag{2.12}
$$

$$
dy/dt = \lambda_2 y. \tag{2.13}
$$

The *flow lines* (also known as integral curves or trajectories or orbits) are given by the curves $(x, y)(t)$ that are solutions of the linear equation. In this case we have

$$(y/y_0)^{|\lambda_2|} - (x/x_0)^{|\lambda_1|} = 0 \qquad \text{for } x_0 \neq 0 \text{ and } y_0 \neq 0. \qquad (2.14)$$

We speak of repulsive (a) or attractive (b) nodes (or fixed points) depending on whether the flow diverges from the node or converges to it for large positive times. Note that cases (a) and (b) can be interchanged by considering negative times.

(c) **Saddle.** We call the fixed point a saddle when one eigenvalue, say λ_1, is smaller than zero and the other (λ_2) is larger than zero. The flow pattern is illustrated in figure 2.1(c) and the trajectories respond to the curves

$$(y/y_0)^{-|\lambda_1|} - (x/x_0)^{|\lambda_2|} = 0 \qquad \text{for } x_0 \neq 0 \text{ and } y_0 \neq 0. \qquad (2.15)$$

(d, e) **Focus.** We call the fixed point a focus when both eigenvalues are complex, $\lambda_1 = \lambda_2^*$ (remember that the matrix A is real). The focus has two subcases, those spiralling inward, when $\text{Re}(\lambda_1) > 0$ (figure 2.1(d)), and those spiralling outward, when $\text{Re}(\lambda_1) < 0$ (figure 2.1(e)). The representative equations are

$$dx/dt = \alpha x + wy \qquad (2.16)$$
$$dy/dt = -wx + \alpha y \qquad (2.17)$$

$$\alpha = \text{Re}(\lambda_1) = \text{Re}(\lambda_2) \text{ and } w = \text{Im}(\lambda_1) = -\text{Im}(\lambda_2). \qquad (2.18)$$

The solutions are easier to present in polar coordinates $\rho = (x^2 + y^2)^{1/2}$, $\theta = \tan^{-1}(y/x)$. It can be verified that the trajectories correspond to logarithmic spirals of the form

$$\rho = e^{\alpha\theta/w}. \qquad (2.19)$$

These three (or five) cases, i.e. node, saddle and focus, are *general* in the sense that the 2×2 real matrices form a four-dimensional space, $\mathbb{R}^4$, and these solutions of equation (2.11) correspond to

$$\text{node if } \Delta > 0 \text{ and } \det(A) > 0, \qquad (2.20)$$
$$\text{saddle if } \Delta > 0 \text{ and } \det(A) < 0, \qquad (2.21)$$
$$\text{focus if } \Delta < 0, \qquad (2.22)$$

where $\Delta = \text{Tr}(A)^2 - 4\det(A)$.

Each one of these conditions is satisfied in a four-dimensional submanifold of the manifold of 2×2 real matrices.

In addition to the general cases, there are a number of special cases that are possible in manifolds of smaller dimension. Associated with three-dimensional manifolds are the following:

(f) **Jordan node.** We call the fixed point a Jordan node when $\lambda_1 = \lambda_2$ (see figure 2.1(f)). After a proper change of coordinates, the matrix A takes the form

$$\begin{pmatrix} \lambda & 1 \\ 0 & \lambda \end{pmatrix}. \tag{2.23}$$

(g) **Centre.** We call the fixed point a centre when $\lambda_1 = -\lambda_2 = \lambda_2^*$ (the real part of both eigenvalues is zero). See figure 2.1(g). In a proper coordinate system, the trajectories are represented by circles (check it!).

(h) **Singular node.** We call a fixed point a singular node when $\lambda_1 = 0$ and $\lambda_2 \neq 0$ (see figure 2.1(h)). If x_1 and x_2 are the associated eigenvectors, the flow occurs along x_2, while there is no motion along x_1. Note that every point in the the one-dimensional subspace $(x_1, 0)$ is a fixed point!

Finally, there are special cases associated with manifolds of dimension two, one and zero. The nilpotent singularities related to the matrices where $\lambda_1 = \lambda_2 = 0$ (bi-dimensional manifold), the singular node where $\lambda_1 = \lambda_2$ related to the one-dimensional manifold of matrices that are multiples of the identity and the (trivial) case when the flow is identically null, $A = 0$.

(i) **Nilpotent singularity.** We call the fixed point a nilpotent fixed point when A is not identically zero but has two zero eigenvalues (see figure 2.1(i)). In such a case A can be transformed into the nilpotent form

$$\begin{pmatrix} 0 & 1 \\ 0 & 0 \end{pmatrix}. \tag{2.24}$$

(j) **Bicritical node.** We call the fixed point a bicritical node when it has an associated normal form (see figure 2.1(j))

$$\begin{pmatrix} \lambda & 0 \\ 0 & \lambda \end{pmatrix} \tag{2.25}$$

which occurs if $b = c = 0$ and $a = d$ in equation (2.11).

2.3.2 Forced flows

We discuss here the case where A or b depend on time. The different levels of difficulty that this case may carry will be related to the character of the time dependences. The name *forced* arises from the fact that in this way one can represent the action of time-varying forces considered *external* to the system.

We call *additively* forced flows those where b depends on time, while A is constant, and *parametrically* forced flows those where A is time dependent.

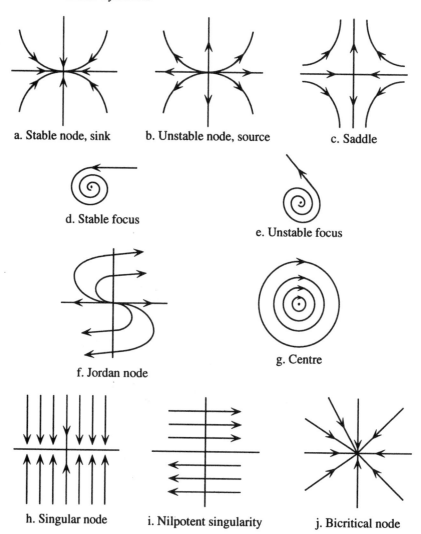

Figure 2.1. Bi-dimensional singularities: (a) and (b) node; (c) saddle; (d) and (e) focus; (f) Jordan node; (g) centre; (h) singular node; (i) nilpotent singularity; (j) bicritical node.

2.3.2.1 *Additively forced flows*

Once again, the general solution can be found as the sum of the solutions of the associated homogeneous equation and a particular solution of the inhomogeneous equation. The latter can be found by method of variation of the constants. The general solution is then written as

$$x(t) = x_h(t) + \int_0^t e^{A(t-s)} b(s) \, ds. \qquad (2.26)$$

x_h solves the associated homogeneous equation:

$$\frac{dx_h}{dt} = Ax_h. \qquad (2.27)$$

The present situation covers the inhomogeneous equation (2.9) as a special case. In particular, the integral in equation (2.26) corresponds exactly to the solution in equation (2.10).

2.3.2.2 Parametrically forced flows

The study of parametrically forced flows is of a more complex nature than the previous ones. One cannot state much more than some formal results without considering the nature of the forcing terms. We propose the reader to consider an example instead, which appears in different areas of physics, namely

$$\frac{d^2x}{dt^2} = Q(t)x \qquad (2.28)$$

or

$$\begin{cases} \dot{x} & = \quad p \\ \dot{p} & = \quad Q(t)x. \end{cases} \qquad (2.29)$$

$Q(t)$ is a real periodic function of (minimal) period 2π, i.e., $Q(t) = Q(t + 2\pi)$.

Equation (2.28) (Hill's equation) appears in several contexts in physics. It was proposed for the study of the lunar perigee by G W Hill [hill77] and it is also used in a model to describe electrons moving in a periodic array [cohe77, kitt76]. Because of the different physical motivation different features of the equation are stressed in each case:

- From the celestial point of view a key question is: Under which circumstances does equation (2.28) present *stable* (in this context, bounded) solutions? Clearly, unstable (unbounded) solutions will imply increasingly large amplitude for the oscillations of the lunar perigee.
- For the solid-state physicist the problem is: Which values of λ (the *energy* in this interpretation) are permitted?
- One may also be interested in the equation itself, regardless of the previous motivations. A natural question may be what the solutions of equation (2.28) look like. In particular, do periodic solutions exist?

In order to partially answer these (classical) questions we first observe that since equation (2.29) is a homogeneous linear equation, every solution

$((x, p)(t))$ of the initial value problem can be written as the superposition of two fundamental solutions, (x_1, p_1) and (x_2, p_2) with fixed (standardized) initial condition:

$$(x_1, p_1)(t_0) = (1, 0)$$
$$(x_2, p_2)(t_0) = (0, 1).$$

A solution with arbitrary initial condition (a, b) evolves in time into

$$(x, p)(t) = (ax_1(t) + bx_2(t), ap_1(t) + bp_2(t)) \tag{2.30}$$

which can be rewritten in the form of a matrix equation:

$$\begin{pmatrix} x \\ p \end{pmatrix}(t) = \begin{pmatrix} x_1 & x_2 \\ p_1 & p_2 \end{pmatrix}(t) \begin{pmatrix} a \\ b \end{pmatrix} = U(t, t_0) \begin{pmatrix} a \\ b \end{pmatrix}. \tag{2.31}$$

The *evolution operator* $U(t, t_0)$ contains all relevant information about the flow. It satisfies the following relations:

$$U(t_0, t_0) = I \tag{2.32}$$
$$U(t', t)U(t, t_0) = U(t', t_0) \tag{2.33}$$
$$U(t', t) = U(t' + 2\pi, t + 2\pi). \tag{2.34}$$

The second property affirms that the evolution does not depend on any intermediate state. The third property is a consequence of the periodicity of the coefficients: the final state depends upon the initial conditions, the phase of the oscillating coefficients (i.e., $t \bmod 2\pi$) and the total interval of the evolution. $U(t, t_0)$ satisfies the following equation:

$$U(t, t_0)^\dagger \begin{pmatrix} 0 & 1 \\ -1 & 0 \end{pmatrix} U(t, t_0) = \begin{pmatrix} 0 & 1 \\ -1 & 0 \end{pmatrix} \tag{2.35}$$

which in turn implies that $\det(U(t, t_0)) = 1$ (this is a consequence of the *Hamiltonian* structure of Hill's equation). The inverse matrix

$$U(t, t_0)^{-1} = U(t_0, t) \tag{2.36}$$

is related to the adjoint matrix $U(t, t_0)^\dagger$ by

$$U(t_0, t) = -\begin{pmatrix} 0 & 1 \\ -1 & 0 \end{pmatrix} U(t, t_0)^\dagger \begin{pmatrix} 0 & 1 \\ -1 & 0 \end{pmatrix}. \tag{2.37}$$

This last relation can be verified realizing that the inverse matrix satisfies the 'adjoint differential equation'

$$\frac{dU^\dagger}{dt}(t_0, t) = -\begin{pmatrix} 0 & 1 \\ Q(t) & 0 \end{pmatrix}^\dagger U^\dagger(t_0, t). \tag{2.38}$$

Exercise 2.11: Find the differential equation that $U(t, t_0)$ fulfills and verify relation (2.38).

Exercise 2.12: Show that $\det(U(t, t_0)) = 1$.

It is of particular interest to examine the evolution operator after a complete period of the forcing. The operator $U \equiv U(t + 2\pi, t)$ receives the special name of *monodromy matrix* and, we anticipate, it is the fundamental object in the study of the stability of periodic orbits. Note that *any* time evolution can be written as

$$U(t', t_0) = U(t', t_0 + 2n\pi)U(t_0 + 2n\pi, t_0) = U(t', t)\,(U(t_0 + 2\pi, t_0))^n \tag{2.39}$$

where $n = [(t' - t_0)/2\pi]$ and $t = t_0 + 2n\pi$. If the time-span of the evolution is large, n becomes large but $(t' - t) \leq 2\pi$.

The stability of the solutions is associated with the eigenvalues of the monodromy matrix, U. The result is stated in a restricted version of Floquet's theorem that we take from Magnus and Winkler [magn79]. In order to lighten the notation, we will consider in what follows the initial time $t_0 = 0$ (without loss of generality).

2.3.2.3 *Floquet's theorem*

Let λ_1 and λ_2 be the eigenvalues of the monodromy matrix U associated with equation (2.29). Then

(i) If $\lambda_1 \neq \lambda_2$, then Hill's equation has two linearly independent solutions

$$f_1(t) = e^{i\alpha_1 t}\, q_1(t)$$
$$f_2(t) = e^{i\alpha_2 t}\, q_2(t)$$

where $q_1(t), q_2(t)$ are periodic functions of period 2π and

$$e^{i 2\pi \alpha_j} = \lambda_j, \quad j = 1, 2. \tag{2.40}$$

(ii) If $\lambda_1 = \lambda_2$, i.e. $\mathrm{Tr}(U) = \pm 2$, then equation (2.28) has one periodic solution. This solution is of period 2π when $\lambda_1 = \lambda_2 = 1$ and of period 4π when $\lambda_1 = \lambda_2 = -1$. There is a second periodic independent solution iff U is diagonal.

The main elements in the proof of the theorem are simple enough to be given here.

(i) Let $\begin{pmatrix} f_j(0) \\ g_j(0) \end{pmatrix}$ be the eigenvectors associated with the eigenvalues λ_j, $j = 1, 2$ considered as initial conditions of equation (2.29). Then

$$U \begin{pmatrix} f_j(0) \\ g_j(0) \end{pmatrix} = \lambda_j \begin{pmatrix} f_j(0) \\ g_j(0) \end{pmatrix} = \begin{pmatrix} f_j(2\pi) \\ g_j(2\pi) \end{pmatrix}. \tag{2.41}$$

Define

$$\begin{pmatrix} f_j(t) \\ g_j(t) \end{pmatrix} = U(t, 0) \begin{pmatrix} f_j(0) \\ g_j(0) \end{pmatrix} \tag{2.42}$$

$$q_j(t) = e^{-i\alpha_j t} f_j(t), \quad j = 1, 2. \tag{2.43}$$

We only need to show that $q_j(t)$ is a periodic function of t, since $e^{i\alpha_j t} q_j(t) = f_j(t)$ is a solution of equation (2.28). The result follows immediately from equation (2.39) since

$$\begin{pmatrix} f_j(t + 2\pi) \\ g_j(t + 2\pi) \end{pmatrix} = U(t + 2\pi, 0) \begin{pmatrix} f_j(0) \\ g_j(0) \end{pmatrix}$$

$$= U(t, 0)U(2\pi, 0) \begin{pmatrix} f_j(0) \\ g_j(0) \end{pmatrix}$$

$$= \lambda_j \begin{pmatrix} f_j(t) \\ g_j(t) \end{pmatrix} \tag{2.44}$$

and therefore

$$q_j(t + 2\pi) = e^{-i\alpha_j(t + 2\pi)} f_j(t + 2\pi)$$
$$= e^{-i\alpha_j(t + 2\pi)} \lambda_j f_j(t) = q_j(t)$$

(ii) The case of degenerate eigenvalues is similar to the previous case except that in general there may be only one proper eigenvector associated with each eigenvalue. Defining $f_1(t)$ as before we find that

$$f_1(t + 2\pi) = \lambda f_1(t) \tag{2.45}$$

and recalling that $\det(U) = 1 = \lambda_1 \lambda_2 = \lambda^2$, the eigenvalue λ must be 1 or -1 and $f_1(t)$ will be periodic of period 2π or period 4π respectively.
Finally, if U is diagonal, i.e., it is a scalar multiple of the identity, there is a second (linearly independent) eigenvector of U and we are able to find a second periodic solution.

The *stability* of the solutions of Hill's equation is directly related to the eigenvalues of the monodromy matrix. Recall that in this context solutions that grow unboundedly are called *unstable* while bounded solutions are called *stable*. There will be no stable solution iff

$$|\lambda_j| \neq 1, \quad j = 1, 2 \tag{2.46}$$

and there will be two stable solutions iff

$$|\lambda_1| = |\lambda_2| = 1 \text{ and } \lambda_1 \neq \lambda_2 \tag{2.47}$$

or

$$\lambda_1 = \lambda_2 = \pm 1 \text{ and } U \text{ is diagonal.} \tag{2.48}$$

To prove these statements we recall that $\det(U) = 1$. Consequently, there is no eigenvalue larger than unity in absolute value iff both eigenvalues have absolute value unity. Whenever there is an eigenvalue larger than one, almost every solution is unbounded.

Note that in discussing Floquet's theorem we made almost no use of the actual form of Hill's equation. The theorem can be generalized to any (finite-dimensional, ordinary) linear differential equation with periodic coefficients. In the case of non-degenerate eigenvalues the proof will be identical to the proof of part (i) above, while the results for degenerate eigenvalues will have to be rephrased in terms of the different possibilities for the Jordan normal form associated with the monodromy matrix.

2.3.2.4 Oscillation theorem—Liapunov–Haupt

Consider Hill's equation in 'standard' form:

$$\mathrm{d}^2 x / \mathrm{d}t^2 = (\rho + Q(t))x. \tag{2.49}$$

Floquet's theorem provides a criterion for the stability of the solutions of homogeneous linear equations with periodic coefficients; the following theorem (oscillation theorem, see [magn79]) states the existence of regions of stability and instability for equation (2.28). In solid-state theory, these regions will be interpreted as permitted and prohibited bands.

The oscillation theorem, which we give without proof, admits a graphic presentation. The eigenvalues of the monodromy matrix evolve continuously as functions of ρ. Considering that their product (the determinant of U) is constant and equal to one, and that the monodromy matrix is real, we have that either $\lambda_1 = \lambda_2^*$ and $|\lambda_j| = 1$ or both of them are real. The transition from two real eigenvalues (one of them larger than one in absolute value, the other smaller) to a pair of complex conjugated eigenvalues occurs when both of them are either one ($\mathrm{Tr}(U) = 2$) or minus one ($\mathrm{Tr}(U) = -2$) (see figure 2.2).

The theorem tells us that the regions with real and negative eigenvalues alternate with those regions with real and positive eigenvalues of the monodromy matrix; that between one case and the other there is a region of stability and that this alternating sequence is semi-infinite.

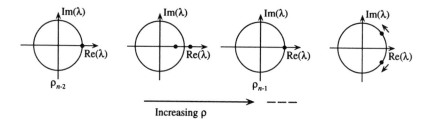

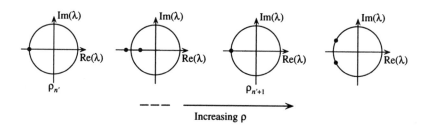

Figure 2.2. The eigenvalues of Hill's equation as a function of ρ.

For every differential equation (2.49), there belong two monotonically increasing finite sequences of real numbers

$$\rho_0 \; \rho_1 \; \rho_2 \; \cdots$$

and

$$\rho_0' \; \rho_1' \; \rho_2' \; \cdots$$

such that equation (2.49) has a periodic solution of period 2π if and only if $\rho = \rho_n$, $n = 0, 1, 2, \ldots$ and a solution of period 4π if and only if $\rho = \rho_n'$, $n = 0, 1, 2, \ldots$. The ρ_n and ρ_n' satisfy the inequalities

$$\rho_0 < \rho_1' \leq \rho_2' < \rho_1 \leq \rho_2 < \rho_3' \leq \rho_4' < \rho_3 \leq \rho_4 < \cdots \qquad (2.50)$$

and the relations

$$\lim_{n \to \infty} \rho_n^{-1} = 0 = \lim_{n \to \infty} \rho_n'^{-1}. \qquad (2.51)$$

The solutions of equation (2.49) are stable in the intervals

$$(\rho_0, \rho_1'), \; (\rho_2', \rho_1), \; (\rho_2, \rho_3'), \; (\rho_4', \rho_3), \; \ldots. \qquad (2.52)$$

At the end points of these intervals the solutions of (2.1) are, in general, unstable. This is always true for $\rho = \rho_0$. The solutions of equation (2.49) are stable for $\rho = \rho_{2n+1}$ or $\rho = \rho_{2n+2}$ if and only if $\rho_{2n+1} = \rho_{2n+2}$, and they are stable for $\rho = \rho_{2n+1}'$ or $\rho = \rho_{2n+2}'$ if and only if $\rho_{2n+1}' = \rho_{2n+2}'$.

For complex values of ρ, equation (2.49) has always unstable solutions.

The ρ_n are roots of the equation $\text{Tr}(U) = 2$ and the ρ'_n are roots of $\text{Tr}(U) = -2$.

Readers familiar with quantum mechanics can notice that the oscillation theorem is directly related to the Bloch theorem describing the possible energies of an electron in a periodic potential. It can be read as:

The permitted energies, $-\rho$, associated with the one-dimensional periodic Hamiltonian $p^2/(2m) + Q(x)$, where $Q(x)$ is 2π periodic, are organized in *permitted* bands that alternate with *forbidden* bands. The eigenfunctions associated with the energies of the boundary of the permitted bands are periodic functions of period 2π and 4π respectively, for the lower and upper limits of the band.

Exercise 2.13: Consider a small-amplitude real pendulum with its equivalent length subject to oscillatory variations in time in the form $l = l_0 + \epsilon f(t)$

$$f(t) = \begin{cases} 1 & \text{for } 0 \le t < \pi \\ -1 & \text{for } \pi \le t < 2\pi. \end{cases}$$

Find the stability regions in the parameter space $\omega = (g/l_0)^{1/2}$ and ϵ (Hint: See [arno73, guck86]). Could you use the result to provide a tentative explanation of the physics involved in the swing?

2.3.2.5 Arnold tongues

To complete our discussion of Hill's equation we quote a result from V I Arnold which has a twin result in the study of nonlinear resonances. The statement is of perturbative nature and can be proven with methods of Rayleigh–Schrödinger perturbation theory. We first rewrite Hill's equation (2.49) in the form

$$\mathrm{d}^2x/\mathrm{d}t^2 = -(w^2 + \epsilon a(t))x. \tag{2.53}$$

Theorem (Arnold tongues) [arno83a]: Assume that the coefficient a in equation (2.53) is an even trigonometric polynomial of degree p:

$$a(t) = \sum_{s=-p}^{p} a_s\, \mathrm{e}^{\mathrm{i}st} \qquad \text{where } a_{-s} = a_s. \tag{2.54}$$

Then the width of the Nth forbidden zone decreases no more slowly than $C\epsilon^r$ as $\epsilon \to 0$, where $r = 1 + [N/p]$ (see figure 2.3).

Note that in the general case, i.e., $p = \infty$, the decrease of the unstable region is linear.

A similar result regarding the width of resonant zones in nonlinear maps of the circle into itself can be proven with identical methods [arno83a].

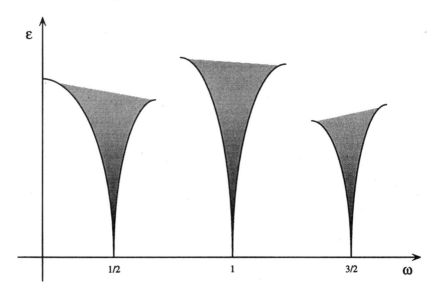

Figure 2.3. Arnold tongues in Hill's equation. Shaded zones represent the unstable (forbidden) zones.

2.4   Summary

This chapter was intended to be a survey of linear and affine dynamical systems. Since most readers are probably familiar with the subject, we tried to shift the stress towards the most unusual topics, such as non-diagonalizable matrices, singular points of two-dimensional flows and some theorems about Hill's equation.

We established what the set of fixed points of linear systems looks like, in two cases: (i) the general case of autonomous homogeneous systems and (ii) the particular case of two-dimensional flows. At this stage, the reader may try to compare the asymptotic motions which are possible in linear systems with the properties inferred from the systems in chapter 1, in order to answer the question: Why are nonlinearities necessary?

2.5   Additional exercise

Exercise 2.14: Find T such that

$$\begin{pmatrix} 0 & 1 \\ 0 & 0 \end{pmatrix} = T \begin{pmatrix} 1 & 1 \\ -1 & -1 \end{pmatrix} T^{-1}. \tag{2.55}$$

Chapter 3

Nonlinear examples

3.1 Preliminary comments

We are about to undertake the study of three algebraically simple systems: a model of a CO_2 laser, the Duffing oscillator and the Lorenz system. Our intention here is twofold. First, we will derive some of the models in order to make more evident their connection with physical problems and (more important) their limitations as tools for understanding nature. Second, we want to focus our attention on the relevant elements for the description of the dynamics, exploring the systems following our (educated) intuition. We will delay the abstraction of the concepts, formal definitions and the sharpening of our tools until the coming chapters.

3.2 A model for the CO_2 laser

3.2.1 The model

There are a number of models for the CO_2 laser with modulated losses mentioned in chapter 1 [tred86]. However, we are here concerned with the basic elements of the dynamics and correspondingly we shall adopt the simplest of the phenomenological models available [arec84, risk89]. See [nard88] for a discussion of the different models in the context of laser instabilities.

The semiclassical approach to lasers [risk89] assumes that the electromagnetic field can be treated by way of Maxwell's equations, while the material (in our case CO_2 gas) is described quantum mechanically (under severe approximations, however). Lowest-order perturbation theory is used for the interaction between field and matter.

Maxwell's equations for a homogeneous, linear, isotropic medium lead to [bald69]

$$\frac{1}{c^2}\frac{\partial^2 E}{\partial t^2} - \frac{\partial^2 E}{\partial z^2} + \frac{2\chi}{c^2}\frac{\partial E}{\partial t} = -\frac{1}{\epsilon_0 c^2}\frac{\partial^2 P}{\partial t^2} \tag{3.1}$$

39

where the electric field vector E is assumed to lie in the x direction, z is the only propagation direction for the (electric) waves and $\chi = \sigma/\epsilon_0$ describes the losses of the electric field (this term arises from the constitutive relation $J = \sigma E$). The matter polarization, $P = D - E = \chi \epsilon_0 E$, acts as the 'driving force' responsible for the sustained oscillations of E. The equation (3.1) has to be solved with appropriated boundary conditions at mirrors delimiting the laser cavity being the simplest resonators the 'ring cavity' (the light travels along the sides of a parallelogram in only one direction) and the Fabry–Perot cavity with opposing mirrors (light travels inside the cavity in counter-propagating directions).

The CO_2 gas is described as a two-state quantum mechanical system (this approximation does not represent how the pumping of the laser is performed except for its main dynamical effects). The Hamilton operator reads $H = H_0 - exE$ (we are assuming that x dependences in E can be neglected in a first approximation). The density operator Γ satisfies Heisenberg's equation:

$$i\hbar \dot{\Gamma} = [\Gamma, H]. \tag{3.2}$$

Consider the unperturbed states as a basis $(H_0|j\rangle = \epsilon_j|j\rangle$; $\hbar\omega_0 = \epsilon_2 - \epsilon_1)$ and assume that the matrix of x in this basis has only one (real, constant, off-diagonal) element $a = \langle 1|x|2\rangle$. Two extra contributions to equation (3.2) will account for the damping and pumping of the state populations. Hence, we have

$$\begin{aligned}
(\dot{\Gamma}_{22} - \dot{\Gamma}_{11}) &= -2i\frac{eEa}{\hbar}(\Gamma_{21} - \Gamma_{12}) + \gamma_1(W_0 - (\Gamma_{22} - \Gamma_{11})) \\
\dot{\Gamma}_{21} &= i\omega_0\Gamma_{21} - i\frac{eEa}{\hbar}(\Gamma_{22} - \Gamma_{11}) - \gamma_2\Gamma_{21},
\end{aligned} \tag{3.3}$$

recalling that Γ_{21} is the lower left element of the 2×2 matrix Γ. W_0 describes a reference population difference and γ_i are the damping constants. These equations were derived by Bloch in a different context (spin resonances).

The description of the field–matter interaction is completed by providing a microscopic model for the gas polarization: $P = ea(\Gamma_{12} + \Gamma_{21})$. The factor a is taken constant since all the atoms were assumed to be identical irrespective of their position in the laser cavity.

Since we are assuming that the field–matter interaction is weak (we are using perturbation theory), the wave amplitudes will change slowly. Thus, in the case of a ring cavity we may represent the electric field by essentially one wave with a slowly varying amplitude:

$$E(z, t) = \sqrt{\frac{\hbar\omega}{2\epsilon_0}}\{b(t)\, e^{i\omega(t-z/c)} + CC\}. \tag{3.4}$$

From equation (3.1), equation (3.3) and equation (3.4), neglecting all terms that do not have the proper space–time exponential factor, we obtain a set of ordinary first-order differential equations for b, $\Gamma_{21} = s\, e^{i\omega(t-z/c)}$ and $W = \Gamma_{22} - \Gamma_{11}$ (g is a suitable constant):

$$\dot{b} = -\chi b + igs \qquad (3.5)$$

$$\dot{s} = -\gamma_2 s + i\theta s - igbW \qquad (3.6)$$

$$\dot{W} = \gamma_1(W_0 - W) + 2ig(s^*b - sb^*). \qquad (3.7)$$

To obtain the first equation one needs to assume that $\partial^2 P/\partial t^2 \sim -\omega_0^2 P$, that $\dot{b} \ll \omega \dot{b}$ and finally that $|\chi| \ll |\omega|$, in other words that the variations due to losses and damping have much larger time-constants than those coming from the characteristic frequency of the CO_2 states.

The quantity $\theta = w_0 - w$ represents the detuning between the atomic frequency, w_0, and the 'cavity frequency' $w = c/k$. For a well tuned laser (a condition often sought in the experiments) $\theta \sim 0$.

Furthermore, when $\gamma_2 \gg \gamma_1$ or χ, the magnitude s decays very fast to the 'slowly varying equilibrium' when compared with the typical time-scale for the variation of W and b. Hence, we can further approximate the system taking $\dot{s} = 0$ and $s = -igbW/\gamma_2$ (this is called *adiabatic elimination*).

The dynamics of this type of laser is thus described by two variables within this approximation: the population inversion, W, that is proportional to the difference between the population of excited and non-excited atoms in the gas; and the intensity $|b|^2$ of the light beam. Using the scaling

$$I = 4\chi\tau|b|^2 \qquad \hat{W} = \frac{2|g|^2\tau}{\gamma_2}(W - \Delta) \qquad \Delta = \frac{\chi\gamma_2}{|g|^2} \qquad (3.8)$$

$$t \to \tau t \qquad \tau = \sqrt{\frac{\gamma_2}{2\gamma_1|g|^2(W_0 - \Delta)}} \qquad (3.9)$$

we arrive at the following set of equations [sola87]:

$$dI/dt = I(\hat{W} - R\,\cos(\omega t)) \qquad (3.10)$$

$$d\hat{W}/dt = 1 - \epsilon_1\hat{W} - (1 + \epsilon_2\hat{W})I \qquad (3.11)$$

where the modulated losses are described by the change $\chi \to \chi + R/(2\tau)\cos\omega t$, $\epsilon_1 = \tau\gamma_1$ and $\epsilon_2 = \frac{|g|^2}{\gamma_2\chi} = 1/\Delta$. In the sequel we will drop the tilde over W.

The physics involved in the simple model given by equation (3.11) is that an increase (decrease) in the population of the upper level above a certain equilibrium value produces an increase in the intensity of the light, while an increase of the intensity produces a depletion of exited atoms. The laser action is sustained by a continuous pumping of atoms from the lower to the upper level (the constant unity in the equation for dW/dt). The remaining terms represent losses of several kinds such as non-radiative decay of the upper level (controlled by ϵ_1) and light leaving the cavity (controlled by ϵ_2).

We can summarize the basic facts described by equation (3.10) and equation (3.11):

(i)   The physical domain of the intensity is $I \geq 0$. The solutions cannot abandon this region.

(ii) The terms proportional to ϵ_1 and ϵ_2 represent linear and nonlinear losses in the various elements of the laser (cavity, mirrors, ...) while the oscillatory term represents the modulated losses.

(iii) Energy is pumped into the system through an electrical discharge here represented by the constant unity.

(iv) The increase in the intensity deploys the excited atoms.

(v)   Excited atoms in excess, relative to a reference value (here represented by 0), will prompt higher-intensity outputs.

(vi) Setting all the terms associated with losses to zero, i.e., $\epsilon_1 = \epsilon_2 = R = 0$, we obtain a conserved quantity $\mathcal{E}$ in the form

$$\mathcal{E} = I - \ln(I) + W^2/2. \tag{3.12}$$

Moreover, if we define $x = \ln(I)$ the equation without losses turns out to be Hamiltonian in the canonical variables (x, W).

3.2.2 Dynamics of the laser

In the absence of the forcing term, i.e., $R = 0$ in equation (3.10), the dynamics of the laser presents two distinguished solutions: the laser-off case, where the output intensity is zero $(I, W) = (0, 1/\epsilon_1)$ and the laser-on solution $(I, W) = (1, 0)$, where a constant output is achieved. The solutions have in common that the system remains for ever at the same values of intensity and population inversion.

There is a second type of distinguished solution: the line $I = 0$ represents the *separatrix* between the physical and unphysical regions of I and contains the trajectories that converge towards the laser-off state for positive times, $(I, W) = (0, 1/\epsilon_1 + A\, e^{-t\epsilon_1})$.

We can further explore the character of the laser-on and laser-off solutions if we consider initial conditions lying 'close' to these solutions. We can tentatively argue that since we are interested in a small region around the stationary solution we can expand the vector field (i.e., the right hand side of equation (3.10) and equation (3.11)) in Taylor series and disregard terms of order higher than one, obtaining a linear vector field of the kind studied in chapter 2.

The linearized equation near the laser-off solution is

$$\begin{pmatrix} di/dt \\ dw/dt \end{pmatrix} = \begin{pmatrix} 1/\epsilon_1 & 0 \\ -(1 + \epsilon_2/\epsilon_1) & -\epsilon_1 \end{pmatrix} \begin{pmatrix} i \\ w \end{pmatrix} \tag{3.13}$$

where i and w are the linear deviations from the laser-off values.

The laser-off state represents a saddle, according to the discussion in chapter 2, with eigenvalues $-\epsilon_1$ and $1/\epsilon_1$. The attractive direction is

$$\begin{pmatrix} 0 \\ 1 \end{pmatrix}$$

and the repelling direction is

$$\begin{pmatrix} 1+\epsilon_1^2 \\ -\epsilon_1(1+\epsilon_2/\epsilon_1) \end{pmatrix}.$$

The laser-on state is a stable focus provided that $0 < (\epsilon_1 + \epsilon_2) < 2$. The linearized equation reads

$$\begin{pmatrix} di/dt \\ dw/dt \end{pmatrix} = \begin{pmatrix} 0 & 1 \\ -1 & -(\epsilon_1 + \epsilon_2) \end{pmatrix} \begin{pmatrix} i \\ w \end{pmatrix}. \qquad (3.14)$$

In order to complete the picture of the model we need to have some global information. This information is provided in this simple case by the following observations:

(i) The flow crosses the $W = 0$ line with increasing population inversion if $I < 1$ and with decreasing population inversion if $I > 1$.

(ii) The energy, equation (3.12) is always *dissipated* ($dE/dt < 0$), except at the stationary solutions.

The result of these two observations is the phase portrait of figure 3.1.

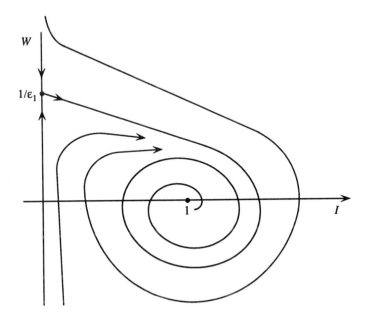

Figure 3.1. Phase portrait for the CO_2 laser without modulation of the losses.

3.3   Duffing oscillator

3.3.1   The model

The original Duffing [duff18] oscillator was introduced in relation to the single (spatial) mode vibrations of a beam subjected to external periodic forces. The model has been extensively used as a prototype of periodically forced systems thanks to its simplicity.

The Duffing equations read

$$
\begin{aligned}
du/dt &= v \\
dv/dt &= -\beta u - u^3 - \mu v + A\ \cos(\omega t)
\end{aligned}
\tag{3.15}
$$

where $\mu \geq 0$ is a damping term and $\beta = \pm 1$. The two values of β correspond to the original problem, $\beta = 1$ [duff18] and to its pedagogical version, $\beta = -1$ [holm79, guck86, wigg90]. Clearly, for $\beta = 1$ dropping the u^3 term we recover the damped–forced harmonic oscillator of standard physics textbooks.

The Duffing oscillator has an important symmetry. If we perform the change of coordinates

$$
\begin{aligned}
u &= -U \\
v &= -V \\
t &= t' - \pi/\omega
\end{aligned}
\tag{3.16}
$$

the equation (3.15) in the new coordinates is identical to the original equation. It follows from this observation that if $(u(t), v(t))$ is a solution of the Duffing equations, then also $(-u(t+\pi/\omega), -v(t+\pi/\omega))$ is a solution of equation (3.15). This 'new' solution may in turn be the same original orbit, if $u(t) = -u(t'+\pi/\omega)$ and $v(t) = -v(t'+\pi/\omega)$, or it may be an entirely different solution. In the first case we will say that the solution is *symmetric* while in the second case we will speak of a pair of *symmetry-related solutions*. Note that a second application of the symmetry transformation equation (3.16) is equivalent to a phase shift of 2π: the point $(u(t + 2\pi/\omega), v(t + 2\pi/\omega))$ belongs to the same original trajectory, i.e., the square of the symmetry transformation gives the identity transformation when full trajectories are considered.

Exercise 3.1:   Verify the existence of the symmetry and also that if $(u(t), v(t))$ is a solution for the Duffing system, then $(-u(t+\pi/\omega), -v(t+\pi/\omega))$ is also a solution.

3.3.2   Dynamics in the unforced model

We consider the case when there is no forcing action exerted onto the beam $(A = 0)$. Then we can analyse the behaviour of the solutions in the (u, v) plane.

Just as in the CO_2 laser system, when $\mu = 0$ we have a Hamiltonian problem with an associated energy

$$E = v^2/2 + (u^4/4 + \beta u^2/2). \qquad (3.17)$$

The flow has one (if $\beta = 1$), or three (if $\beta = -1$), *equilibrium points* (equilibrium point is just another name for fixed point!), given by

$$(u, v) = (0, 0)$$
$$(u, v) = (\pm 1, 0) \qquad \text{iff } \beta = -1.$$

For $\beta = 1$ the point $(0, 0)$ represents the minimum of the energy. For $0 < \mu < 2$, all the orbits spiral down, i.e., with decreasing energy, towards the equilibrium point.

For $\beta = -1$ and $\mu = 0$ the symmetric solution $(0, 0)$ is a relative maximum of the energy and is unstable, of saddle type. The pair of symmetry-related equilibrium points $(\pm 1, 0)$ are stable foci or nodes associated with the minimum of the energy.

Exercise 3.2: Verify the stability type of the equilibrium points by linearizing the flow around them.

The global picture of the Duffing oscillator without forcing is represented in figure 3.2. The flow spirals towards one of the foci for almost all trajectories. There are two special trajectories, *separatrixes*, that terminate at the saddle point and divide the basins of attraction of the stable foci: they are associated with the stable directions of the saddle.

3.4   The Lorenz equations

The celebrated Lorenz equations [lore63] are related to the convective movements of the atmosphere and to weather forecasting. However, their fame does not come from their success as a meteorological model but rather from historical reasons: the Lorenz model made completely clear that the possibilities of forecasting depend not only on the physical (biological, economical, ...) quality of the model and the precision of the initial data but also on qualitative aspects of the dynamics. The rediscovery of the type of motion that we now call *chaotic* initiated a new era of studies in nonlinear dynamics[†].

[†] The Lorenz model has reappeared in the context of laser physics. There is a correspondence between the equations describing certain lasers working in a single spatial mode and the Lorenz equations [hake75]. Since the laser is described by five variables, $(E, P, W) \in \mathbb{C}^2 \times \mathbb{R}$, the correspondence is established only for a subset of initial conditions (those with E and P real) and for perfectly tuned lasers. See [abra95] for a recent discussion of the model and [tang91, weis95] for experimental results.

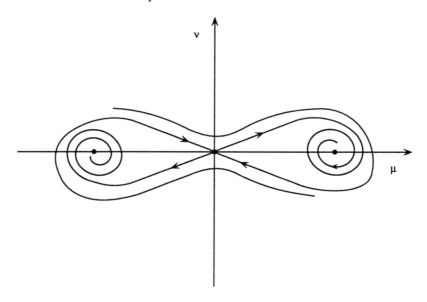

Figure 3.2. Phase portrait of the unforced Duffing oscillator for $\beta = -1$.

3.4.1 The model

The approximations to Navier–Stokes equations leading to the Lorenz system remind us of the Bénard experiment. However, the Lorenz equations are insufficient to describe all the dynamical features observed in the experiment.

Let us consider a thin layer of fluid enclosed between two horizontal surfaces kept at constant temperature, with full translational symmetry in one horizontal direction and periodicity length l in the other. We choose a coordinate system so that the system has the translational symmetry along the y axis (i.e., all dynamical variables are y independent). The lower surface ($z = 0$) is kept at temperature $T_0 + \Delta T$ while the upper one ($z = h$) has (cooler) temperature T_0. Recalling Bénard's experiment, we want to describe the rotating motion of the xz blocks (rolling motion).

The fluid is described with the Navier–Stokes and continuity equations, while the diffusion–convection heat equation is used for the temperature profile:

$$\frac{\partial v}{\partial t} + (v \cdot \nabla)\, v = \frac{1}{\rho}\left(F - \nabla p + \mu\, \nabla^2 v\right)$$

$$\frac{dT}{dt} = \frac{\partial T}{\partial t} + v \cdot \nabla T = \kappa\, \nabla^2 T \tag{3.18}$$

$$\frac{\partial \rho}{\partial t} + \nabla \cdot (\rho v) = 0.$$

Following Saltzman [salt62] we assume that the density of the fluid will vary only linearly with T: $\rho = \bar{\rho}(1 - \gamma(T - T_0))$. Furthermore, we will approximate $\rho \to \bar{\rho}$ everywhere except in the force term $\mathbf{F} = -\rho g \hat{\mathbf{z}}$. We also define the variable θ describing the deviation of temperature from a (vertical) linear profile:

$$T(x, z, t) = T_0 + \Delta T \left(1 - \frac{z}{h}\right) + \theta(x, z, t). \tag{3.19}$$

The continuity equation (under the approximation $\rho = \bar{\rho}$) is automatically fulfilled by defining a scalar function ψ such that

$$v_x = -\frac{\partial \psi}{\partial z}$$

$$v_z = \frac{\partial \psi}{\partial x}.$$

Taking the curl of the Navier–Stokes equation, we obtain a pair of scalar equations for θ and ψ:

$$\frac{\partial}{\partial t} \nabla^2 \psi = -\frac{\partial(\psi, \nabla^2 \psi)}{\partial(x, z)} + v \nabla^2(\nabla^2 \psi) + g\gamma \frac{\partial \theta}{\partial x} \tag{3.20}$$

$$\frac{\partial \theta}{\partial t} = -\frac{\partial(\psi, \theta)}{\partial(x, z)} + \frac{\Delta T}{h} \frac{\partial \psi}{\partial x} + \kappa \nabla^2 \theta \tag{3.21}$$

where $v = \mu/\bar{\rho}$ (it was called v in chapter 1) and

$$\frac{\partial(a, b)}{\partial(x, z)} = \frac{\partial a}{\partial x} \frac{\partial b}{\partial z} - \frac{\partial b}{\partial x} \frac{\partial a}{\partial z}. \tag{3.22}$$

Expressed in terms of ψ and θ the boundary conditions describing the rolling motion of the Bénard blocks require that $\theta = 0$ in the upper and lower surfaces as well as in the centre of the block, while the velocities (derivatives of ψ) at the edge of the block have to be tangent to the edges (no flux of fluid to the neighbouring blocks). These conditions are fulfilled by the following low-order Fourier expansions [lore63]:

$$\psi(x, z, t) = \frac{\kappa(1 + a^2)}{a} X(t) \sin \frac{\pi a x}{h} \sin \frac{\pi z}{h}$$

$$\theta(x, z, t) = \frac{\pi R}{R_c \Delta T} \left(Y(t) \cos \frac{\pi a x}{h} \sin \frac{\pi z}{h} - \frac{Z(t)}{2} \sin \frac{2\pi z}{h} \right)$$

where $R = g\gamma h^3 \Delta T/(\kappa v)$ is the Rayleigh number, $R_c = \pi^4(1 + a^2)^3/a^2$ and $a = h/l$ is the aspect ratio.

Substituting the above expressions into the differential equations and neglecting all higher-order terms in the trigonometric functions, one obtains

$$dX/dt = \sigma(Y - X)$$
$$dY/dt = rX - Y - XZ \tag{3.23}$$
$$dZ/dt = XY - bZ$$

where $\sigma = \nu/\kappa$ is the Prandtl number (in chapter 1 we used P instead), $r = R/R_c$ and $b = 4/(1 + a^2)$ (in Lorenz' paper $\sigma = 10$ and $b = 8/3$). Time has been rescaled with a factor $\pi^2 \kappa (1 + a^2)/h^2$.

3.4.1.1 *Properties*

The Lorenz model also presents a reflection symmetry. The equations (3.23) are invariant under the transformation

$$X = -x \qquad (3.24)$$
$$Y = -y \qquad (3.25)$$
$$Z = z \qquad (3.26)$$

which leaves the Z axis invariant. This symmetry constrains the trajectories either to lie completely on the Z axis or to have no point belonging to this axis. The solutions for trajectories lying on the Z axis are

$$X = Y = 0 \qquad (3.27)$$
$$Z = e^{-bt} z_0. \qquad (3.28)$$

The Lorenz flow always contracts volumes in the following sense: if we consider an infinitesimal volume $V = (dX \, dY \, dZ)$ at X, Y, Z and let this volume evolve following the flow during an infinitesimal time dt, the resulting change in volume will be

$$dV/dt = -V(t)(\sigma + 1 + b)) \qquad (3.29)$$

which is always a decreasing function of time.

Exercise 3.3: Show that for a system of first-order ordinary differential equations

$$dx_i/dt = f_i(x_1, ..., x_n) \qquad i = 1, ..., n \qquad (3.30)$$

infinitesimal volumes change at a rate

$$dV/dt = V(t) \, \text{Tr}(Df) \qquad (3.31)$$

where Tr is the trace operator and Df is the matrix of derivatives with elements $(Df)_{ij} = df_i/dx_j$.
Hint: (i) Use that the volume of a slanted cube is the determinant of vectors determining the cube sides. (ii) The exercise is discussed in [arno89, pp 68–70].

3.4.2 Simple dynamics in the Lorenz model

The study of the number, stability and relations of the fixed points in the Lorenz equations will provide the basic ground for further studies. It is convenient to perform this study for increasing values of the control parameter r.

There is one symmetric fixed point present for all values of r and a pair of symmetry-related fixed points for $r > 1$

$$(X, Y, Z) = \begin{cases} (0, 0, 0) \\ (x_0, x_0, x_0^2/b), \quad x_0 = \pm(b(r - 1))^{1/2}. \end{cases} \tag{3.32}$$

The symmetric fixed point changes stability at $r = 1$, being a stable sink for $r < 1$ and a saddle for $r > 1$. For $r = 0$ its stability is not defined in the linearized flow. In this case, the equilibrium point is degenerate since the three fixed points present for $r > 1$ coalesce into one point at this parameter value. No further changes affect the symmetric fixed point.

The stability of the pair of symmetry-related equilibrium points is described by the matrix of derivatives

$$D = \begin{pmatrix} -\sigma & \sigma & 0 \\ 1 & -1 & -x_0 \\ x_0 & x_0 & -b \end{pmatrix} \tag{3.33}$$

with $x_0 = \pm(b(r - 1))^{1/2}$. Note that the eigenvalues of D depend on x_0^2, being thus the same for both fixed points. The stability of these fixed points changes with the value of r. For $\sigma = 10$ and $b = 8/3$ the change occurs as follows [spar82]: for $1 < r < 1.3456...$, D has three (real) negative eigenvalues; for $1.3456... < r < 470/19 = 24.7368...$ there is one real, negative eigenvalue and two complex conjugate eigenvalues with negative real parts; for $470/19 < r$ there is one negative eigenvalue and two complex conjugate eigenvalues with positive real parts. The flow types are illustrated in figure 3.3.

Although the flows illustrated in figure 3.3 can be deduced analytically, at this point of the exposition we are not prepared to perform this study (we have not presented the necessary tools yet). For the time being, the additional information can be considered as *de facto* information provided by a 'numerical experiment'.

3.5 Summary

The systems presented in this chapter are intrinsically nonlinear. By this we mean that it is not possible to transform any of them into a linear system with a suitable change of coordinates. The simplest verification of this statement relies on the existence of *several* isolated fixed points for all three systems, a feature impossible to achieve with linear systems, as pointed out in the previous chapter.

Basically, the three problems presented, despite their algebraic simplicity, cannot be easily classified in any form allowing to have an *a priori* knowledge

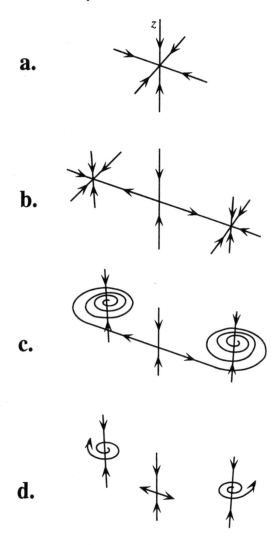

Figure 3.3. Simple Lorenz flows. (a) $r < 1$; (b) $1 < r < 1.246$; (c) r slightly above 1.246; (d) $24.74 < r$.

of the details of the dynamics. For example, while a harmonic oscillator can be well understood by a physicist just by 'looking' at its dynamical equation (the oscillation frequency is displayed in the equation), it is a much harder task to identify the wild oscillations that the Lorenz system may display (we will discuss the Lorenz oscillations in chapters 8 and 16).

The reader can now give a more elaborated answer to the question at the end of chapter 2: Why are nonlinearities necessary?

The examples in this chapter are built from the most basic building blocks, the very same building blocks that make up all low-dimensional dynamical systems. To identify these blocks and to make the most of them is the task of the coming chapters. We mention some topics that will deserve further attention:

- What is the relationship between stability change and the appearance of new equilibrium points in the Lorenz system? Is there a general rule?
- What is the relationship between the loss of stability of the focus at $r = 470/19$ in the Lorenz system and the disappearance of periodic orbits?
- What is the validity of the procedure we have called 'linearization'?
- What is the nature of separatrixes (between basins of attraction)? Are they always made of a countable number of (complete) trajectories?

Exercise 3.4: Compose a list of your own doubts and spend some time trying to answer them.

Moreover, some of the specific features of our systems will be the starting point to develop fundamental tools of the theory of nonlinear dynamics. For example, the Duffing oscillator and the Lorenz model have a reflexion symmetry that more sooner than later will prove to be relevant. The laser system is closely related to a Hamiltonian problem in which the period of the oscillations increases with the energy (this feature, however, is not as evident as the symmetries of the other problems). We will address symmetries in general in chapter 15, some features of the Duffing oscillator and laser will be considered in chapter 14 and the symmetry of the Lorenz system will be addressed in chapter 16.

In short, we can say that for every model representing a natural phenomenon (physical systems, for example), knowledge of the nature (physical) of the problem and its particularities can (and should) be transformed in a path towards a deeper understanding of the phenomenon.

3.6 Additional exercises

Exercise 3.5: Let

$$x' = \mu x - (x^2 + 2y^2)x + by \cos(z + \theta) \tag{3.34}$$
$$y' = \mu y - (2x^2 + y^2)y + bx \cos(z - \theta) \tag{3.35}$$
$$z' = \alpha(x^2 - y^2) - b(x/y \; \sin(z - \theta) + y/x \; \sin(z + \theta)) \tag{3.36}$$

with the variables $(x, y, z) \in \mathbb{R}^+ \times \mathbb{R}^+ \times \mathbb{S}^1$, and the parameters $(b, \theta) \in \mathbb{R}^+ \times \mathbb{S}^1$.

(a) Find the fixed points of these equations located at $x = y$.
(b) Linearize the vector field around these fixed points. For which parameter values does the linearized vector field have two zero eigenvalues?

Exercise 3.6: As we have discussed throughout this chapter, ordinary differential equations in physics usually arise after a *sensible* truncation of the number of active modes of a partial differential equation (see the discussion of the Lorenz example). In this exercise (adapted from a discussion in [mann91], chapter 5) we will train our intuition in this 'truncation routine'. Let

$$v_t + vv_x = av - (v_{xxxx} + 2v_{xx} + v) \tag{3.37}$$

for v a scalar field (i.e., a function of x and t) such that $v(t,0) = v_{xx}(t,0) = v(t,l) = v_{xx}(t,l) = 0$ (subindices denote partial derivatives), and a a real parameter. We are interested in studying the dynamics of v for a range of parameters such that v is small.

(a) Find the eigenfunctions of the linear part of this equation, looking for solutions of the form

$$v_i = e^{\lambda_i t} f(k_i x). \tag{3.38}$$

For which values of k_i do the solutions satisfy the boundary conditions?
(b) Let us pay attention to the temporal eigenvalue associated to a *mode* v_i. Find the curves in the (a, l) parameter space in which λ_i is zero. Find the regions in the parameter space in which λ_i is greater than zero and smaller than zero. Repeat this procedure for $i = 1, 2, 3, 4$.
(c) Assume that we can expand v as

$$v = \sum_{i=1}^{i=\infty} X_i f(k_i x). \tag{3.39}$$

Find the set of ordinary differential equations ruling the behaviour of X_i.
(d) Notice that there is a curve in parameter space where the real part of the eigenvalue for the mode $i = 1$ is zero. For parameter values slightly above this curve (called the *curve of marginal stability*), find a *sensible* truncation of the equations. Find the fixed points of the truncated system.
Hint: Keeping just two modes, the system reads as follows:

$$A_1' = \epsilon A_1 + \tfrac{1}{2} A_1 A_2 \tag{3.40}$$

$$A_2' = (\epsilon - 9) A_2 - A_1^2. \tag{3.41}$$

(e) Repeat the study with the boundary conditions changed to $v(t,0) = v_{xx}(t,0) = 0$ and $v(t,l) = v_{xx}(t,l) = \epsilon$.

Chapter 4

Elements of the description

'Force, for example, is not a metaphysical conatus of an unknown kind which hides behind its effects (accelerations, deviations, etc.); it is the totality of these effects. Similarly an electric current does not have a secret reverse side; it is nothing but the totality of the physical–chemical actions which manifest it (electrolysis, the incandescence of a carbon filament, the displacement of the needle of a galvanometer, etc.). No one of these actions alone is sufficient to reveal it. But no action indicates anything which is *behind itself*; it indicates only itself and the total series.

[...] That is why we can equally reject the dualism of appearance and essence. The appearance does not hide the essence, it reveals it; it *is* the essence.

[...] But the appearance, reduced to itself and without reference to the series of which it is part, could be only an intuitive and subjective plenitude, the manner in which the subject is affected. If the phenomenon is to reveal itself as *transcendent*, it is necessary that the subject himself transcend the appearance toward the total series of which it is a member.'

Jean-Paul Sartre [sart66]

4.1 Introduction

In the previous chapters, we have introduced some of the elements pertaining to the analysis of a dynamical system mostly from an intuitive point of view. Although intuition will always be essential for our job, it will play a lesser role when we enter less familiar regions of dynamics and we will have to rely increasingly on pure reason (the reader can generalize this statement and make a rule of it: for example our (regular) intuition is of little help—and even the source of trouble—in areas like relativity, since we do not have an intuition fed with relativistic facts).

The guiding line of this chapter is to convert our gained experience and intuition into less informal elements that will be at the basis of our reasoning.

The formalization will often be done defining the important elements of analysis. Note that the full process of abstraction will be concentrated in the art–craft–science of the definitions.

4.2   Basic elements

4.2.1   Phase space

We have so far represented the dynamics by means of differential equations relating the variables that describe the system with time and with the parameters that are kept constant during the evolution (we leave aside for a moment the time-dependent coefficients).

The *phase space* is the space of definition of the (dependent) variables that describe the system. To fix ideas, in the dynamical system

$$\dot{x} = f(x) \tag{4.1}$$

x is a point in phase space whose time-evolution is described by the *vector field* f that defines the time-derivatives of x.

We will assume that phase space is locally equivalent to an Euclidean space $\mathbb{R}^n$ of dimension n, i.e., that for every point in phase space it is possible to find a neighbourhood and a map that carries the neighbourhood of the point into an open set of $\mathbb{R}^n$. In other words, the phase space is a *manifold*.

Not only the local structure of the phase space is relevant but also the global structure, i.e., the manifold as a whole.   Consider for example a differential equation representing the dynamics on a circle and the same equation representing the dynamics on the real line. The two spaces are locally isomorphic but the global behaviour of the system will be rather different. In the case of the circle the time-asymptotic (long-time) behaviour could be a fixed point or a periodic orbit while in the real line it would be a fixed point or an unbounded evolution towards $\pm\infty$.

The case in which the coefficients depend on time can be cast into the same terms introducing an additional variable with constant derivative. For example, the Duffing equations can be written as

$$du/dt = v \tag{4.2}$$
$$dv/dt = -\mu v + u - u^3 + R\,\cos(\omega t) \tag{4.3}$$

or

$$du/dt = v \tag{4.4}$$
$$dv/dt = -\mu v + u - u^3 + R\cos(\phi) \tag{4.5}$$
$$d\phi/dt = \omega. \tag{4.6}$$

In this example the variable ϕ can be considered modulo 2π since the dynamics only depends on ϕ mod (2π). The phase space in this case is $\mathbb{R}^2 \times \mathbb{S}^1$.

4.2.2 Flow

The solutions of a system of differential equations can be regarded as a function that given a time, t, and a point in phase space, $x \in M$, produces a unique point x_t in phase space (the uniqueness of the solution of differential equations is guaranteed if the system fulfills some physically reasonable smoothness condition, such as being *locally Lipschitz*[†]). Formally, the *flow*, F, is defined as the continuous function

$$F : M \times \mathbb{R} \to M \tag{4.7}$$

$$F(x, t) = x_t \qquad \text{for all } x \in M \text{ and } t \in \mathbb{R}. \tag{4.8}$$

x can be regarded as the initial condition at e.g. $t = 0$, and x_t is the time evolution of this initial condition after a period of time t.

Some differential equations are *not* associated with flows. They produce unbounded images for some finite initial condition and time. In such a case it is said that the equation *blows up* in finite time. The standard example for this class of equations is

$$\mathrm{d}x/\mathrm{d}t = x^2. \tag{4.9}$$

The solution $x(t) = x_0/(1 - tx_0)$ diverges for $t = 1/x_0$.

This kind of equation is rare in natural sciences where the proper nature of the problems prevents such 'pathological' behaviour. In the rest of this book we are not going to be concerned with this possibility and will assume the existence of a flow associated with all the equations of physical (or natural) interest.

It is clear that for *zero time* an initial condition 'evolves' into itself:

$$F(x, 0) = x. \tag{4.10}$$

Also, the evolution of an initial condition x, associated with an autonomous equation, during a time t can be split into two parts (and these into as many parts as we want) by composing the evolution during times t_1 and t_2 (such that $t_1 + t_2 = t$). The result is stated in the following property:

$$F(x, t) = F(F(x, t_1), t_2) = F(F(x, t_2), t_1). \tag{4.11}$$

In the special case when $t = 0$ the last equation reads $F(x, 0) = x = F(F(x, t'), -t') = F(F(x, -t'), t')$ which shows that $F(\cdot, -t)$ is the *time-reversed* or *inverse* flow.

Exercise 4.1: Show that the properties (2.32), (2.33), (2.34), (2.39) of the evolution operator U, defined for Hill's equation in chapter 2, are direct consequences of the general properties of flows.

[†] A Lipschitz function f satisfies $\|f(x) - f(y)\| < K\|x - y\|$ for some finite K. K is called the Lipschitz constant for f.

Note that we will quite often abuse notation and apply the flow to a set. The expression $F(S,t)$, with S a set, simply denotes shorthand for the set $F(x,t)$, for all $x \in S$.

Exercise 4.2: Show that

$$F(A,t) \cup F(B,t) = F(A \cup B,t) \qquad (4.12)$$
$$F(A,t) \cap F(B,t) = F(A \cap B,t). \qquad (4.13)$$

4.2.3 Invariants

4.2.3.1 Orbits

Our study of dynamical systems in chapter 2 was based on the idea of identifying special trajectories or orbits in the flow. We call the set

$$O(x) = \{y \in M \text{ such that } y = F(x,t) \text{ for some } t \in R\} \qquad (4.14)$$

an *orbit*, $O(x)$, through the point x.

Exercise 4.3: Orbits divide phase space into classes of equivalence according to the relation $x \equiv y$ iff $O(x) = O(y)$. Show that this relation is reflexive, symmetric and transitive.

The most relevant property of the orbits is their invariance with respect to the flow. To be precise, we say that a set, $S \subseteq M$, is *invariant*—with respect to the flow F, but this part usually goes without saying and is clear from the context—when for all $x \in S$ and all $t \in R$, $F(x,t) \in S$, i.e. the image of S by the flow coincides with S, $F(S,t) = S$.

The union and intersection of invariant sets are invariant sets. We verify immediately that all orbits are invariant and that every invariant set either contains complete orbits, $O(x) \cap S = O(x)$, or no point of an orbit, $O(x) \cap S = \emptyset$.

There are special classes of orbits that are very useful in the discussion of the dynamics and deserve special attention. In what follows we will focus on the simplest and most relevant cases.

Fixed points. We have already encountered in the examples the notion of fixed points or equilibrium points as trajectories that evolve into themselves. In the present context we can say that a *fixed point* is an orbit that has a single element.

Exercise 4.4: Show that the two definitions are equivalent.

Periodic orbits. An orbit is said to be *periodic* if, after a finite time of evolution, the trajectory goes back to the initial point. In symbols, $O(x)$ is periodic if $F(x, T) = x$ for some $T > 0$. T is called the *period* of the orbit. Clearly, if after a time T the position in phase space equals the starting point, the same will happen after a time nT, for n integer. If an orbit is periodic with period T it is also periodic with period nT. We usually associate the word period with the *minimum period*, i.e., the minimum $T > 0$ such that $F(x, T) = x$.

According to the definition, the period of $O(x)$ seems to depend on x. In fact, the period characterizes the orbit regardless of the chosen point x. This intuitive idea can be verified with the definition.

Exercise 4.5: Show that if $y \in O(x)$ and $F(x, T) = x$ then $F(y, T) = y$.

Quasiperiodic orbits. The kind of orbit which we call quasiperiodic is frequent in physics. Everybody familiar with completely integrable systems in classical mechanics [gold80, land69] has been in contact with this type of motion. The simplest example that can be given consists of two, sort of, uncoupled harmonic oscillators

$$dx_1/dt = \Omega_1 p_1 \tag{4.15}$$
$$dp_1/dt = -\Omega_1 x_1 \tag{4.16}$$
$$dx_2/dt = \Omega_2 p_2 \tag{4.17}$$
$$dp_2/dt = -\Omega_2 x_2. \tag{4.18}$$

We can let the frequencies depend on the energies, thus moving out of the linear realm:

$$\Omega_1 = \Omega_1(x_1^2 + p_1^2, x_2^2 + p_2^2) \tag{4.19}$$
$$\Omega_2 = \Omega_2(x_1^2 + p_1^2, x_2^2 + p_2^2). \tag{4.20}$$

In order to have a closed, periodic orbit both oscillators must have the same period (not necessarily the same minimum period). Given the relation of the period with the frequency, the requirement reads

$$n\Omega_1 - m\Omega_2 = 0 \tag{4.21}$$

where $n \neq 0$ and $m \neq 0$ are integers. In such a case the frequencies are said to be *commensurate*.

Orbits with incommensurate frequencies densely fill a two-dimensional torus imbedded in the four-dimensional Euclidean space. Such a motion is the perfect image of *quasiperiodicity*.

In general, a quasiperiodic orbit will fill a bi-dimensional surface isomorphic to a torus (i.e., a deformed torus).

4.2.4 Attractors

4.2.4.1 Attracting sets

For the CO_2 laser, the Duffing oscillator and the Lorenz system we found that after sufficiently long time, most orbits approached some invariant sets (fixed points or periodic orbits). We can call these asymptotic states 'attractors' in the sense that they seem to 'attract' other trajectories.

Two elements are essential to an attracting set: first the attracting set is a closed invariant set; second, there has to be *something* that is attracted: we expect at least orbits starting nearby the attracting set to be attracted to it. The simplest definition of *attracting set* is as follows: we call a set, A, an attracting set if it is a closed invariant set and in addition there is a neighbourhood $U(A)$ such that for every $x \in U(A)$, $F(x, t) \in U(A)$ for positive times and $\lim_{t \to \infty} F(x, t) \to A$.

This definition would be fine if only we knew what $\lim_{t \to \infty} F(x, t) \to A$ meant. The expression has a clear sense if A is a point, but what does it mean for more complex attracting sets? We can make sense of the expression if we say what the *distance to a set* is.

The distance from a point x to a set S is the minimum distance from x to the points in S, $d(x, S) = \inf_{y \in S} d(x, y)$. For a sequence x_n to have a set S, as limit it is required that the distance from points in the sequence to the set goes to zero as n goes to infinity, i.e., $\lim_{n \to \infty} d(S, x_n) = 0$.

Exercise 4.6: What would you say a *repelling set* is? (Hint: see [guck86, p 34]).

Exercise 4.7: Show that the union and intersection of attracting sets are attracting sets.

A classical example shows that the idea of attracting set, although useful, does not contain all the elements that most of us intuitively expect. Consider the bi-dimensional flow of figure 4.1. There is a region U that can be recognized as the neighbourhood of the definition above. There are also three fixed points, one of saddle type and two attractive foci (stable nodes would do as well). Individual trajectories end up in the vicinity of one of the foci while there are two trajectories that end up at the saddle. We can recognize several attracting sets in this flow, namely, both foci (each by itself), their union and the line going from one focus to the other passing through the saddle.

4.2.4.2 Attractors

Intuitively, only the foci in the previous example constitute what most people would call an 'attractor', showing that our definition of attracting set fails to

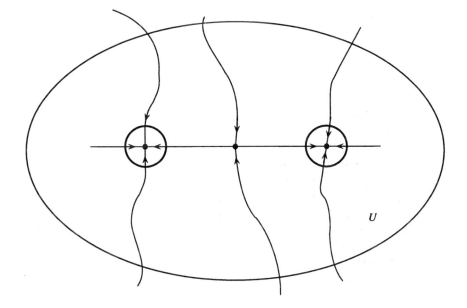

Figure 4.1. Attracting sets and attractors.

meet all the (unstated) intuitive requirements.

What is needed is to incorporate into our formalization of attractor the idea that *every individual orbit should get infinitesimally close to every point in the attractor.* This property can be formalized as follows: a closed invariant set, A, is said to be *topologically transitive* if for any two open[†] sets, $U \subseteq A$ and $V \subseteq A$, there exists a time $t \in R$ such that $F(U, t) \cap V \neq \emptyset$. An *attractor* is a topologically transitive attracting set.

Exercise 4.8: Show that in the previous example the only attractors are the two foci.

4.2.4.3   Basin of attraction

Each attractor, A, posses an associated *basin of attraction* which is the set of points with orbits eventually entering the attracting neighbourhood of the attractor. As such, the basin of attraction can be built as the infinite union of pre-images of the neighbourhood (basin of $A = \bigcup_{t=0}^{\infty} F(U, -t)$).

[†] Note that we are using the *relative topology*, relative to the invariant set A. i.e., the open set U—in the relative topology—is the intersection of an open set of the whole phase space with A.

Exercise 4.9: Show that the basin of attraction is an invariant set.

4.2.4.4 Boundaries of the basin of attraction

In cases like the Duffing oscillator, the Lorenz system and the forced CO_2 laser, in which we find more than one attractor, there exists an invariant set that includes those points in phase space that are in the limit between the two basins of attraction. We call those points the *boundary between the basins of attraction*.

A distinction should be made between the boundary *of* a basin of attraction and the boundary *between* basins of attraction. The former refers to a single basin of attraction. The basin boundary does not necessarily limit with two (or more) different basins. For example, in the CO_2 laser without modulation, there is only one attractor, i.e., the stable focus (for appropriate ϵ_1, ϵ_2). If we consider the physical phase space $(\mathbb{R}^+ \cup \{0\}) \times \mathbb{R}$, the line associated with zero intensity output $I = 0$ is *not* in the basin of attraction of the focus but is an accumulation point of elements in the basin, i.e., it is in the basin boundary.

On the other hand, the boundary between two or more basins of attraction is the common boundary of these attractors, i.e., the set of points that simultaneously belong to the closure of all the considered basins of attraction.

Although this example can be somehow challenged by extending phase space to the non-physical region and introducing an attractor at $-\infty$ (i.e., by 'constructing' a new system having the line $I = 0$ in (the closure of) another basin of attraction), we will find instances of complex chaotic transients where the distinction between 'boundary of a basin' and 'boundary between basins' cannot be circumvented [esch89].

4.2.5 Trapping regions

The vicinity of an attractor is an example of a *trapping region*: the closure U of a (bounded) connected open set, of phase space such that $F(U, t) \subset U$ for non-negative times.

A trapping region always contains at least one attracting set. The set

$$\bigcap_{t>0} F(U, t) = A \qquad (4.22)$$

is the *maximal attracting set* in U, i.e., the only attracting set in U that is not a part of a larger attracting set.

Note that the maximal attracting set has to be connected [hale88]. The region U is compact. The continuity of the flow assures that all forward images $F(U, t)$ are compact and connected. Hence, A is also compact. If it were not connected, one could cover A with two open disjoint sets $V, W \in U$. Clearly, for any $t > 0$ one can find points $x_t \in F(U, t)$ which are *not* in $V \cup W$. Any accumulation point of these x_t is still not in $V \cup W$, but has to be in A. Hence, A is connected.

4.2.6   Stable and unstable sets

An orbit of saddle type has no basin of attraction simply because it is not an attractor (it has an associated repelling direction!). However, we have seen that any orbit initiated with zero intensity in the example of the CO_2 laser approaches (for positive times) the saddle point that represents the laser-off state, as close as we are willing to wait. The same is true for the z axis and the symmetric fixed point in the Lorenz system.

Consider the unforced Duffing oscillator. Where do the points in the boundary between the basins of the two attractors go asymptotically? They cannot end up arbitrarily close to one of the attractors since they do not belong to the basin of attraction, and they cannot move towards infinity because that will represent an increase of energy along a trajectory. Finally, by the same energy argument they have to go asymptotically towards an equilibrium point where the energy no longer changes. The only such point available apart from the attractors is the saddle. We conclude that the saddle is necessarily in the common boundary between the two basins of attraction and is the limit point for orbits belonging to the boundary.

These examples motivate the definition of stable and unstable sets.

Stable set of an invariant set or *inset*: is the set of orbits that have the invariant set as their infinite-time limit for positive times ($t \to \infty$).

Unstable set or *outset*: is the set of orbits that have the invariant set as their infinite-time limit for negative times ($t \to -\infty$).

Exercise 4.10: Show that the inset and outset are invariant sets.

It is worth noticing that an attractor has the empty set as its unstable set and the basin of attraction as its stable set.

The stable set of saddle type orbits is frequently associated with basin boundaries as in the preceding example.

Exercise 4.11: *Stable, centre and unstable subspaces.* Consider an autonomous linear system in the form $dx/dt = Ax$. Show that the stable and unstable sets are in fact subspaces expanded by the generalized eigenvectors of the matrix A with real part smaller and larger than zero respectively. Note that when A has an eigenvector with associated eigenvalue having real part equal to zero, then there exists also a centre subspace.

4.3 Poincaré sections

4.3.1 Stroboscopic section

Studying Hill's equations we have noticed that it is enough to know the monodromy matrix U in order to establish the long-time behaviour of the solutions. This result has two relevant elements: first, the information can be collected in a matrix only because of the linear character of the flow; second, the necessary information can be collected in the form of an evolution over one period of the forcing. The reason for this second property has to do with the periodicity of the vector field (the fact that the vector field is periodic does not necessarily imply that the individual orbits are periodic!).

In the case of periodically forced systems, like the general case of the CO_2 laser and the Duffing equation, the behaviour of the flow can be studied with a *stroboscopic map*.

Given a time-periodic flow F,

$$x(t) = F(x, \phi, t) \tag{4.23}$$

$$\phi(t) = \phi + t \in \mathbb{S}^1 \tag{4.24}$$

$$F(x, \phi, t) = F(x, \phi + 2\pi, t) \tag{4.25}$$

the long-time evolution of the orbits is essentially dictated by the map

$$x' = F(x, \Phi, 2\pi) \equiv \hat{F}(x) \tag{4.26}$$

for an arbitrary, but fixed, phase Φ. For every time t and phase ϕ_0 we can divide the evolution in three steps: the first step goes from time zero to the first intersection with the surface at $\phi = \Phi$ ($\Delta t = (\Phi - \phi_0) \mod (2\pi)$); the second step consists of the evolution for a time $n2\pi$ with $n = [(t - \Delta t)/(2\pi)]$; the last step consists of the remainder of the evolution time, starting at the surface at $\phi = \Phi$ and evolving for a time $t' = t - \Delta t - n2\pi$

$$x(t) = F(\hat{F}^n(F(x, \phi_0, \Delta t)), \Phi, t') \tag{4.27}$$

where $\hat{F}^n = \hat{F}(\hat{F}(\ldots \hat{F}()\ldots))$, n times. Apart from the (somewhat limited) effects in the first and third steps, it is clear that on increasing the time n will increase and thus the iterations of the map $\hat{F}$ will determine the asymptotic behaviour of the solutions. The map $\hat{F}$ is called the *stroboscopic map* and represents the evolution of an initial condition on the surface $\phi = \Phi$ arriving to a new point on the same surface after a *fly time* of 2π in normalized units (one period of the forcing term).

Both the Duffing equations and the laser equations can be treated in this form. For example, periodic orbits of the systems become periodic orbits of the stroboscopic map, figure 4.2, with a period that can be defined with respect to the stroboscopic section as the minimum integer, n, such that $\hat{F}^n(x) = x$. The point x will be called a *periodic point*, i.e., an intersection of the periodic orbit with the stroboscopic section. The case of period one sometimes is referred to as a *fixed point*. Note that fixed points of maps and flows are certainly not equivalent invariant sets, and they are usually distinguished by context.

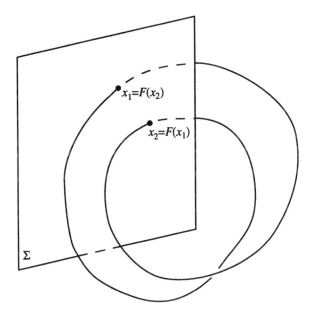

Figure 4.2. Periodic orbits and the Poincaré map.

4.3.2 Transverse section

Stroboscopic sections are not the only way of associating a map with a flow. Let us go back to the laser without forcing and study the characteristics of the $W = 0$ surface. It is easy to realize from the equations that the orbits always cross this surface with non-zero velocity in the interval $I \in [0, 1)$, since

$$dW/dt = 1 - I > 0 \tag{4.28}$$

on the surface. We say that the flow intersects transversely the $W = 0$ plane.

In general, we say that two manifolds *intersect transversely* at a point when the tangent planes to the manifolds expand the full space.

Exercise 4.12: Show that a curve tangent to a plane at $x = 0$ is not a transversal intersection. Show that two planes that intersect along a line have a transversal intersection.

Exercise 4.13: Show that the stroboscopic section $\phi = \Phi$ is a transverse section.

Exercise 4.14: Show that if $\hat{n}(x)$ is the normal to a surface at the point x, the condition of transversality reads $\hat{n}(x) \cdot f(x) > 0$ for all x in the surface (or $\hat{n}(x) \cdot f(x) < 0$ for all x in the surface). Recall that the *vector field* $f(x)$ is the set of derivatives defining the flow ($\mathrm{d}y/\mathrm{d}t = f(y)$), where y belongs to phase space).

4.3.3 Poincaré sections and Poincaré first-return maps

4.3.3.1 *Global Poincaré section*

Can we generalize the stroboscopic section? Which are the properties that a section of phase space should have in order to generate a map containing all the important information of the dynamics? Would a transverse section have all of them?

The answer to these questions is simple. For a *control section* to be a *global Poincaré section* there are three essential requirements ([nite71]):

- Every orbit with initial condition on the control section returns infinitely many times to the control section for both positive and negative times.
- Every orbit intersects the control section.
- The control section is transverse to the orbits.

Note that the transversality condition implies that the dimension of the section is one less than the dimension of the phase space (hence the name Poincaré *surface*, sometimes used even in dimensions other than three).

These particular properties generate a 'well behaved' *Poincaré* or *first-return map*, i.e., a map that continuously sends each point on the Poincaré section to the point where the flow crosses the section in the next return. When a Poincaré surface is not available, a common practical procedure is to use a *control surface* that does not satisfy all the required conditions. It is a useful exercise to observe how each of these conditions affects the return map.

The infinite-return condition assures that the domain and the image of the first return map are the complete section. The second condition guarantees that every orbit of the flow will be represented as an orbit of the map. The third condition is more subtle.

If the transversality condition is not satisfied, we expect to find orbits crossing the section in both directions. As a limit case, we will have orbits that are tangent to the section. The return map at the point of tangency cannot be expected to be continuous as illustrated in figure 4.3.

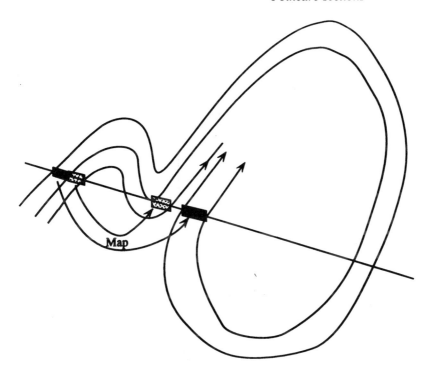

Figure 4.3. Transverse condition and the continuity of the Poincaré map.

The existence of Poincaré sections for arbitrary flows is an open problem. However, a flow may admit one or several *inequivalent* Poincaré sections and the corresponding Poincaré maps. The simplest example of an inequivalent Poincaré surface corresponds to a quasiperiodic flow on a bi-dimensional torus, the (inequivalent) Poincaré surfaces being the surfaces obtained by keeping one or the other angle constant. Such a flow can be obtained, for example, in the case of an integrable Hamiltonian system with two degrees of freedom [gold80, land69].

4.3.3.2 Local Poincaré section

There are many circumstances under which we are mainly interested in a small region of phase space, typically a region containing an invariant set. In such a case we may not need to worry about capturing a complete image of the dynamics and consequently we can drop the first two conditions in the definition of a Poincaré section, keeping the transversality condition. The map associated to this local section will represent the dynamics in the chosen region.

The main example of application for this procedure is the local Poincaré map in the vicinity of a periodic orbit. In such a case, it suffices to use as control surface a small enough section of a hyperplane with an orientation perpendicular to the periodic orbit (see figure 4.4).

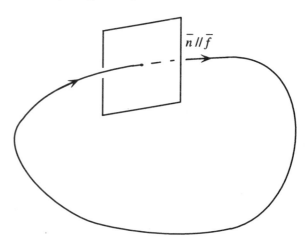

Figure 4.4. Local Poincaré section at a periodic orbit.

4.4 Maps and dynamics

Maps are relevant in dynamics *per se*, not only in the form of stroboscopic 'sampling' or first-return (Poincaré) maps. The most famous examples come from population dynamics [may 76]. For example, some insects are born in spring, grow during the summer, lay their eggs and die. Two generations of these species never coexist (this is called 'seasonally breeding populations in which generations do not overlap'). The fact that populations do not overlap in time and moreover that different generations appear (and disappear) in 'bursts' intuitively discourages the use of differential equations. It seems more natural to describe the system in terms of discrete equations, i.e. by a map that relates populations of consecutive periods [may 75]:

$$x_{n+1} = F(x_n). \tag{4.29}$$

The most naive approach to population dynamics [may 76] is to assume that the magnitude of the population of the new generation is proportional to that of the parent generation (no parents, no children; the more parents, the more children):

$$x_{n+1} = bx_n. \tag{4.30}$$

Although the model fits many observed populations for 'small' values of x, it leads to exponential unbounded growth for large n. It seems reasonable to assume that the proportionality in equation (4.30) is 'modulated' by a factor depending on the size of the population: the larger the population, the less it grows. Choosing adequate units, we get

$$x_{n+1} = ax_n(1 - x_n). \tag{4.31}$$

This discrete version of the *logistic equation* has been widely used in a number of population dynamics models [murr89] of the kind mentioned above. However, it must be recalled that the goal of modelling is *dictated* by the specific problem (be it biological, physical, ...) [murr89]. It is equally—if not more— important to provide sound observational grounds for the model assumptions as to attempt to understand its mathematical intricacies.

4.4.1 Properties of maps

We say that a map, $F(x)$, is *orientation preserving* if $\det(DF(x)) > 0$ for all x in the space of definition, where $DF(x)$ is the matrix of derivatives of the map at the point x. A map is said to be *orientation reversing* if $\det(DF(x)) < 0$ for all x in the space of definition.

Poincaré maps are always invertible and orientation preserving as a consequence of the uniqueness of the solutions of differential equations and the initial condition $F(\cdot, 0) = I$. The flow is a continuous deformation of the identity and so is the resulting Poincaré map. If the map were to change continuously from orientation preserving to orientation reversing, the determinant would have to change from positive to negative, i.e., it would be zero at some point y_0 and there would exist y_1 and y_2 arbitrarily close to y_0 such that $\det(DF(y_1)) > 0$ and $\det(DF(y_2)) < 0$. Such a map would no longer be invertible, thus violating the uniqueness result. There are however many cases in biology where one finds direct applications of non-invertible maps [may 75, may 76] not mediated by differential equations.

Exercise 4.15: Show that all Poincaré maps are orientation preserving. Hint: Use that $F(\cdot, t)$ can be viewed as a family of maps that depends continuously of t and that $\det(F(\cdot, 0)) = 1$.

Exercise 4.16: Extend the ideas of attracting set, attractor and trapping region to dynamical problems represented by maps.

4.4.1.1 Conjugated maps

Two different maps do not necessarily represent different dynamics. They might be the 'same' map written in different coordinates. The notion of change of coordinates coincides with the mathematical notion of *conjugation*.

We say that two maps $F_1 : X \to X$ and $F_2 : Y \to Y$ are *conjugated* if there is an invertible transformation $\phi : X \to Y$ such that $F_2 \circ \phi = \phi \circ F_1$ (the notation $\circ$ means composition of functions, $F \circ G(x) = F(G(x))$). Clearly, the function ϕ plays the role of a change of coordinates.

Exercise 4.17: Show that if F_1 and F_2 are conjugated, any orbit of F_1 is transformed by ϕ into an orbit of F_2.

If in addition we demand that special features of the map F_1 are preserved in changes of coordinates, we will have to reduce the class of allowed transformations to those that preserve the desired feature. For example if F_1 is continuous with n continuous derivatives, $F_1 \in C^n$, then we may want to restrict ϕ to C^n invertible transformations.

Exercise 4.18: Show that if F_1, F_2 and ϕ are C^1 then $DF_2 D\phi(x) = D\phi DF_1(x)$.

The notion of *equivalence* between Poincaré sections can be introduced in identical form. Consider two different sections immersed in phase space, Σ_0 and Σ_1, and a diffeomorphism (i.e., an invertible, two-way differentiable function) $\phi : \Sigma_0 \to \Sigma_1$ such that the respective Poincaré maps, F_0 and F_1, are conjugated by ϕ, i.e.

$$\phi^{-1}(F_1(\phi(x))) = F_0(x) \tag{4.32}$$

for $x \in \Sigma_0$.

The simplest example of equivalent Poincaré sections is given by a Poincaré section Σ_0 and its images with respect to the flow, $F(\cdot, t)$.

4.5   Parameter dependence

4.5.1   Families of flows and maps

One of the themes of nonlinear dynamics is the description of the changes observed in flows and maps depending on parameters. The discussion of the changes in simple attractors of the Lorenz system is illustrative of this point. We will frequently be considering dynamical systems that depend continuously on a set of real numbers or parameters: in other words, we will be considering *continuous families of dynamical systems*.

In this context, the *bifurcation set* is the set of points in *parameter space* where the dynamics presents qualitative changes; the *bifurcations* are the associated changes in the dynamics.

The bifurcation set separates regions of parameter space characterized by inequivalent dynamics. Consequently, we expect the bifurcation set to divide parameter space into connected regions, each region with a characteristic dynamic at least in those aspects that we are explicitly considering.

Going back to the example of the Lorenz system and considering only the number of fixed points and their stability, we can distinguish two bifurcation surfaces. The first one corresponds to the loss of stability of the symmetric fixed point and the simultaneous appearance of the pair of symmetry-related stable equilibrium points, $r = 1$. The second surface corresponds to the loss of stability of the pair of stable fixed points and the appearance of a conjugated (by the symmetry) pair of periodic orbits. This surface intersects the r axis at $r = 470/19$ for $\sigma = 10$ and $b = 8/3$.

The study of bifurcations is one of our central goals. We will repeatedly come back to this subject.

4.6 Summary

We have presented most of the set-theoretical tools which are central to nonlinear dynamics. The identification of invariant sets, basins of attraction and different kinds of limit set is among the strongest predictive tools that nonlinearity lets survive. More material will be provided in the coming chapters.

Further, we presented the connection between maps and flows. Although flows seem more natural to the physicist (being used to study continuous phenomena governed by differential equations), a great deal of the theoretical work in dynamical systems has been developed on maps instead. The use of Poincaré sections (when possible) will allow us to switch back and forth between map and flow representations, so that one can take advantage of both physical and mathematical intuition on the same problem.

4.7 Additional exercise

Exercise 4.19: Let us consider the following C^1 ODE [cecc93]:

$$r' = \mu f(r) \tag{4.33}$$
$$\theta' = 4\pi/r + 1 \tag{4.34}$$

with $f(r) = -r + r^2/r_c$ if $r < r_c$, $f(r) = (r - r_c) - (r - r_c)^2/(1 - r_c)$ if $r_c < r < 1$ and $f(r) = 1 - r$ otherwise.

(i)   Find the solutions (fixed points and periodic orbits) of this system.
(ii) Let us *kick* this system periodically, i.e., $(x, y) \rightarrow (x, y + F_e)$ at intervals of time τ. Find the stroboscopic map $M(r, \theta) = M_2 \circ M_1$, where M_1 accounts for the kicking and M_1 for the evolution of the system between consecutive kicks.
(iii) Study the fixed points of M, M^2 and M^3 for different values of F_e and τ.

Chapter 5

Elementary stability theory

5.1 Introduction

What happens with trajectories close to an equilibrium point (or an invariant set)?
Do they approach equilibrium? Do they depart from equilibrium? Under which
conditions should we expect one or the other behaviour? Can these trajectories
neither approach nor depart from equilibrium? Can some trajectories approach
equilibrium while others depart from equilibrium?

These questions can be addressed by considering the stability of equilibrium
points. The answers are of extreme relevance in all areas of science and its
applications. Consider the following examples:

- Le Châtelier–Braun principle: any system in chemical equilibrium
  undergoes, as a result of a variation of the factors governing the equilibrium,
  a compensating change in a direction such that, had this change occurred
  alone, it would have produced a variation of the factor considered in the
  opposite direction [glan71, pp 64].
- Supply and demand: '... The effects, which are produced, in practice, by
  the want of adaptation in the parts of demand and supply, are familiar. The
  commodity, which happens to be in superabundance, declines in price; the
  commodity, which is defective in quantity, rises. This is the fluctuation
  of the market, which every body sufficiently understands. The lowness
  of the price, in the article which is superabundant, soon removes, by the
  diminution of profits, a portion of capital from that line of production: The
  highness of price, in the article which is scarce, invites a quantity of capital
  to that branch of production, till profits are equalized, that is, till the demand
  and supply are adapted to one another...' [mill26, pp 235].
  Modern texts in economy consider this behaviour not as a 'natural law',
  but rather as hypothetical conditions for approaching equilibrium in ideal
  markets [hend71].
- Lenz's law: In the context of electromagnetism, and specifically dealing
  with electromotive forces, the law states '... if a current should flow in the

direction of the induced electromotive force, this current itself would create some flux through the loop in a direction to *counteract* the assumed flux change. That is an essential physical fact [···] It is a manifestation of the tendency of the system to resist change.' [purc65, pp 236].

These laws are particular cases of the general characterization of *stable* equilibria.

In chapter 3 we have addressed intuitively some of these questions, 'linearizing' the vector field around the fixed points. We would like to know the validity and the restrictions of this procedure. In which sense is the linearized equation equivalent to the original one?

Is the trajectory of the moon stable? Will the moon some day be hit by a meteorite and fall towards the earth? Will it go away instead? We could try to answer these questions using Kepler's equations, disregarding the influence of the sun and all other celestial bodies except the earth. In doing so, we would find that the trajectory of the moon is stable, but we would not be relieved because the following question will haunt us: would the stability of the moon change if we reconsidered the neglected influence of the rest of the universe?

This is a different, but related, class of problems, namely the problem of *structural stability*, i.e., whether the solutions and features we find with a given model equation will be present in all the equations that can be regarded as *perturbations* of the one studied. Surprisingly for some, naturally for others, there are relations between stability in phase space and structural stability; the ideas and methods of this chapter will also be relevant in the study of structural stability.

These are the kinds of question that we are going to address in the following sections. The chapter is actually a recollection of classical results in the theory of differential equations. We will not present the proofs of the theorems but rather intuitive and brief sketches of them. Complete proofs can be found in the references.

5.2   Fixed point stability

For a fixed point to deserve the qualification of *stable* we have to require that orbits with initial conditions close to the fixed point remain nearby for positive times. The formalization of this property requires a notion of proximity. We can assume that a notion of norm is well (and often naturally) defined for the system.

Fixed point stability: A fixed point, x, associated with the flow $F(\cdot, t)$ of the vector field f is said to be *stable* if for every neighbourhood $\epsilon(x) = \{y : |y-x| < \epsilon\}$ there exists a neighbourhood $\alpha(x) = \{y : |y-x| < \alpha\}$, such that for every $y \in \alpha(x)$ and every positive time t the orbit $y(t) = F(y, t)$ is in $\epsilon(x)$, i.e. $y(t) \in \epsilon(x)$ for all $t \geq 0$. This type of stability is known as *Liapunov stability*.

Exercise 5.1: Show that a fixed point is stable, if and only if, we can interchange the limits in time and initial conditions:

$$\lim_{y \to x} \limsup_{t \to \infty} |F(y, t) - x| = 0 = \limsup_{t \to \infty} \lim_{y \to x} |F(y, t) - x|.$$

Exercise 5.2: Extend the notion of Liapunov stability to an arbitrary orbit that is well defined for all positive times (i.e. it does not blow up in finite time).

An example of stable systems can be given in the form of a Hamiltonian problem with an attractive potential having one degree of freedom. Let $H = p^2/2 + V(x)$ with $V(0) = 0$ and $dV/dx(0) = 0$. The equations of motion are [gold80, land69]:

$$\dot{x} = \frac{\partial H}{\partial p} = p \tag{5.1}$$

$$\dot{p} = -\frac{\partial H}{\partial x} = -\frac{dV}{dx}. \tag{5.2}$$

Consider the neighbourhood $\alpha(0) = \{(x, p) : H(x, p) < \delta\}$. Given $\epsilon(0)$, thanks to the attractive character of the potential, we can always find a δ that is small enough so that $\alpha(0) \subset \epsilon(0)$. From the invariance (conservation) of the energy it follows that the orbits cannot leave the region $\alpha(0)$. In fact, orbits move on the curves $H(x, p) = E$. Within $\alpha(0)$ we have $E \le \delta$. Hence, the fixed point is (Liapunov) stable.

The idea of stability just formalized is as much as we can hope for in Hamiltonian systems but may not match our intuition, especially if our intuition has been nurtured in thermodynamics or dissipative systems. For example, consider the potential $V(x) = x^2/2$, i.e., the ideal harmonic oscillator. The potential fulfills the conditions of the example above. Hence, the point $(0, 0)$ is Liapunov stable. However, we know that any orbit starting outside the origin will *never* approach equilibrium. The system will oscillate around the origin *in æternum*. Sticking to our (dissipative) intuition, we may like to see that trajectories starting near equilibrium approach it asymptotically in time.

Asymptotic stability: We say that a fixed point is *asymptotically stable* if it is Liapunov stable and in addition all the trajectories in a sufficiently small neighbourhood of the fixed point tend to the equilibrium point as time tends to infinity.

5.2.1 Liapunov stability criteria

There are two fundamental stability criteria introduced by Liapunov. We used these criteria from an informal and intuitive point of view in chapter 3. In the discussion of the Duffing equations (3.15) and the laser system (3.10) and (3.11) we argued that since the system permanently loses energy (because of the dissipative terms) the orbits will have to follow paths of decreasing energy until they eventually reach a fixed point. We can generalize this argument realizing that there is nothing special about the energy from a dynamical point of view, since none of these two systems is Hamiltonian. Therefore, we can expect that any positive function which is everywhere decreasing except at the fixed point will have the same sort of behaviour. This is the content of *Liapunov's first stability criterion.*

A function H defined on a suitable neighbourhood, $U \subset M$, of the fixed point x associated with the vector field f is called a *Liapunov function* iff

$$H(x) = 0 \text{ and } H(y) > 0 \qquad \text{if } y \neq x \qquad (5.3)$$

$$\frac{dH}{dt}(y) = \sum_{i=1}^{dimM} \frac{\partial H}{\partial y_i} f_i(y) \leq 0 \qquad \text{for } y \in U - \{x\}. \qquad (5.4)$$

Theorem (Liapunov): A fixed point for which a Liapunov function exists is Liapunov stable. Furthermore, if $dH(y)/dt < 0$ then x is asymptotically stable [hirs78].

Exercise 5.3: Prove the theorem. Hint: Consider the sets $U_n = \{y : H(y) < 1/n\}$.

The *second stability criterion* introduced by Liapunov states that:

Theorem (linear stability) [arno88a]: A fixed point, x, is (asymptotically) stable if all the eigenvalues of the linearized system at the point x have negative real parts (note that one usually says 'stable' instead of 'asymptotically stable').

In other words, let

$$\frac{dy}{dt} = f(y) \qquad (5.5)$$

be our dynamical problem, and

$$\frac{dy}{dt} = Ay$$

the linearized problem at x with

$$\{A\}_{ij} = \frac{\partial f_i}{\partial x_j}.$$

The fixed point will be stable if (but *not* only if)

$$\text{Re}(\lambda) < 0 \qquad (5.6)$$

for all the eigenvalues λ of A.

The proof of this statement uses the notion of a Liapunov function. First of all, we can consider the stability of the point $y = 0$ for the linear system performing a suitable change of coordinates if necessary. Our knowledge of linear systems, chapter 2, allows us to state that

$$y(t) = \exp(At)y(0) \tag{5.7}$$

since all the eigenvalues of A have negative real parts

$$|y(t)| < \exp(\mu t)|y(0)| \tag{5.8}$$

for some negative real number $\mu > \mathrm{Re}(\lambda_1) \cdots > \mathrm{Re}(\lambda_n)$ and t sufficiently large (we recall that all the matrix elements in $\exp(At)$ are sums of polynomials times exponentials of the eigenvalues).

We can look for a Liapunov function as a quadratic expression

$$T(y) = y^\dagger T y. \tag{5.9}$$

The time derivative of $T(y)$ is

$$dT/dt = y^\dagger (A^\dagger T + T A) y \tag{5.10}$$

which is again a quadratic form. Let us see that the linear mapping between matrices defined by

$$L(T) = A^\dagger T + T A \tag{5.11}$$

is bijective.

If L were not bijective there would be an element T_0 such that $L(T_0) = 0$, then $dT_0(y)/dt = 0$ and $T_0(y)$ would be constant along trajectories, then $T_0(y(0)) = T_0(y(t))$ and, since $y(t) \to 0$ equation (5.8) and $T_0(y) = 0$ for all y we have that $T_0 = 0$.

The Liapunov function we are looking for can be obtained in the following form:

- Pick V a negative definite matrix, at will (for example $V = -I$)
- Find the unique T such that $L(T) = V$
- We claim that T is positive definite and, correspondingly, $T(y)$ is a Liapunov function.

In order to verify the last point assume that there exists an initial condition $y(0) \neq 0$ such that $T(y(0)) \leq 0$. Since $dT(y)/dt = y^\dagger V y < 0$ we have that $\lim_{t \to \infty} T(y(t)) \neq 0$ but $\lim_{t \to \infty} T(y(t)) = T(0) = 0$ which is a contradiction; hence T is positive definite.

The extension of the stability property from the linear case to the nonlinear situation is straightforward. It suffices to consider the same quadratic form for the nonlinear system, provided we restrict ourselves to a sufficiently small neighbourhood of the fixed point so as to be able to disregard higher-order corrections to the time-derivative.

We close this section with a geometrically intuitive criterion of *instability* for a fixed point (i.e., how to determine that a fixed point is *not* stable). Consider the following example of a system with a fixed point of saddle type:

$$\dot{x} = -x + f(x, y) \tag{5.12}$$

$$\dot{y} = y + h(x, y). \tag{5.13}$$

f and h are smooth functions vanishing at the origin together with their derivatives f' and h'. For any small positive ϵ, consider the closed region $W_a = \{0 \leq y \leq a, |x| \leq y/(1 + \epsilon)\}$. The function $g(x, y) = x^2 + y^2$ is positive in W_a. For sufficiently small a (so that f and h are much smaller than the linear part) dg/dt is positive in W_a outside the origin. Moreover, the flow crosses the lines $|x| = y/(1 + \epsilon)$ inwards. This guarantees that the intersection with W_a of any small circle centred at the origin will be expanded by the flow. In other words, any initial condition in W_a outside the origin will be pushed further away and hence the fixed point cannot be stable. Once again notice that the fact that the linear part of the vector field is dominant near the fixed point is the crucial point of the example. The function g is called a *Chetaev function*. We proceed with the formalization.

A differentiable function, g, is called a *Chetaev function* for the fixed point $x = 0$ of the vector field f if [arno88a]

- g is defined on a domain W whose boundary contains 0.
- The part of the boundary of W strictly contained in a sufficiently small ball with its centre 0 removed is a piecewise-smooth, C^1 hypersurface along which f points into the interior of the domain.
- $g(x) \to 0$ as $x \to 0$, $x \in W$; g is positive and $\sum_i \partial g/\partial x_i f_i > 0$, everywhere in W (see figure 5.1).

Theorem (Chetaev): A fixed point of a C^1 vector field for which a Chetaev function exists is unstable.

5.3   The validity of the linearization procedure

The validity of the linearization procedure in the vicinity of a fixed point goes beyond Liapunov's linear stability theorem but, necessarily, cannot be used in all situations. In fact, there is only hope for the procedure to be meaningful when we can 'disregard' the nonlinear terms in favour of the linear ones. If the matrix of the linearization has eigenvalues with $\text{Re}(\lambda) = 0$ we cannot hope for the linear terms to dominate in any neighbourhood no matter how small it is. With this in mind we introduce the notion of a hyperbolic fixed point.

A fixed point is called *hyperbolic* if $\text{Re}(\lambda) \neq 0$ for all the eigenvalues λ of the *Jacobian matrix*. The Jacobian matrix, or just the Jacobian, is simply another name for the matrix of first-order derivatives defining the linearized equation.

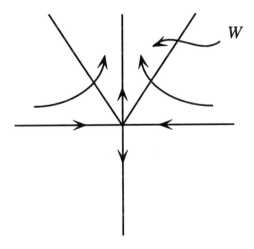

Figure 5.1. Chetaev function for a saddle point.

The following theorem states the relation between the linearized flow and the full flow at a hyperbolic point:

Theorem (Hartman–Grobman): For every hyperbolic fixed point, x, of the flow $F(\cdot, t)$ of $dy/dt = f(y)$ there exists a homeomorphism H defined on some neighbourhood U of x in $\mathbb{R}^n$ locally taking orbits of the nonlinear flow $F(\cdot, t)$ to those of the linear flow, $G(\cdot, t) = e^{tJ}$ (associated to the linearized system $dy/dt = Jy$, where J is the Jacobian matrix Df at x). The homeomorphism preserves the sense of orbits and can also be chosen to preserve the parameterization in time [guck86]. The homeomorphism H is of the *near-identity* type, i.e., $H(x) - x = o(|x|)$.

Proofs of the Hartman–Grobman theorem can be found in most books on ordinary differential equations, such as [arno88b, chow82].

The Hartman–Grobman theorem can only assure the equivalence up to homeomorphisms, but not up to smoother transformations. The difficulty actually occurs only for fixed points of saddle type.

Since the structure of orbits near the fixed point can be regarded as a homeomorphic image of the orbits in the linearized problem, the stable, centre and unstable subspaces of the linear problem will be present in the nonlinear problem as *local stable, centre* and *unstable manifolds* (recall that the *stable—unstable, centre—subspace* is the subspace of the linear problem expanded by the eigenvectors associated with the eigenvalues having real part smaller—greater, equal—than zero). The stable and unstable manifolds are also the insets and outsets defined from the set theoretical point of view in chapter 4. The following theorem advances on the characteristics of these manifolds.

Theorem (centre manifold theorem for flows) [guck86]: Let f be a C^r vector field on $\mathbb{R}^n$ vanishing at the origin ($f(0) = 0$) and let $A = Df(0)$. Divide the spectrum σ of A into three parts, σ_s, σ_c and σ_u according to

$$\text{Re}(\lambda) \begin{cases} < 0 \\ = 0 \quad \lambda \in \sigma. \\ > 0 \end{cases} \tag{5.14}$$

Let the (generalized) eigenspaces of σ_s, σ_c and σ_u be E_s, E_c and E_u, respectively. Then there exist C^r stable and unstable invariant manifolds W_u and W_s tangent to E_s and E_u at 0 and a C^{r-1} centre manifold W_c tangent to E_c at 0. The manifolds W_u, W_s and W_c are all invariant for the flow of f. The stable and unstable manifolds are unique, but W_c need not to be.

The following example taken from Guckenheimer and Holmes [guck86] illustrates the difficulties with the centre manifold.

The dynamical system

$$dx/dt = x^2 \tag{5.15}$$
$$dy/dt = -y \tag{5.16}$$

has an infinite number of centre manifolds. The system is the Cartesian product of two dynamics, one of them, the x equation, is 'pathological' (for our purposes) since it blows up in finite time. However, we are interested in the local behaviour at $(x, y) = (0, 0)$ and in that regard we will see later that the equation is rather general. Direct integration gives the solutions of 5.16 with initial condition (x_0, y_0) in the form

$$x(t) = x_0/(1 - tx_0) \tag{5.17}$$
$$y(t) = y_0 \, e^{-t}. \tag{5.18}$$

The y axis is the stable manifold. The centre manifold associated with the zero eigenvalue coming from the x equation has to be tangent to the x axis at the origin.

All trajectories not lying on the stable manifold can be written as $y(x) = y_0 \, e^{(1/x - 1/x_0)}$, since $dy/dx = -y/x^2$. The limit of these trajectories for $x \to 0^\pm$ is rather different in the plus and minus cases. If $x \to 0^-$ the trajectory will tend to $(0, 0)$ and, moreover, will be tangent to the x axis, but if $x \to 0^+$ the trajectory will diverge unless $y_0 = 0$, i.e., unless the trajectory is *on* the x axis.

Thus, for positive x the centre manifold coincides with the x axis and for negative x it can be chosen to coincide with any of the trajectories: all of them are invariant sets which are tangent to the x axis at the origin. We have thus an infinite number of centre manifolds and all of them are C^∞ (!), since equation 5.16 is analytic and hence C^∞. Finally, there is only one analytic centre manifold, namely the x axis, which would certainly have been our *a priori* candidate for a centre manifold.

5.4 Maps and periodic orbits

5.4.1 Maps

We established in chapter 4 that periodic orbits can be considered as fixed points of a suitably defined (local) Poincaré map. It is then natural to study the stability of periodic orbits and that of fixed points in maps simultaneously. Although we are going to consider throughout this section the case of period-one orbits (or fixed points of a map), the case of higher periodicities can be reduced to the present one considering the pth iteration of the map, F^p, in the case of a period p orbit of the map F. This point of view is useful in the present context but may be less useful in different contexts, for example developing numerical algorithms for the location of periodic orbits.

The idea of stability for fixed points of maps is the same as that of fixed points of flows.

Fixed point stability (maps): A fixed point, x, associated with the map F is said to be Liapunov stable if for every neighbourhood $\epsilon(x) = \{y : |y - x| < \epsilon\}$ there exists a neighbourhood $\alpha(x) = \{y : |y - x| < \alpha\}$ such that for every $y \in \alpha(x)$ and every positive integer n the orbit $\{y_n\} = \{F^n(y)\}$ is in $\epsilon(x)$, i.e. $y_n \in \epsilon(x)$ for all $n \geq 0$.

Asymptotic stability (maps): A fixed point, x, associated with the map F is said to be (asymptotically) stable if it is Liapunov stable and in addition there exists a sufficiently small neighbourhood $U_\epsilon(x)$ such that for all $y \in U_\epsilon(x)$ we have that $F^n(y) \to x$ for $n \to +\infty$.

The linear stability analysis is equally valid in the vicinity of a periodic orbit. Let us consider once again the fixed point 0 of a differential equation, $dx/dt = f(x)$, with linearization $dx/dt = Ax$. If we consider any arbitrary discrete time, t_0, we obtain a map $F(\cdot, t_0)$ and the corresponding linear map $L = e^{t_0 A}$. According to the Liapunov linear stability criterion, if the flow $F(\cdot, t_0)$ is stable, then all the eigenvalues of A have real part smaller than zero. Correspondingly, since the matrix L is the exponential of $t_0 A$, its eigenvalues will have absolute value smaller than unity. Thus, we have:

Stability criterion for maps: A fixed point x of a map F is stable when all the eigenvalues of the associated linear map at x, $DF(x)$, have absolute value smaller than unity.

There is also a 'map' version of the Hartman–Grobman theorem. In fact, the Hartman–Grobman theorem is usually proved for maps and then extended to flows, following a procedure which is exactly the inverse of our discussion [arno88b, chow82].

Theorem (Hartman–Grobman) [arno88b]: Let $L : \mathbb{R}^n \to \mathbb{R}^n$ be a linear transformation without eigenvalues equal to unity in absolute value. Every local diffeomorphism $B : (\mathbb{R}^n, x) \to (\mathbb{R}^n, x)$ with linear part L at the fixed point x is topologically equivalent (conjugated) to L in a sufficiently small neighbourhood of x.

A fixed point of a map is said to be *hyperbolic* if the absolute value of the eigenvalues of the linearized map is different from unity.

In addition, the stable, unstable and centre manifold theorem translates from the flow case in the same form. We repeat it here for completeness.

Theorem (centre manifold theorem for maps) [guck86, carr81]: Let F be a C^r diffeomorphism on $\mathbb{R}^n$ vanishing at the origin ($F(0) = 0$) and let $A = DF(0)$. Divide the spectrum of A into three parts, σ_s, σ_c and σ_u according to

$$|\lambda| \begin{cases} < 1 \\ = 1 \quad \lambda \in \sigma \\ > 1. \end{cases} \tag{5.19}$$

Let the (generalized) eigenspaces of σ_s, σ_c and σ_u be E_s, E_c and E_u, respectively. Then there exist C^r stable and unstable invariant manifolds W_u and W_s tangent to E_s and E_u at 0 and a C^{r-1} centre manifold W_c tangent to E_c at 0.

5.4.2 Periodic orbits of flows, Floquet stability theory

In order to be able to apply the stability analysis outlined in the previous subsection to a periodic orbit we have to be able to compute the local Poincaré map associated with the orbit or at least the linear part of this map.

Let $x^0(t) = x^0(t + T)$ be a T-periodic solution of a differential equation of the form $dx/dt = f(x)$. Our goal is to study the stability of this orbit. To begin with, we can linearize the flow in a neighbourhood of the periodic solution. Writing $x(t) = x^0(t) + y(t)$ and retaining only the first terms in the Taylor expansion of $f(x^0(t) + y(t))$ for small y we obtain

$$\frac{dy_i}{dt} = \frac{\partial f_i}{\partial x_j^0} y_j, \tag{5.20}$$

which is a linear equation with periodic coefficients of the kind discussed in chapter 2.

The flow associated with this kind of equation is described by a linear operator in the form of a matrix $U(t, t_0)$ that satisfies (5.20). Its asymptotic properties are described by the monodromy matrix $U(t_0 + T, t_0)$ (see chapter 2). The time t_0 is irrelevant for all our concerns since the monodromy matrices obtained by changing t_0 are all conjugated by $U(t_0, t_0')$. The monodromy matrix has in this case an eigenvalue equal to unity since the velocity dx^0/dt is a periodic solution of (5.20).

Exercise 5.4: Verify that the velocity is a solution with eigenvalue unity.

The eigenvalue unity is only the indication that the distance between two points on the periodic orbit will not increase nor decrease after a period T. We are however interested in all the other eigenvalues of the monodromy matrix since we shall see immediately that it is closely related to the Poincaré map.

The Poincaré plane or section is conveniently chosen perpendicular to the periodic orbit. We can then reformulate the question of the stability of the periodic orbit $x^0(t)$ as follows: we are interested in knowing whether displacements *on* the Poincaré plane (but not perpendicular to it) around the fixed point z_0 corresponding to the periodic orbit $x^0(t)$ will approach (or depart from) z_0.

The Poincaré map will be of the form

$$F(z) = \bar{U}(t_0 + T, t_0)z + \text{ nonlinear terms} \tag{5.21}$$

with z in the Poincaré plane. We see that $\bar{U}(t_0 + T, t_0)$ is the monodromy matrix projected onto the Poincaré plane.

Finally, to establish that the linear stability of z_0 (when applicable, i.e., for hyperbolic orbits[†]) is governed by the monodromy matrix, it remains to confirm that this result is independent of the point $x^0(t_0)$ used as reference.

Exercise 5.5: Using the Hartman–Grobman theorem show that if the periodic orbit is hyperbolic the Poincaré maps obtained by sectioning the periodic orbit with planes perpendicular at different points are (locally) conjugated.

The local Poincaré maps obtained choosing different sections are actually conjugated by the flow. This fact is a direct consequence of the already mentioned fact that the image of a Poincaré section is a Poincaré section. Note however that in order to be rigorous a little extra care has to be exercised since the image of a plane will not be a plane in general, not even locally. The image of one of our Poincaré planes will be in general a surface tangent to the local Poincaré plane at the orbit.

Exercise 5.6: Propose a map that relates the image by $F(\cdot, t)$ of the Poincaré plane at x to the Poincaré plane at $F(x, t)$. Sketch the proof of the equivalence between Poincaré maps given by different planes perpendicular to the periodic orbit.

The stability theory of periodic orbits outlined in this section is traditionally known as part of Floquet's theory [ioos80, arno88b] and can be cast entirely in the same terms as Floquet's theorem presented in chapter 2.

Exercise 5.7: Note that the version of Floquet's theorem presented in chapter 2 is restricted to the case of conservative flows where $\det(U) = 1$. Extend the theorem to the general case.

[†] The notion of hyperbolic orbit extends immediately from that of hyperbolic fixed point of maps.

5.5   Structural stability

It is convenient to introduce at this stage another notion of stability. The situation so far studied in this chapter aimed at establishing criteria for deciding whether a solution of a dynamical system would be 'tolerant' (stable) under perturbations of the initial conditions. As mentioned in the introduction, we can also inquire whether a solution is 'tolerant' to small changes in the dynamical system itself. This is the idea behind *structural stability*, i.e., to assess when a system (in the end a *family of systems*) retains its structure despite small changes in the vector field.

The simplest examples of *structurally unstable* systems are taken from Hamiltonian mechanics. Consider for example a harmonic oscillator with and without energy losses. If the rate of energy loss is small, we can regard the dissipative problem as a perturbation of the conservative one; actually, the orbits of the dissipative harmonic oscillator will remain close to those of the conservative case for finite time. However, the orbits of the dissipative system will decay towards the fixed point given sufficient time. Although they may be similar for finite time, the orbits of these systems are qualitatively different and this difference is *intrinsic*. The orbits of the Hamiltonian case are all closed (circles or ellipses) while the orbits of the dissipative case are open (spirals going towards the equilibrium point). We conclude that the conservative harmonic oscillator is structurally unstable since no matter how small the amount of dissipation introduced it will be enough to change the characteristics of the solutions. The present discussion can be reproduced for any Hamiltonian problem and its dissipative counterpart.

This section has been partially inspired by the discussion found in V I Arnold's *Geometrical Methods in the Theory of Ordinary Differential Equations* [arno88b, p 89]. We recommend Arnold's discussion as further reading and as a fine example of the process of inducing a definition.

5.5.1   Orbital equivalence

The proper formulation of the concept of structural stability requires a notion of distance in the space of dynamical systems as well as criteria for comparison of the dynamics generated by different systems. Moreover, whatever criteria we introduce they would have to make sense when applied to the oscillator example.

If two different dynamical systems are to be considered equivalent and yet different we will have to admit some tolerance in our way of looking at them. The concept of *conjugation* introduced in chapter 4 to get rid of arbitrary changes of coordinates is again useful in this context. We say that two systems are *topologically equivalent* if they are conjugated, i.e., if there exists a change of coordinates carrying the solutions of one system into the solutions of the other.

Since the concept of conjugation establishes classes of equivalence, there is a corresponding classification of dynamical systems according to this criterion. The harmonic oscillators with and without dissipation clearly belong to different classes since there is no way in which we can transform a circle into a spiral with a change of coordinates.

The classification just achieved may be challenged as too narrow. For example, in our harmonic oscillator example two conservative oscillators with slightly different frequencies will be considered inequivalent since the orbits of one cannot be carried into orbits of the other, since the period of the orbits are slightly different, despite the geometrical fact that the orbits differ only in the parameterization by time.

The idea of orbital equivalence bridges this problem. Two systems are said to be *topologically orbitally equivalent* if there exists a change of coordinates carrying oriented orbits of the first into oriented orbits of the second [arno88b].

Note that the change of coordinates may depend on time. In other words, the transformations are more general that just mapping solutions $x(t)$ of one system onto solutions $y(t)$ of the other system.

5.5.2   Structural equivalence

Before stating the definition of structural stability we have to find the measure of distance in the space of dynamical systems. In this context it is usual to require not only that the difference between vector fields defining the differential equation (or the homeomorphism defining the map) is small, but also that the difference between the derivatives is small. The space of vector fields is then provided with a norm that measures both the 'size' of the vector field and that of its r first derivatives [wigg90].

A system is said to be *structurally stable* if there exists a neighbourhood of the vector field (map) such that every vector field (map) in the neighbourhood is topologically orbitally equivalent to the initial vector field (map).

Note that since topological equivalence only guarantees continuity, there is no point in demanding the *perturbed* vector fields to be C^r even if the starting one is so (however, the C^1 requirement is frequently encountered, since we want to take derivatives of the vector fields).

The question of how to define an 'appropriate' norm is far from elementary. Two vector fields defined in unbounded spaces (such as $\mathbb{R}^n$) may be arbitrarily apart far away from the fixed point(s). For the harmonic oscillator, the difference of the vector fields in the cases with/without dissipation is just the dissipation:

$$|f(x, p) - f'(x, p)| = |\nu p|$$

which is unbounded unless we restrict our considerations to a small compact submanifold in phase space where for example the supreme norm would do the job.

5.6 Summary

Stability theory is the natural continuation of the set-theoretical analysis started in chapter 4. It provides tools to both identify some invariant sets and to assess the behaviour of the flow (or map) around these sets.

Now we can improve our predictive tools by saying not only how certain invariant sets look, but also under what conditions the system will end up in one of these sets.

Structural stability will be a useful tool when we try to answer questions such as the one raised in chapter 3: can the stability of a fixed point change when we alter the parameters in the system? (Remember Lorenz in chapter 3.) Is there a systematic way of understanding the possible changes?

5.7 Additional exercise

Exercise 5.8: Let us consider the stability of the solutions of the autonomous system described in exercise 4.19. Show that this 2-d system has a fixed point at the origin and two limit cycles: an unstable one at $r = r_c$ and a stable one at $r = 1$. Notice that the frequency of the inner limit cycle is larger than the frequency of the outer one.

Chapter 6

Bi-dimensional flows

There is an important reason for discussing flows in two-dimensional phase spaces. Beyond their relevance as simple models of dynamical phenomena and beyond their usefulness in the study of higher-dimensional problems stands their intrinsic beauty.

The results of this chapter are of geometrical and topological nature and to a good extent they are a consequence of the special limitations imposed by the dimension of the phase space upon the possible flows. For example, the fact that a closed Jordan curve divides the plane into two regions will play an important role in the discussion. These results cannot be extended to higher-dimensional systems, their beauty and their limitations spawning from the very same topological nature.

The goal of this chapter is to characterize the possible structures of the attractors in bi-dimensional flows. In this process we will attempt to make sense of the popularized statement 'there is no chaos in bi-dimensional flows'. This characterization can be achieved in the form of a theorem (Poincaré–Bendixson theorem). Before presenting the theorem we will refine the notions developed in chapter 4 and 5.

We will restrict our discussion to the case of flows in $\mathbb{R}^2$, in order to achieve a presentation without additional difficulties. In this chapter we refer to a flow $F(\cdot, t)$ associated with a smooth vector field f defined on a submanifold of $\mathbb{R}^2$ that can be thought of as a disc D^2. The case of other bi-dimensional manifolds can be treated in similar terms.

6.1 Limit sets

In chapter 4 we introduced several kinds of invariant set and related notions. Among them we introduced the notion of an attractor and that of an attracting set as well as the notions of a *positively invariant region* (just a different name for *trapping region* as you have guessed) and a maximal attractor.

The notions introduced in chapter 4 are far from exhausting the different kinds of invariant set, although they are, arguably, the most important ones. In the discussion of the Poincaré–Bendixson theorem (see below) the notion of limit sets will play an important role.

We say that y belongs to the ω-*limit* $\omega(x)$ of an initial condition x if and only if there exists a sequence of positive times t_i, $\lim_{i \to \infty} t_i = \infty$, such that $\lim_{i \to \infty} x(t_i) = y$.

In the same way, but changing positive times by negative times, we can define the α-limit of x, $\alpha(x)$. We say that $y \in \alpha(x)$ if and only if there exists a sequence of negative times t_i, $\lim_{i \to \infty} t_i = -\infty$, such that $\lim_{i \to \infty} x(t_i) = y$. Figure 6.1 illustrates the limit-set idea.

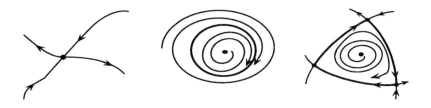

Figure 6.1. ω-limit sets (thick lines and saddle fixed points).

The following properties of the limit sets are somewhat intuitive[†] or require little proof [hirs78]:

- If x and y are on the same trajectory, i.e., $x = F(y, t)$ for some finite time t, then $\omega(x) = \omega(y)$ and $\alpha(x) = \alpha(y)$.
- The ω and α limit sets are closed.
- $\omega(x)$ and $\alpha(x)$ are invariant sets.
- $\omega(x) = \cap_{s \geq 0} \mathrm{cl}\{x_t : t \geq s\}$ where $\mathrm{cl}(A)$ denotes the closure of a subset A.
- For any fixed time t, $F(\omega(x), t) = \omega(F(x, t))$.
- If M is a closed positively invariant set and $x \in M$, then $\omega(x) \subset M$.
- If M is a closed positively invariant set and $x \in M$, then $\omega(x)$ is connected.
- A closed invariant set contains the α- and ω-limit of every point in the set.

Exercise 6.1: Find the attracting sets, the attractors and the ω-limit sets for the following flows:

$$(1) \quad \begin{aligned} dx/dt &= x(1 - x^2) \\ dy/dt &= -y \end{aligned}$$

[†] As usual, they are intuitive for all those that share our background. Those that find no immediate reason for these properties to hold will do well in testing them in the examples of figure 6.1.

$$(2) \quad \begin{aligned} dx/dt &= x(1-(x^2+y^2)) - \Omega y \\ dy/dt &= y(1-(x^2+y^2)) + \Omega x. \end{aligned}$$

We emphasize that, in contrast to the rest of this chapter, the idea of limit sets carries to higher-dimensional spaces without change; however they should not be expected to be, in general, as simple as those in figure 6.1.

6.2 Transverse sections and sequences

The second observation needed to construct the Poincaré–Bendixson theorem has to do with transverse sections. Consider a transverse section Σ, of the flow $F(\cdot, t)$ under consideration (see figure 6.2). We observe that if the orbit of $p \in \Sigma$ intersects the transverse section at a second point, p', this segment of the orbit between the two intersections, together with the segment $[p, p'] \subset \Sigma$, determine a positively (or negatively) invariant region which is also a simply connected region of $\mathbb{R}^2$. The invariance comes from the perceived fact that any orbit with initial condition in the region will not be able to leave it for positive (negative) times. In other words, no orbit with initial condition in the region can intersect the trajectory of p (except p's orbit itself). Also, because of the transversality property, the segment $[p, p']$ on Σ cannot be crossed 'outwards' from the region, i.e., in the opposite direction to that of the orbit of p at p.

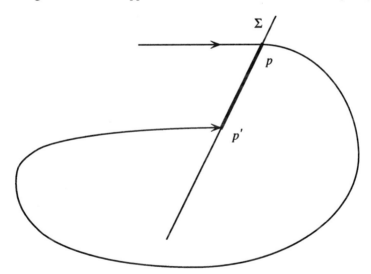

Figure 6.2. Intersections with a transverse section and trapping regions.

Although the above discussion is intuitive we note by passing that it requires several properties of $\mathbb{R}^2$. The argument with the transverse section would be dead wrong if the manifold were not orientable, like a Möbius band. Also, the Jordan curve may or may not divide the space in two regions for other bi-dimensional manifolds such as, for example, a torus.

As a direct consequence of the observed properties of transverse sections with regard to orbits, we have that the sequence of the intersections of the orbit of p with the section Σ, $\{x_i = x(t_i), t_{i+1} > t_i\}$, ordered according to the time of the intersection is a monotonic sequence with respect to the natural order in the interval $[p, p']$. This sequence needs not to be infinite however. The reader can convince himself/herself of this statement by playing with orbits for figures like figure 6.2 or, in case she/he feels that a proof is necessary, working it out as an exercise (the solution can be found in [hirs78, wigg90]).

In what follows we restrict our attention to the case of a positively invariant region; the case of a negatively invariant region can be treated in a completely analogous form by exchanging negative and positive times.

Lemma: $\omega(x)$ intersects any transverse section in no more than one point.

The reason is the following. For every $p \in \omega(x) \cap \Sigma$ we will consider local orbits $O(p) = \{F(p, t)$ for $-\epsilon < t < \epsilon\}$. We can choose ϵ small enough so that $O(p) \cap \Sigma = p$. For every $q \in \omega(x) \cap \Sigma$ there is a sequence $\{x_i^q\}$ of points in the positive orbit of x that converges to q. Moreover, the segments of orbits $O_i^q = \{F(x_i^q, t)$ for $-\epsilon < t < \epsilon\}$ converge to the segment $O(q)$. Then for i large enough, the segments O_i^q have to intersect the transverse section. Let $\{o_i\}$ be the sequence of all these intersections (for all possible q) ordered after increasing intersection time, regardless of the limit point q. This sequence is monotonic, and it is defined on a compact region of $\mathbb{R}^2$ (i.e., Σ or a closed segment in Σ). Hence, we have that the limit $\lim_{i \to \infty} o_i = p$ is unique. Therefore $\omega(x) \cap \Sigma$ consists of only one point (see figure 6.3).

6.3 Poincaré–Bendixson theorem

Using the results shown in the previous section regarding the intersection of ω-sets with transverse sections we can characterize the possible ω-sets for bi-dimensional flows. The following theorem achieves part of this characterization:

Theorem (Poincaré–Bendixson) [hirs78]: A non-empty compact limit set of a C^1 planar dynamical system, which contains no equilibrium point, is a closed orbit.

The theorem is proved in two steps. In the first step we show that $\omega(y)$ for $y \in \omega(x)$ is a closed orbit; in the second step we show that $\omega(x)$ itself is a closed orbit, see figure 6.4.

(i) To see that $\omega(y)$ for $y \in \omega(x)$ consists of a periodic orbit, consider a transverse section Σ at $z \in \omega(y)$. There is a sequence of times t_i such that $F(y, t_i)$ tends to z (by definition of ω-set). Moreover, following the reasoning in the previous lemma, the points in the sequence can be assumed to belong to Σ. Finally, all these points belong to $\omega(x)$, since they are forward orbits of y. Since $\omega(x)$ intersects Σ at only one point we have

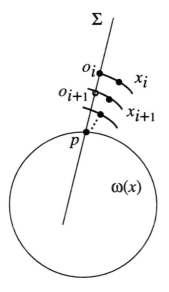

Figure 6.3. Intersection of $\omega(x)$ with a transverse section.

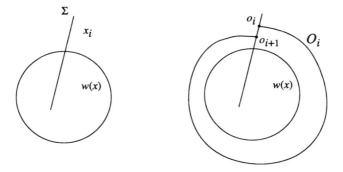

Figure 6.4. The periodic orbit theorem.

that $F(y, t_i) = F(y, t_{i+1})$ and hence the orbit is periodic, i.e., $\omega(y) = O(y)$ is a periodic orbit.

(ii) To realize that $\omega(x) = O(y) = \omega(y)$, we have to prove only that $\omega(x) \subseteq \omega(y)$ since the reciprocal is immediately true. Let us consider again a transverse section Σ that is crossed by $\omega(x)$ at some point y_0 in $O(y)$. The piece of the orbit of x lying between successive intersections with this transverse section,

$$O_i = \{F(x, t), \ t_i \leq t \leq t_{i+1}\}$$

delimits, together with the segment $[F(x, t_i), F(x, t_{i+1})] \in \Sigma$, a positively invariant region which contains the (connected) limit set $\omega(x)$. The forward crossings of both endpoints of O_i on Σ build convergent sequences tending to the unique point of $\omega(x)$ in Σ, namely y_0. Hence the forward iterates of O_i tend to $O(y) = \omega(y)$ (we refer the reader to [hirs78] for a detailed proof along these lines).

Alternatively, let us assume that $\omega(x)$ contains other orbits apart from $O(y)$ (recall that $\omega(x)$ does not have fixed points). Consider an open band of size ϵ around $\omega(y)$. Since $\omega(x)$ is connected, there has to be some point $z \in \omega(x)$ in the band. Choose a point $y_1 \in O(y)$ such that z is inside a ball of size ϵ centred on y_1. A transverse section containing y_1 will be crossed by the orbit of z for ϵ sufficiently small, thus contradicting the previous lemma. Hence, $\omega(x) = O(y) = \omega(y)$.

The second result, also known as the Poincaré–Bendixson theorem [wigg90], completes the characterization of the possible ω-sets for bi-dimensional flows.

Theorem (Poincaré–Bendixson) [wigg90]: Let M be a positively invariant region for a vector field containing a finite number of fixed points. Let $p \in M$, and consider $\omega(p)$. Then, one of the following possibilities holds:

(i) $\omega(p)$ is a fixed point;
(ii) $\omega(p)$ is a closed orbit;
(iii) $\omega(p)$ consists of a finite number of fixed points $p_1, \ldots, p_n$ and orbits γ with $\alpha(\gamma) = p_i$ and $\omega(\gamma) = p_{i+1}$[†].

The theorem is proved considering the following observations, see also figure 6.5:

(i)   If $\omega(p)$ consists only of fixed points, then it has in fact only one point, since the fixed points are isolated (its number is finite) and $\omega(p)$ is a connected set.
(ii) If there are no fixed points, then by the previous theorem $\omega(p)$ is a periodic orbit.
(iii) If $\omega(p)$ contains both fixed points and other points, the orbits of these other points, y, have to tend to the fixed points both in past and future. Otherwise, we would be in the case of the previous theorem and $\omega(y)$ as well as $\omega(p)$ would be a periodic orbit, which is a contradiction.

The Poincaré–Bendixson theorem shows that the possible ω-sets in $\mathbb{R}^2$ are very limited in their topology. Only fixed points, limit circles (periodic orbits) and *heteroclinic* cycles (see footnote †). Everything is pretty much ordered and there is nothing in sight that might eventually be called 'chaotic'. There is no chaos in bi-dimensional flows (whatever chaos might exactly be!).

---

† These orbits γ 'starting' in a (necessarily non-stable) fixed point and 'finishing' in another fixed point are called *heteroclinic orbits*. Similarly, a *homoclinic orbit* starts and finishes in the same (non-stable) fixed point.

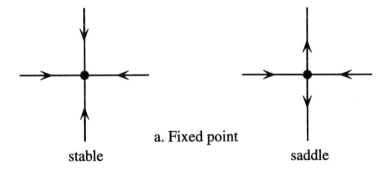

a. Fixed point

stable   saddle

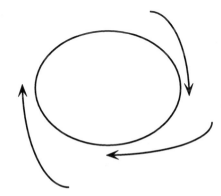

b. Periodic orbit

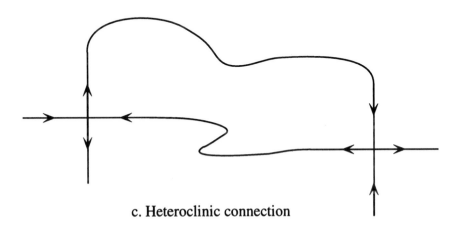

c. Heteroclinic connection

Figure 6.5. $\omega(p)$ is: (a) a fixed point; (b) a periodic orbit; (c) a heteroclinic connection.

A pair of results that we will not discuss in detail give even more force to the theorem.

6.4 Structural stability

The Pontryagin–Andronov theorem establishes conditions that are equivalent to structural stability for flows confined to a bi-dimensional disc D^2.

Theorem (Pontryagin–Andronov) [hirs78]: Suppose f, the vector field, points inward on ∂D^2 (the *border* of the disc). Then, the following conditions taken together are equivalent to structural stability on D^2:

(a) the equilibrium points in D^2 are hyperbolic;

(b) each closed orbit in D^2 is either a periodic attractor or a periodic repeller;

(c) no trajectory in D^2 goes from saddle to saddle.

The situation is illustrated in figure 6.6.

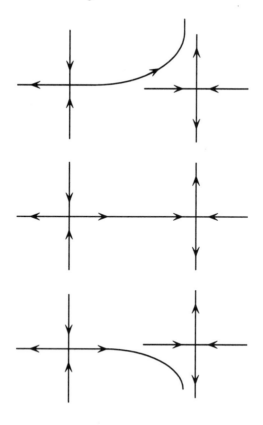

Figure 6.6. Structurally unstable flows and their perturbed counterparts.

Finally, Peixoto's theorem establishes that almost all (in a probabilistic sense) flows on D^2 are structurally stable.

Theorem (Peixoto) [hirs78]: Let $\mathcal{V}_0(W)$ be the set of C^1 vector fields on $W \subset \mathbb{R}^2$ that point inward on ∂D^2. The set

$$\mathcal{S} = \{f \in \mathcal{V}_0(W) \ f \text{ is structurally stable on } D^2\}$$

is dense and open.

In a certain sense, the combination of the theorems by Poincaré–Bendixson, Pontryagin–Andronov and Peixoto characterize completely the ω-sets, and then the attractors, of the bi-dimensional flows that we are likely to find.

6.5 Summary

This chapter concludes the 'elementary' part of the analysis of dynamical systems started in chapter 4. In all, we have given special names to the sets of points in phase space bearing the asymptotic properties of the flow, studied the properties of these sets and considered some of the conditions in which the flow will *tend* to these sets (be it for positive or negative times).

The results of this chapter in particular indicate that for the special case of bi-dimensional systems we can determine which are the possible kinds of limit set that are to be expected as well as when the properties of a flow will be tolerant to small perturbations.

Chapter 7

Bifurcations

7.1 The bifurcation programme

7.1.1 Families of flows and maps

The experiments of chapter 1 had in common, among other properties, the dependence of the dynamics with the value of certain parameters (or external controls[†]) of the experiment. The Reynolds number in turbulence, the Prandtl and Rayleigh numbers in Bénard convection, the amplitude and frequency of the modulation of losses in the CO_2 laser are examples of these 'control' parameters. In general we can say that each set of possible parameters characterizes a dynamical system as a member of a broader *family of dynamical systems* (labelled by the values of these parameters). Furthermore, we expect in general a smooth (C^r) dependence of the equations with respect to the parameters, i.e., a continuous (and possibly differentiable) family of dynamical systems.

As an example, the Lorenz system, for constant Prandtl number $\sigma = 10$ and aspect ratio $b = 8/3$, constitutes a one-parameter family of dynamical systems depending on the (normalized) Rayleigh number r: $\dot{x} = f_r(x)$, $x \in \mathbb{R}^3$. We will come back to this system several times along this chapter.

For some critical values of the parameters the characteristics of the dynamics change somewhat abruptly. We shall presently specify what is meant by this. The study of the changes in the dynamics undergone by systems as a response to the changes in the environment is one of the fundamental themes of nonlinear dynamics.

We can consider that the parameters live in a *parameter space* or *configuration space*. We will use the word *bifurcations* to denote the qualitative changes of the dynamics undergone by the flow and *bifurcation set* the set of points in *parameter space* where the dynamics presents qualitative changes. Quite often we will focus on a particular kind of orbit, X, of the problem (X

[†] We note on passing that with the expression *control parameters* we mean that the parameter values determine the dynamics and **not** that the researcher has control over the parameters, something that can never be fully achieved outside the universe of paper and pencil of the theorist.

might be for example fixed points) and speak of bifurcations of X, meaning the qualitative changes that affect these solutions. Much of this chapter will deal with bifurcations of fixed points of flows and maps.

The bifurcation set separates regions of parameter space characterized by inequivalent dynamics. Consequently, we expect the bifurcation set to divide the parameter space into connected regions, each region with a characteristic dynamics at least in those aspects that we are explicitly considering.

For example, in the Lorenz system, regarding only the number of fixed points, their stability and their changes as a function of the control parameter r, we can distinguish two bifurcation surfaces. The first one corresponds to the loss of stability of the symmetric fixed point and the simultaneous appearance of a pair of symmetry-related stable equilibrium points, $r = 1$. The second surface corresponds to the loss of stability of the pair of stable fixed points and the disappearance of a conjugated (by the symmetry) pair of periodic orbits. This surface intersects the r axis at $r = 470/19$ (again, for $\sigma = 10$ and $b = 8/3$).

The bifurcation set can be quite complex when all the aspects of the flow are considered.

7.1.2 Local and global bifurcations

The simplest cases of bifurcations are associated with the study of the number and stability of fixed points in flows and maps (periodic orbits). These types of orbit are localized in phase space and their transformations, with respect to changes of parameters, can be studied restricting the analysis to a small neighbourhood (in phase space) of the fixed point. Because of this property we can name them *local bifurcations* emphasizing the confinement of the events to an arbitrarily small region of the space.

In partial opposition to local bifurcations are *global bifurcations*, in this case the analysis cannot be restricted to a small region of phase space and requires some knowledge of the dynamics in an extended region.

The distinction between global and local bifurcations is not a perfect one, some global bifurcations can be packed (under special circumstances) in local bifurcations [guck86, pp 117, 118].

The rest of this chapter develops the bifurcation programme for local bifurcations and studies the fundamental (and simplest) cases.

7.1.3 Local bifurcations: the programme

Now that we know (or so we believe!) what bifurcations are, we can pose a few more questions: When (or why) do bifurcations occur? How do they happen? How do we go about studying them?

The three questions are tied together and we will have to answer them almost in parallel. Fortunately, we already have some ideas to push forward: to what degree can the Jordan normal form method be extended to nonlinear problems? Can the flow be linearized (in the same lines as in the Hartman–Grobman theorem) at least in some directions?

One of the central ideas in the Jordan normal form, discussed in chapter 2, was the decomposition of the flow in independent movements associated with the different eigenspaces of the linear equation. We shall come back to this idea with the method of reduction to the centre manifold.

In chapter 5 we learned that in the case of hyperbolic (fixed) points, the behaviour of the flow was equivalent to the linearized flow in a small enough neighbourhood of the fixed point (see below). We shall further explore the relation between the linearized and the full problem, aiming to recognize the 'persistent' terms in the nonlinear equations, i.e., those (nonlinear) terms that *cannot* be removed with eventually nonlinear changes of coordinates. This is the cornerstone of the reduction to *normal forms* (see below).

7.2 Equivalence between flows

In order to give some rigorous content to the concept of bifurcation, we must specify what is meant by 'qualitative changes in the flow'. This is closely related to the concept of equivalence between flows, i.e.: when are two flows 'qualitatively' equivalent? This problem was addressed in chapter 5. Let us revisit it yet another time.

Two flows are *linearly equivalent* if there is a linear change of coordinates mapping one onto the other [arno83b]. For *linear* flows this amounts to demanding the coefficient matrices to be linearly equivalent. The classification of two-dimensional linear flows of chapter 2 rests upon this concept. Note that matrices with different eigenvalues are *not* linearly equivalent.

Exercise 7.1: Verify that the *nonlinear* transformation $w = x^{1/|\lambda|}$ maps the flow $\dot{x} = \lambda x$ onto $\dot{w} = \mathrm{sg}(\lambda)w$, where λ is a real, non-zero number (sg() means sign of, i.e. $\mathrm{sg}(\lambda) = \lambda/|\lambda|$).

In chapter 5 however, along with the discussion of linear stability analysis and the Hartman–Grobman theorem, we developed a more intuitive and less restrictive concept of equivalence, namely that two flows are *topologically orbitally equivalent* if there is a continuous, invertible change of coordinates (probably involving time) that maps one onto the other preserving time-orientation. As the previous exercise may suggest, topological orbital equivalence is a broader concept than linear equivalence. Let us educate our intuition with the following exercises:

Exercise 7.2: Show that the linear flow

$$\dot{x} = -ax - wy$$

$$\dot{y} = wx - ay$$

(a and w positive real numbers) is topologically orbitally equivalent to $\dot{X} = -X$, $\dot{Y} = -Y$. Proceed as follows:

- Take a point $p_i = (X, Y)$ and follow it forward (or backward) in time during a time $t = \ln R \equiv \ln(X^2 + Y^2)$ until it crosses the unit circle. You are now at the point $p_0 = (X/R, Y/R)$.
- Take the identity mapping between the unit circles in (X, Y) and (x, y) spaces.
- Follow back the point p_0 in (x, y) space during a time $t = -\ln R$. You arrive at a point p_f (compute its coordinates!).
- Verify that the mapping $p_i \mapsto p_f$ realizes the topological orbital equivalence between both flows.

Exercise 7.3: Try a similar procedure for the flow $\dot{x} = -ax + y$, $\dot{y} = -ay$.

The summarizing ideas are formulated in the following theorem:

Theorem (Ladis' theorem) [ladi73]: Two linear differential equations $dx/dt = Ax$ and $dy/dt = By$, $x, y \in \mathbb{R}^n$, are topologically (orbitally) equivalent if and only if the number of eigenvalues with negative (and positive) real parts of the operators A and B are equal, and if the restrictions of these operators to the invariant subspaces corresponding to the purely imaginary eigenvalues are linearly equivalent (in other words, the zero-real-part subspaces must have the *same* Jordan blocks).

Exercise 7.4: Show that the flows corresponding to two harmonic oscillators of different frequency are topologically orbitally equivalent but *not* linearly equivalent.

Now we can give content to the original phrasing of the bifurcation problem. A bifurcation point μ_0 in parameter space will be such that in any open neighbourhood of μ_0 we can find flows ϕ_μ which are *not* topologically orbitally equivalent to ϕ_{μ_0}. In other words, the flow at $\mu = 0$ in the parametric family is not *structurally stable*.

7.3 Conditions for fixed point bifurcations

In the discussion of the Lorenz model (chapter 3) we observed that the appearance of new fixed points for the system coincides with the change of stability of the original solution.

We would like to show that this is in fact the general case. The fixed points of a parametric family of flows, $dx_i/dt = f_i(x; a)$, are given by the zeros of the vector field

$$f_i(x; a) = 0, \ i = 1, \cdots, n \qquad (7.1)$$

(n the dimension of the phase space).

The location of the fixed points is given by the intersection of n hypersurfaces in phase space. Whenever this intersection is *transverse*, i.e., $\det(Df(x; a_0)) \neq 0$, the solutions of equation (7.1) form a unique continuous curve $x = x(a)$ in a small enough parameter interval around a_0, according to the implicit function theorem. It immediately follows that the necessary condition for the existence of degenerated fixed point solutions is the vanishing of the determinant, i.e., the existence of a zero eigenvalue of the fixed point. Moreover, if two (or more) fixed points are to coalesce into one fixed point for a given parameter value, say $a = a_0$, it is necessary that the curve $x = x(a)$ be multivalued with branching point at a_0, thus the coalescence of fixed points also implies the existence of at least a zero eigenvalue.

A similar reasoning can be followed concerning fixed points in maps (and hence periodic orbits) replacing the condition equation (7.1) by $x_i - f_i(x; a) = 0$. The non-degeneracy condition translates accordingly to demanding the eigenvalues of $Df(x; a)$ to be different from one in modulus.

The second kind of behaviour observed for fixed points in the Lorenz system appeared when the determinant of $Df(x; a)$ had a pair of imaginary, complex conjugated eigenvalues, with the disappearance of a periodic orbit. Here again we can argue that the change of stability is a necessary condition for the disappearance of the periodic orbit. Without loss of generality we can consider the fixed point to be at $x = 0$ for the bifurcation value of the parameter $a = 0$. If the matrix of the linearization $Df(0; 0)$ corresponds to a hyperbolic fixed point, then by the Hartman–Grobman theorem the nonlinear flow is equivalent to the linearized flow. Hence, there are no periodic orbits present and the fixed point $x = 0$ is either a sink, a source or a saddle. Moreover, for parameter values sufficiently near $a = 0$, the fixed point remains hyperbolic, since $Df(0; a)$ changes continuously. We conclude that if a periodic orbit is to coalesce into a fixed point, the fixed point cannot be hyperbolic at the bifurcation value.

In the coming sections we will develop the standard method of bifurcation analysis in two steps: reduction to the centre manifold and reduction to a *normal form*.

Exercise 7.5: Can the number of fixed points of a linear system $\dot{x} = A(a)x + b(a)$ *change* when a is varied? Under what conditions?

7.3.1 Unfolding of a bifurcation

As we mentioned above, if the determinant $\det(Df(x, a))$ is to be zero, some conditions will have to be imposed on a general flow. If we consider the space of all possible vector fields, those fields having a fixed point with $\det(Df(x, a)) = 0$ constitute a hypersurface of our space. In finite-dimensional spaces, the hypersurface would have dimension $d - 1$ (with d the original dimension). Since the space of vector fields has infinite dimensions, rather than speaking about the dimension of the hypersurface (which is also infinite) one speaks about the *co-dimension*, i.e., roughly speaking the dimension(s) 'lost' when imposing a restriction (in this case $\det(Df(x, a)) = 0$).

If we consider a one-parameter family of vector fields, this family would be described by a curve. This curve will intersect transversely the hypersurface $\det(Df(x, a)) = 0$ except when very specific conditions are fulfilled. Note that a small perturbation can destroy the specificity and restore the transversal crossing. The transversal intersection, by contrast, will persist in spite of small perturbations of the family of flows.

It is in this sense that we say that a bifurcation has *co-dimension* one. It is enough to consider a one-parameter family of flows to describe the bifurcation point and the modifications of the flow in the surroundings of the bifurcation.

This argument can be extended in a natural way. There may exist two hypersurfaces describing zero eigenvalues, and they may intersect along some curve (despite infinite dimensionality, let the dimension $d = 3$ to help us visualize the discussion). The simplest parametric family of flows that necessarily will include this intersection must be a two-parameter family, i.e., it has to look like a 'plane' in order to assure that the intersection curve will be crossed for some parameter values (in a way which is persistent in front of small perturbations).

A *bifurcation diagram* is a plot (in parameter space) indicating for which parameter values there is a bifurcation point. The previous example would be a 2-d plot having two curves (each one describing the intersection of our two-parameter family with one of the hypersurfaces $\det(Df(x, a)) = 0$) that intersect at one point. The bifurcation diagram is usually accompanied with a qualitative description of the flow in the dynamically different regions resulting in parameter space.

In general, the problem of *unfolding* a bifurcation is that of giving the simplest and most general parametric family of flows that under all perturbations will present the same flow types and the same (distorted) bifurcation diagram. We could say that the family is structurally stable in the 'space of families of vector fields'. The *complete* unfolding of a bifurcation is interesting from the mathematical point of view. In applications, one is usually satisfied with the (part of the) unfolding that is dictated by the specific physical, biological, etc problem.

7.4 Reduction to the centre manifold

7.4.1 Adiabatic elimination of fast-decaying variables

We have established that the topological changes associated with a bifurcation of fixed points in a parametric family of dynamical systems occur when the fixed points change stability. Similar arguments using the implicit function theorem can convince us that these changes are restricted to the centre manifold associated with the neutral directions at the bifurcation, i.e., the directions associated with eigenvalues with zero real part. In this direction, this section presents the methods that allow us to decouple (up to a certain degree) the dynamics in the centre manifold with respect to the dynamics in the stable and unstable manifolds.

Let us begin with an example. We can consider the problem

$$dx/dt = ax^2 + bxy \qquad (7.2)$$
$$dy/dt = -y + x^2. \qquad (7.3)$$

Let us consider the dynamics in a small neighbourhood of the origin, i.e., where x and y are sufficiently smaller than one. We can argue that since the coordinate x has derivatives of order[†] $O(x^2, y^2)$, while the y coordinate evolves with derivatives of order $O(|y|)$, the variable y will evolve faster and decay towards the 'equilibrium value' obtained by setting $dy/dt = 0$, i.e. $y \approx x^2$. This procedure is known as *adiabatic elimination of the fast variables* and has been widely used in physics [hake83, risk89, sola94].

The result is an equation for the 'slow' variable x in the form

$$dx/dt = ax^2 + bx^3 \qquad (7.4)$$

valid for x sufficiently small. The procedure results in a reduced set of (approximated) equations for the slow variable and can be recast in terms of changes of variables.

Considering again the original example, equations (7.2) and (7.3), we propose the following change of variables:

$$x' = x + bxy \qquad (7.5)$$
$$y' = y - x^2. \qquad (7.6)$$

Note that the previous adiabatic elimination would now consist of setting $y' = 0$. In the new coordinate system the equations (7.2) and (7.3) read

$$dx'/dt = ax'^2 + O(3) \qquad (7.7)$$
$$dy'/dt = -y' + O(3) \qquad (7.8)$$

---

† We will often use the shorthand notation O(2), etc, indicating only the power involved, while the coordinates are implicitly understood.

which advances one step further towards separating the system into two decoupled components. In the new coordinates, x', y', the equations are separated up to second order. Thus, the adiabatic elimination can be regarded as the *asymptotic dynamics* of equations (7.7) and (7.8) (loosely speaking, when y' can for practical purposes be equated to zero).

The method of adiabatic elimination of fast variables is the first step in a reduction process where the original equations are transformed by way of nonlinear changes of variables into approximated equations valid up to order $O(n)$ (i.e., in a sufficiently small neighbourhood of the fixed point), where the dynamics on the centre manifold is decoupled from the rest of the system. We will formalize the general procedure in the next subsection.

7.4.2 Approximation to the centre manifold

We will consider fixed points of differential equations of the form

$$\begin{aligned}
\mathrm{d}x/\mathrm{d}t &= Ax + f(x, y, z) \\
\mathrm{d}y/\mathrm{d}t &= B_+ y + g(x, y, z) \\
\mathrm{d}z/\mathrm{d}t &= B_- z + k(x, y, z)
\end{aligned} \tag{7.9}$$

possibly after shifting the fixed point to zero and transforming the linear term to the corresponding Jordan normal form. f, g and k contain no constant or linear terms, the eigenvalues of B_+ (B_-) all have positive (negative) real part, and the eigenvalues of A all have zero real part.

Recalling the Hartman–Grobman theorem, the fact that (locally) the only *essential* nonlinearities are associated with the centre manifold should by now be natural. All other nonlinearities can be dealt with via changes of coordinates (again *locally*). We give without proof the following splitting theorem:

Theorem (A N Šošitaĭšvili) [arno88a, sosi73]: Suppose that a differential equation with C^2 right hand side has a singular point 0 and a linear part as in equation (7.9). Let E^s, E^u and E^c be the invariant subspaces corresponding to the (linear) flow of B_-, B_+ and A. Then, in a neighbourhood of the singular point 0, the equation under consideration is (topologically) equivalent to the direct product of two equations: the restriction of the original equation to the centre manifold, on one side, and on the other side the saddle:

$$\begin{aligned}
\mathrm{d}y'/\mathrm{d}t &= B_+ y' \\
\mathrm{d}z'/\mathrm{d}t &= B_- z'.
\end{aligned} \tag{7.10}$$

It follows from the theorem that only the dynamics in the centre manifold requires further study involving the nonlinear terms. Hence the idea of reduction of the equation to the centre manifold.

In what follows we will group the unstable and stable manifolds in a single manifold since they can be treated in a unified way in the reduction to centre manifold. We simply drop the $+$ and $-$ subindices and use the variable y for (y, z) to lighten the notation.

In order to make practical use of the theorem we need approximate expressions for the dynamics in the centre manifold. These expressions can be obtained as follows:

- The centre manifold (W^c) is tangent to the centre subspace at 0, hence, we can write for $(x, y) \in W^c$, $(x, y) = (x, h(x))$ with $h(0) = 0$ and $\lim_{|x| \to 0} |h(x)|/|x| = 0$.
- The centre manifold is invariant, hence

$$dx/dt = Ax + f(x, h(x)) \tag{7.11}$$
$$dy/dt = dh/dx\, dx/dt = Bh(x) + g(x, h(x)). \tag{7.12}$$

- Equation (7.11) provides the dynamics in the centre manifold while equation (7.12) provides a functional equation for $h(x)$.

Solving the above equations is equivalent to solving the original set of equations. This task is expected to be as difficult as the original problem. However, equation (7.12) can be solved iteratively in the form

$$h_1 = 0 \tag{7.13}$$
$$h_{n+1}(x) = h_n(x) + \Delta_{n+1}(x) \tag{7.14}$$
$$L[\Delta_{n+1}] = d\Delta_{n+1}/dx\, Ax - B\Delta_{n+1}(x) \tag{7.15}$$
$$= Bh_n(x) + g(x, h_n(x))$$
$$- dh_n/dx(Ax + f(x, h_n(x))) + O(n + 2). \tag{7.16}$$

The following exercise gives a hint towards what Δ_{n+1} is.

Exercise 7.6: Verify that the right hand side of the last equation is indeed of order $(n + 1)$ in x. Hint: Recall that f and g are of order two or higher. Use the induction principle.

We can hence devise a systematic iterative procedure that makes it easier to keep track of the different orders:

- At each iteration m, expand the right hand side of equation (7.16) in power series and *truncate* it at order $m + 1$.
- Choose Δ_{m+1} as the homogeneous polynomial of degree $m + 1$ in x that solves the truncated equation. In other words, take Δ_{m+1} of order $m + 1$ such that the following equation is satisfied:

$$L[\Delta_{n+1}] + O(n + 2) = \begin{aligned} &Bh_n(x) + g(x, h_n(x)) \\ &-dh_n/dx(Ax + f(x, h_n(x))). \end{aligned} \tag{7.17}$$

In this way, all contributions to h of degree $m+1$ or lower have been determined at step m or earlier. The right hand side of equation (7.16) can be written as $(Mh_n)(x)$ with

$$(M\Phi)(x) = B\Phi(x) + g(x, \Phi(x)) - d\Phi/dx(Ax + f(x, \Phi(x))). \qquad (7.18)$$

The following theorem establishes the validity of the iterative approximation to the centre manifold just constructed.

Theorem (Approximation to the centre manifold) [carr81]: Suppose that $\Phi(0) = 0$, $d\Phi/dx(0) = 0$ and that $(M\Phi(x)) = O(|x|^q)$ where $q > 1$. Then, as $x \to 0$,

$$|h(x) - \Phi(x)| = O(|x|^q). \qquad (7.19)$$

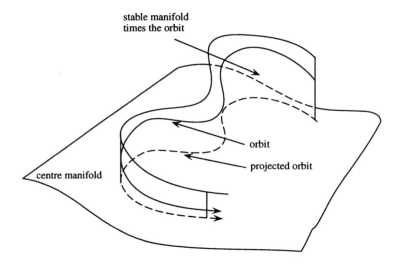

Figure 7.1. Sketch of the projection onto the centre manifold.

At this point we feel the urgent need of presenting a few examples. In figure 7.1 we depict the idea behind the projection onto the centre manifold. As a first example, consider the system

$$
\begin{aligned}
dx/dt &= -xy \\
dy/dt &= -y + zx \\
dz/dt &= y^2.
\end{aligned} \qquad (7.20)
$$

The y axis is the stable subspace while the x–z plane is the centre subspace. We seek a centre manifold of the form $(x, h(x, z), z)$. The $n + 1$ correction to h will satisfy

$$\Delta_{n+1} = -h_n + zx + \partial h_n/\partial x \, xh_n - \partial h_n/\partial z \, h_n^2. \qquad (7.21)$$

The lowest orders in this approximation are

$$h_1 = 0$$
$$h_2 = \Delta_2 = zx$$
$$h_3 = h_2 \quad \text{since } \Delta_3 = 0$$
$$h_4 = h_3 + \Delta_4 = zx(1 + zx).$$

Our second example is related to what in the coming sections we will learn to recognize as a *Hopf bifurcation*. The equation reads

$$
\begin{aligned}
dx/dt &= y + xz \\
dy/dt &= -x + yz \\
dz/dt &= -z + xy.
\end{aligned}
\tag{7.22}
$$

The centre subspace is the (x, y) plane and the stable subspace the z axis. The centre manifold reads $(x, y, h(x, y))$ and the equation for the successive corrections to h_n reads

$$
\begin{aligned}
&y \, \partial\Delta_{n+1}/\partial x - x \, \partial\Delta_{n+1}/\partial y + \Delta_{n+1} \\
&= -h_n + xy - (y + xh_n)\partial h_n/\partial x - (-x + yh_n)\partial h_n/\partial y.
\end{aligned}
\tag{7.23}
$$

To lowest order (two) we can write $\Delta_2 = ax^2 + by^2 + cxy$. Keeping terms up to second order only in equation (7.23) we find an algebraic equation

$$
\begin{aligned}
&2axy + cy^2 - 2byx - cx^2 + ax^2 + by^2 + cxy \\
&= (a - c)x^2 + (b + c)y^2 + (2a - 2b + c)xy = xy
\end{aligned}
\tag{7.24}
$$

which solves in $a = c = -b = 1/5$ and $h_2 = (x^2 - y^2 + xy)/5$.

7.4.3 Centre manifold for maps

The application of the centre manifold reduction to maps is relatively straightforward. The theorems on the splitting of the dynamics and the approximation to the centre manifold, in particular, hold also in the case of maps after adequate re-stating. We review in this subsection the main highlights of the matter.

The general map under consideration is of the form

$$x_{n+1} = Ax_n + f(x_n, y_n) \tag{7.25}$$
$$y_{n+1} = By_n + g(x_n, y_n) \tag{7.26}$$

where A is a matrix having all eigenvalues of absolute value unity and B having all eigenvalues of absolute value different from unity. As in the case of flows,

- The centre manifold is tangent to the centre subspace (which now is the subspace associated with eigenvalues of absolute value unity) and has the expression $(x, h(x))$, $h(0) = 0$ and $Dh(0) = 0$.
- The centre manifold is invariant,

$$x_{n+1} = Ax_n + f(x_n, h(x_n)) \qquad (7.27)$$

$$h(x_{n+1}) = Bh(x_n) + g(x_n, h(x_n)). \qquad (7.28)$$

The crucial difference with the case of flows lies in the form in which we have to solve the functional equation for the centre manifold in the recursive procedure.

In order to solve equation (7.28) up to order $m + 1$ with $h_{m+1} = \Delta_{m+1} + h_m(x)$, we require Δ (we drop for simplicity the index $m + 1$) to satisfy

$$\Delta(x_{n+1}) + h_m(x_{n+1}) = B(h_m(x_n) + \Delta(x_n)) + g(x_n, h_{m+1}(x_n)) \qquad (7.29)$$

$$= \Delta(Ax_n + f(x_n, h_{m+1}(x_n)))$$

$$+ h_m(Ax_n + f(x_n, h_{m+1}(x_n))) \qquad (7.30)$$

or

$$\Delta(Ax_n) - B\Delta(x_n) = Bh_m(x_n) + g(x_n, h_m(x_n))$$

$$- h_m(Ax_n + f(x_n, h_m(x_n))) \qquad (7.31)$$

up to order $n + 1$. Again, we have dropped higher-order terms. We see that the main difference between equation (7.31) and equation (7.16) consists in the form of the linear operator acting on Δ, but both cases (maps and flows) follow the same general idea.

Let us work through an example. Consider the fixed point at $(0, 0)$ of the map

$$x_{n+1} = -x_n + 2(x_n - y_n)^2$$

$$y_{n+1} = -y_n/2 + (x_n - y_n)^2.$$

In this case $A = -1$ while $B = -1/2$. Observe also that the map is orientation preserving as could be expected of a Poincaré map. We look for a centre manifold of the form $(x, h(x))$. To the lowest non-trivial order, the centre manifold at $(0, 0)$ is $(x, 2x^2/3)$ and the map restricted to the centre manifold results in $u_{n+1} = -u_n + 2u_n^2 - 8u_n^3/3$ which is an orientation reversing map. From this example we conclude that orientation reversing maps may arise when we study the reduction to the centre manifold of Poincaré maps (recall the discussion of chapter 4).

7.4.4 Bifurcations

We have discussed, so far, how to describe the dynamics near a fixed point of a flow at the bifurcation set, when the linear dynamics is not equivalent to the full dynamics no matter how small is the neighbourhood (around the bifurcation set) in parameter space where we perform the comparison.

Suppose that we have solved the dynamical problem at the bifurcation; it still remains to be shown how the dynamics of this family of dynamical systems is deformed when the control parameters are varied within this neighbourhood of the bifurcation set.

This task is facilitated by considering the parameters of the family as variables belonging to the centre manifold which are constant in time. To fix ideas, let $dx/dt = f(x; \mu)$ be the family in study. $x = 0$ is a fixed point for $\mu = 0$, i.e., $(f(0; 0) = 0)$. At $\mu = 0$ this fixed point has undetermined linear stability, or, equivalently, some of the eigenvalues of $Df(0; 0)$ have their real parts equal to zero. We consider instead the equivalent problem:

$$\dot{x} = f(x; \mu) \tag{7.32}$$
$$\dot{\mu} = 0. \tag{7.33}$$

The centre subspace becomes the direct sum of the centre subspace of $dx/dt = f(x; 0)$ and the one-dimensional subspace given by $(x; \mu) = (0; \mu)$, and we can apply the previous method, reducing the dynamics to something of the form

$$\dot{x}_c = \Phi(x_c, \mu) \tag{7.34}$$
$$\dot{\mu} = 0 \tag{7.35}$$

where x_c are the variables on the centre manifold.

7.5 Normal forms

The final step in producing prototypes for the different classes of bifurcations consists in making use of the linear approximation to the dynamics on the centre manifold. If the linear part is identically zero we are done, at least for the moment. If the linear part is non-zero we can still make some changes of coordinates trying to remove as many terms as possible. We will elaborate on this idea and become more precise in this section. The 'final' system (i.e., what is left after the coordinate changes) is called the *normal form* of a bifurcation.

7.5.1 Example: Jordan block

There is a good amount of algebraic manipulations in the reduction to the normal form. We will illustrate the procedure in one of the simplest (non-trivial) examples before enunciating the general form.

Consider the bifurcation to a pair of (complex conjugated) imaginary eigenvalues of the two-dimensional double-zero Jordan block:

$$\dot{u} = v + f(u, v) \tag{7.36}$$
$$\dot{v} = 0 + g(u, v) \tag{7.37}$$

or considering the vectors $x = (u, v)^t$, $k = (f, g)^t$ and the matrix $A = \begin{pmatrix} 0 & 1 \\ 0 & 0 \end{pmatrix}$

$$\dot{x} = Ax + k(x). \tag{7.38}$$

We seek a 'near-identity' change of coordinates, $x = y + h(y)$ that simplifies the vector field as much as possible in a sense that will become clear soon. In the new coordinate system the equation equation (7.38) reads

$$\dot{y} = Ay + F(y) \tag{7.39}$$

and h satisfies (we specify F—together with h—below):

$$(I + dh/dy)(Ay + F(y)) = Ay + Ah(y) + k(y + h(y)). \tag{7.40}$$

We can consider, as a first approximation, the Taylor expansion up to second order of equation (7.40), thus obtaining

$$F_2(y) = Ah_2(y) - \frac{dh_2}{dy} Ay + k_2(y) \tag{7.41}$$

where F_2, h_2 and k_2 are vectors of the form (recall that we are considering a two-dimensional problem)

$$F_2, h_2, k_2 = a_i \begin{pmatrix} u^2 \\ 0 \end{pmatrix} + b_i \begin{pmatrix} uv \\ 0 \end{pmatrix} + c_i \begin{pmatrix} v^2 \\ 0 \end{pmatrix}$$
$$+ d_i \begin{pmatrix} 0 \\ u^2 \end{pmatrix} + e_i \begin{pmatrix} 0 \\ uv \end{pmatrix} + f_i \begin{pmatrix} 0 \\ v^2 \end{pmatrix} \tag{7.42}$$

letting $i = F_2, h_2, k_2$.

The expression $Ah(y) - dh/dy\, Ay$ can be thought of as a linear operator acting on the space of vector fields. Moreover, it preserves the degree of polynomials. Therefore, it maps the subspace of vector fields having homogeneous polynomials of degree n as components, into itself. Because of linearity, using equation (7.42) the expression $L(h_2) = Ah_2(y) - dh_2/dy\, Ay$ resolves into

$$a_h L \begin{pmatrix} u^2 \\ 0 \end{pmatrix} + b_h L \begin{pmatrix} uv \\ 0 \end{pmatrix} + c_h L \begin{pmatrix} v^2 \\ 0 \end{pmatrix}$$
$$+ d_h L \begin{pmatrix} 0 \\ u^2 \end{pmatrix} + e_h L \begin{pmatrix} 0 \\ uv \end{pmatrix} + f_h L \begin{pmatrix} 0 \\ v^2 \end{pmatrix} \tag{7.43}$$

which can be further resolved into

$$a_h \begin{pmatrix} -2uv \\ 0 \end{pmatrix} + b_h \begin{pmatrix} -v^2 \\ 0 \end{pmatrix} + c_h \begin{pmatrix} 0 \\ 0 \end{pmatrix}$$
$$+ d_h \begin{pmatrix} u^2 \\ -2uv \end{pmatrix} + e_h \begin{pmatrix} uv \\ -v^2 \end{pmatrix} + f_h \begin{pmatrix} v^2 \\ 0 \end{pmatrix}. \tag{7.44}$$

The coefficients $\{a_h, b_h, c_h, d_h, e_h, f_h\}$ can then be chosen so that F_2 depends on the minimum possible number of coefficients, or, what is exactly the same, h_2 is chosen *to cancel as many terms of k_2 as possible*. The terms which cannot be cancelled will be part of the vector field written in its *normal form*, which in this case reads

$$\begin{pmatrix} \dot{u} \\ \dot{v} \end{pmatrix} = A \begin{pmatrix} u \\ v \end{pmatrix} + a \begin{pmatrix} 0 \\ u^2 \end{pmatrix} + b \begin{pmatrix} 0 \\ uv \end{pmatrix} + \text{HOT}. \qquad (7.45)$$

The election of the vector field in the normal form is far from being unique. The student should verify that there are various (and perhaps more obvious) choices. All these choices, however, have only two independent coefficients and are equivalent under changes of coordinates.

Exercise 7.7: Verify the previous statement.

The process of determining the normal form continues by repeating the procedure with a third-order (and later higher-order) truncation of the equations and the selection of the proper change of coordinates at each level. The algebraic complexity (and the tedium!) increases geometrically. Fortunately, only the lower-order terms are usually necessary to determine the dynamics.

Exercise 7.8: When is it sensible to stop the power expansion of the normal form?

7.5.2 The general case

Let

$$\dot{x} = Ax + k(x) = Ax + \sum_{n \geq 2} k_n(x) \qquad (7.46)$$

be the reduced equation in the centre manifold. Let

$$x = y + h(y) = y + \sum_{n \geq 2} h_n(y) \qquad (7.47)$$

be the near-identity change of coordinates given in an implicit (or inverse) form. Since the nonlinearity is *intrinsic* in the centre manifold we can have no hope of simplifying the vector field as far as to eliminate all the nonlinear terms. We will aim, then, to obtain a normal form of the general type

$$\dot{y} = Ay + F(y) = Ay + \sum_{n \geq 2} F_n(y) \qquad (7.48)$$

where F will have to be determined simultaneously with the change of coordinates. As stated above, we attempt to choose h_n in such a way to render F_n as simple as possible.

Here k_n, h_n and F_n stand for vector fields having homogeneous polynomials of degree n as components.

Replacing the expression of the change of coordinates equation (7.47) and the normal form equation (7.48) in the original equation (7.46), we obtain

$$(I + dh/dy)(Ay + F(y)) = A(y + h(y)) + k(y + h(y)) \qquad (7.49)$$

(cf equation (7.40)). Note the occurrence again of the linear operator

$$L(h) = Ah(y) - dh/dy \, Ay, \qquad (7.50)$$

acting on the space of vector fields. Since L acting on vector fields that have as components homogeneous polynomials of degree n, H_n, returns another element belonging to the same subspace H_n, we can consider the action of L as the direct sum of its action on each of these subspaces. Hence, we can solve the equation above with an iterative scheme attempting to determine F and h order by order:

$$h_1 = 0 \qquad (7.51)$$

$$F_{n+1}(y) = \left(k\left(y + \sum_{i=1}^{n} h_n(y) \right) \right)_{n+1} + Ah_{n+1}(y) - \frac{dh_{n+1}}{dy} Ay$$

$$= k_{n+1}(y) + L(h_{n+1}) \qquad (7.52)$$

where at each order $n+1$ ($n \geq 1$), we choose our change of coordinates h_{n+1} in order to cancel as many terms as possible in k_{n+1}, the $(n+1)$th order expansion of f evaluated with the transformed coordinates of step n. F_{n+1} is computed at each step as whatever terms of k_{n+1} are left uncancelled. Keep in mind that the goal is to make the dynamical system as simple as possible, i.e., to choose h such as F becomes as simple as possible.

The terms that can be removed from f are precisely those that are in the image of the linear operator L. The subspace H_n can hence be decomposed into $H_n = L(H_n) + G_n$; $L(H_n)$ is the image by L of H_n and G_n an arbitrarily fixed complement of $L(H_n)$ in H_n so that each element $h_n \in H_n$ can be written in a unique form as $h_n = l_n + g_n$ with $l_n \in L(H_n)$ and $g_n \in G_n$. Clearly, F_n, i.e., those elements that cannot be removed with a change of coordinates, belong to G_n. The following theorem states the general result:

Theorem (normal forms) [guck86]: Let $dx/dt = f(x)$ be a C^r system of differential equations with $f(0) = 0$ and $Df(0) = A$. Choose a complement G_k for $L(H_k)$ in H_k, so that $H_k = L(H_k) + G_k$. Then, there is an analytic change of coordinates in a neighbourhood of the origin which transforms the system $dx/dt = f(x)$ to $dy/dt = g(y) = Ay + g_2(y) + \ldots + g_r(y) + R_r$ with $g_k \in G_k$ for $2 \leq k \leq r$ and $R_r = O(|y|^{r+1})$.

7.5.3 Normal forms for maps

The reduction to the normal form for maps can be achieved following the same principles (but not the same formulæ) used for flows. Once the system is reduced (in a neighbourhood of a fixed point) to the centre manifold, we have a problem of the form

$$x_{n+1} = Ax_n + f(x_n) \tag{7.53}$$

where A is a matrix having all eigenvalues of modulus one and $f(0) = 0$, $Df(0) = 0$.

We seek near-identity changes of coordinates, $x = y + h(y)$, $h(0) = 0$, $Dh(0) = 0$, that eliminate as many elementary terms from the map as possible. We have

$$y_{n+1} + h(y_{n+1}) = Ay_n + Ah(y_n) + f(y_n + h(y_n))$$
$$y_{n+1} = Ay_n + F(y_n) \tag{7.54}$$

where F represents, again, the (eventual) impossibility of eliminating all non-linear terms of the original equation by changes of coordinates.

As the reader may at this point suspect, we solve the equation in successive attempts, describing h and F as sums of homogeneous polynomials of order m, $m \geq 2$. Using the formal definition of y_{n+1} above, we have

$$h_m(y_{n+1}) = h_m(Ay_n + O(|y|)) \approx h_m(Ay_n) + O(|y|^{m+1})$$

$$y_{n+1} + \sum_p h_p(y_{n+1}) = Ay_n + A \sum_p h_p(y_n) + \sum_p \left(f\left(y_n + \sum_q^{p-1} h_q(y_n) \right) \right)_p$$

$$= Ay_n + \sum_p F_p(y_n). \tag{7.55}$$

Note that the truncation of the sum inside f is possible since the higher-order polynomials h_r, $r \geq p$ do not contribute to order p. Putting together the three equations above, we finally have to solve for each $p \geq 2$ a problem of the form

$$F_p(y_n) = \left(-h_p(Ay_n) + Ah_p(y_n) \right) + \left(f\left(y_n + \sum_q^{p-1} h_q(y_n) \right) \right)_p. \tag{7.56}$$

The relevant linear operator is then

$$L(h_p) = Ah_p(y_n) - h_p(Ay_n). \tag{7.57}$$

Again F_p will be that part of f_p that cannot be eliminated with a change of coordinates governed by L. In other words, the normal form theorem reads now exactly as the normal form theorem for flows except that the linear operator L has the meaning given above.

Exercise 7.9: For any flow $\phi(x,t)$, we can define associated maps $F_s(x) = \phi(x,s)$, by choosing any fixed time s. If 0 is a fixed point of the flow, then 0 is a fixed point of the map F (by definition of fixed point). For small t the flow can be approximately described by $\phi(x,t) = x + f(x)t + O(t^2)$ (this induces an approximate description for F_t). Show that the normal form theorem for maps reduces to the normal form theorem for flows in the limit $t \to 0$.

7.5.4   Resonances

The case of a fixed point of a flow (map) in R^n with n different eigenvalues allows us to get a better idea of the nature of the vector fields appearing in the normal form. Consider the system described by

$$dx_i/dt = \lambda_i x_i + f_i(x_1, \ldots, x_n). \tag{7.58}$$

The linear operator L, equation (7.57), equation (7.50), both leaves the subspace H_k of vector fields with polynomials of degree k as components invariant, and does not mix different components within each k, i.e. *monomials* of degree k are left invariant by L:

$$L \begin{pmatrix} p_1 \\ \cdots \\ p_n \end{pmatrix} = \begin{pmatrix} L_1 p_1 \\ \cdots \\ L_p p_n \end{pmatrix}. \tag{7.59}$$

Let us represent a polynomial of degree k by a column vector, where each entry contains a different monomial (the polynomial would be, then, the sum of the column entries). The action of L on a vector with all components equal to zero, except the jth which reads $\prod_{k=1}^{n} x_k^{q_k}$, results in a new vector of the same type ($v_i = 0$, $i \neq j$) where the non-zero component reads $v_j = (\prod_{k=1}^{n} x_k^{q_k})(\lambda_j - \prod_{i=1}^{n} \lambda_i^{q_i})$ in the case of maps and $(\prod_{k=1}^{n} x_k^{q_k})(\lambda_j - \sum_{i=1}^{n} q_i \lambda_i)$ in the case of flows.

The terms of f that cannot be removed are those that are in *resonance* with the frequency of the corresponding coordinate, i.e., when the powers q_k are such that $\prod_{k=1}^{n} \lambda_k^{q_k} = \lambda_j$ in the case of maps and $\sum_{k=1}^{n} q_k \lambda_k = \lambda_j$ in the case of flows.

For example, in the system

$$\begin{aligned} dx/dt &= x + f(x, y) \\ dy/dt &= -y + g(x, y), \end{aligned} \tag{7.60}$$

which is a regular saddle (not a fixed point at a bifurcation), the normal form will read

$$\begin{aligned} du/dt &= u + au^2v + \text{HOT} \\ dv/dt &= -v + bv^2u + \text{HOT}. \end{aligned} \tag{7.61}$$

7.5.4.1 Small denominators

The reader might feel that there is something odd about the last statements. If we slightly modify equation (7.60) so that the ratio between the positive and negative eigenvalues becomes irrational, all terms of any order could be removed. Thus, a small (as small as we like) change in the original equation would result in a very relevant change in the normal form.

This fact raises itself a second objection. Even if the quotient between the eigenvalues is irrational, the expression $\lambda_1 n_1 + \lambda_2 n_2$ can be arbitrarily close to zero with an adequate choice of positive integers n_1, n_2, provided the eigenvalues are of opposite signs. Hence, arbitrarily small changes in the dynamical equations may turn the system into resonance or near-resonance conditions. This problem is known as the *problem of small denominators* [mars76].

A key to understanding these apparent contradictions is to realize that the proposed changes of coordinates leading to the normal form of a dynamical system are *formal* in nature. The question of whether the series of changes of coordinates converges or not has not even been touched yet. If all the eigenvalues have negative (or positive) real part, a positive answer for the question of convergence is easy to find (in the case of flows, it amounts to determining whether there exist positive integers such that $\sum_{k=1}^{n} q_k \lambda_k = \lambda_j$).

If the fixed point in question is of saddle type the problem is far more complex. At each change of coordinates in the normal form expansion, we introduce coefficients with factors $\left(\prod_{k=1}^{n} \lambda_k^{q_k} - \lambda_j \right)^{-1}$ in the case of maps and $\left(\sum_{k=1}^{n} q_k \lambda_k - \lambda_j \right)^{-1}$ in the case of flows, which arise from the action of L. For near-resonant terms the factors are large. Consequently, we can expect that the domain of validity (in phase space) of the change of coordinates will be smaller than for more regular terms.

Related to this problem of convergence is the difficulty in the Hartman–Grobman theorem in realizing the equivalence with the linearized flow using a transformation smoother than C^0.

7.6 Simplest local bifurcations of fixed points

The bifurcation of a fixed point at a zero eigenvalue of the linearization matrix is the simplest bifurcation we can think of. The centre manifold is one dimensional and, in addition, the linear operator associated with the normal form, L, equation (7.57), is identically zero, i.e., no term can be removed!

In the following subsections we discuss the general form of this bifurcation problem (intuitively, the form we are most likely to find if we choose an arbitrary dynamical system), the *saddle-node* bifurcation, and some particular cases (*non-generic* cases) which are found when additional conditions are imposed to a general flow (the concept of *genericity* is subtler than this intuitive presentation. The curious reader may consult [wigg90, pp 98]).

7.6.1 Saddle-node bifurcation (flows)

The bifurcating equation reads (after truncation and re-scaling of the variables)

$$\dot{x} = \mu \pm x^2. \tag{7.62}$$

The saddle-node bifurcation rests on two assumptions: first we assume that the second order term does not vanish, and second we assume that the parameter enters already at the zeroth order of the vector field. In general this will be the case. Note that this is one of the features of the (undiscussed) concept of genericity: unless a sharp limiting condition is imposed on a family of flows, there exists a flow in which the bifurcation parameter enters at the zeroth order, 'arbitrarily close' to any given flow. Observe that the inclusion of a second small parameter multiplying a linear term would have no effect on the qualitative behaviour of the flow.

Exercise 7.10: Reduce the equation $\dot{x} = \mu + \nu x - x^2$ with ν small to the form of equation (7.62).

Let us consider the minus sign only in equation (7.62). For $\mu < 0$ there are no fixed points in the vicinity of $x = 0$ where the normal form is valid. For $\mu > 0$ there are two fixed points that vary with the square root of the parameter $x_{\pm} = \pm \mu^{1/2}$; for $\mu = 0$ both fixed points coalesce. Furthermore, for $\mu > 0$ the fixed point $\mu^{1/2}$ is stable and $-\mu^{1/2}$ is unstable (see figure 7.2). Note in figure 7.2 that the flow remains unchanged *outside* the region between the fixed points.

Exercise 7.11: Study the case with the plus sign in equation (7.62).

7.6.2 Pitchfork and transcritical bifurcations

The intuitive idea behind the study of non-generic cases of the bifurcation at a single zero eigenvalue is to consider which special restrictions to the flow will substantially modify the behaviour described by equation (7.62).

We will consider two subcases that occur often in applications, namely the *pitchfork* bifurcation and the *transcritical* bifurcation.

7.6.2.1 Pitchfork bifurcation

The most common reason for the second-order term in equation (7.62) to be zero is the presence of a symmetry. When a vector field is antisymmetric upon reflexion, i.e., when $f(x) = -f(-x)$, no terms of even order appear in the normal form. In such a case, the first non-zero term is of order three and the bifurcation parameter can only enter through the linear term (the symmetry forces

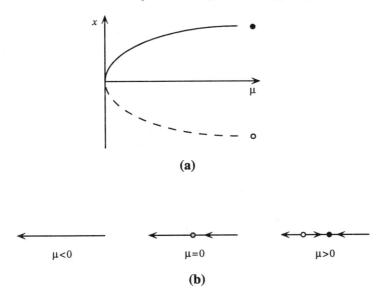

(a)

$\mu<0$ $\mu=0$ $\mu>0$

(b)

Figure 7.2. Saddle-node bifurcation for flows. (a) Fixed points (dashed line for unstable and solid line for stable fixed point). (b) Flow for μ smaller, equal and larger than zero.

also the zeroth order to vanish). The normal form becomes (after truncation and rescaling)

$$\dot{x} = \mu x \pm x^3. \tag{7.63}$$

Taking the minus sign in equation (7.63) we observe that for $\mu < 0$ there is one fixed point at $x = 0$, which is stable. For $\mu > 0$ there are three fixed points: two stable at $x = \pm\mu^{1/2}$ and one unstable at $x = 0$. Taken as a block (see figure 7.3) the stability of a sufficiently large neighbourhood of zero remains unchanged (recall the local character of this bifurcation).

Exercise 7.12: Identify a pitchfork bifurcation in the Lorenz example of chapter 3. Try to perform a centre manifold and normal form analysis of the flow.

7.6.2.2 *Transcritical bifurcation*

The transcritical bifurcation occurs when the control parameter does not enter as a coefficient for the zero power of x in the vector field but rather for the first power. The difference with the previous subcase is that we do not impose the antisymmetry restriction to the flow. Only the (weaker) constraint of vanishing

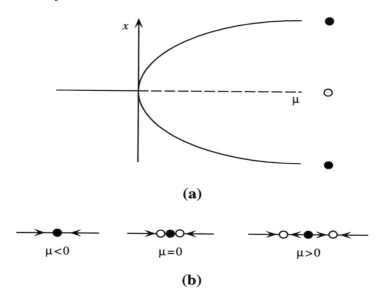

(a)

(b)

Figure 7.3. Pitchfork bifurcation. (a) Fixed points. (b) Flow for μ smaller, equal and larger than zero.

zeroth order is active. The normal form reads

$$dx/dt = \mu x \pm x^2. \tag{7.64}$$

Taking the minus sign in equation (7.64) we see that in this case we have fixed points at $x = \mu$ and $x = 0$ that coalesce for $\mu = 0$. For $\mu < 0$ the fixed point at $x = 0$ is stable and the one at $x = \mu$ unstable while for $\mu > 0$ the situation has been reversed (see figure 7.4).

7.6.2.3 *Structural stability*

The unfolding of the pitchfork and transcritical bifurcation is structurally unstable.

If we add a small term ϵx^2 to the normal form of the pitchfork bifurcation equation (7.63) we observe that it breaks into a saddle-node plus a fixed point not involved in the bifurcation (figure 7.5).

If we add a constant term ϵ to the transcritical bifurcation equation (7.64) we observe that either no bifurcation occurs or two saddle-node bifurcations (at the parameter values $\mu = \pm 2\sqrt{|\epsilon|}$) occur depending on the relation between the sign of the nonlinear term and that of ϵ (figure 7.6).

Exercise 7.13: Verify these statements.

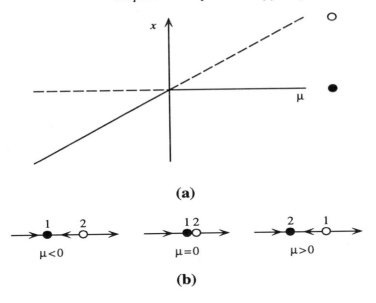

Figure 7.4. Transcritical bifurcation. (a) Fixed points. (b) Flow for μ smaller, equal and larger than zero.

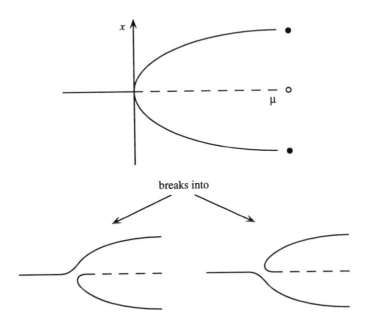

Figure 7.5. Perturbed pitchfork bifurcation.

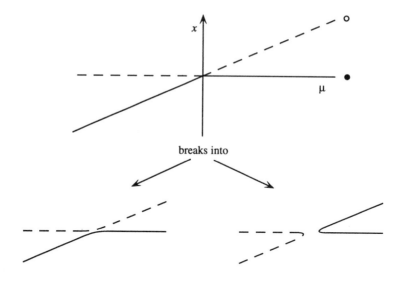

Figure 7.6. Perturbed transcritical bifurcation.

If one is presented with the centre manifold reduction of a flow, $\dot{x} = f_\mu(x)$, with f e.g. an analytic function having the bifurcating fixed point $x = 0$ for $\mu = 0$, there is a simple way of checking to which of the above bifurcation cases it belongs, by studying the derivatives of f at the bifurcating fixed point. For example, the saddle-node satisfies $(\partial f / \partial \mu)_{x=0} \neq 0$, in contrast to the other two cases.

The discussion of the special cases has probably suggested to the reader how to devise an unending list of special cases, just by proposing different restrictions to the general saddle-node flow. Rather than continuing on these (less interesting) lines, we will consider more general types of bifurcation.

7.6.3 Hopf bifurcation (flows)

Increasing the complexity of the bifurcation problem, the 'next' step is to consider the case when two eigenvalues of the linearized flow at a fixed point simultaneously have zero real part for a given parameter configuration. The centre manifold will now be necessarily two dimensional. We will consider the most relevant case, namely the bifurcation at a pair of (complex conjugated) imaginary eigenvalues for a one-parameter family of flows. This is known as the *Hopf* bifurcation[†].

Let us use the coordinates x and y to describe the centre manifold problem.

[†] Arnold [arno88b, pp 274] observes that this problem was already known to Poincaré and it was formulated and proved by Andronov in 1929, while Hopf's original work is from 1942. This would confirm the popular belief that mathematical formulas do not carry the name of their original author.

We can form the complex variable $z = x + iy$ and write

$$\dot{z} = iwz + f(z, z^*)$$
$$\dot{z^*} = -iwz^* + (f(z, z^*))^*. \tag{7.65}$$

Since the linearized flow has eigenvalues of different sign, a discussion in terms of *resonances* will be proper. Operating with L on terms of the form $z^n(z^*)^m$ will produce a factor $iw - (niw + m(-iw))$. Hence, the unremovable terms (in the sense of the normal form theorem) correspond to $n = m + 1$, i.e., $z^{m+1}(z^*)^m$. To lowest order, the normal form reads

$$\dot{z} = (\mu + iw)z + a|z|^2 z \tag{7.66}$$

where we have already restored the term vanishing at the bifurcation.

Transforming the normal form equation (7.66) with $z = \rho \exp(i\theta)$ (polar coordinates), we obtain

$$\dot{\rho} = \mu\rho + \mathrm{Re}(a)\rho^3 \tag{7.67}$$
$$\dot{\theta} = w + \mathrm{Im}(a)\rho^2. \tag{7.68}$$

Taking $\mathrm{Re}(a) = -1$ to fix ideas, we observe that as we sweep the parameter μ from negative to positive the stable focus at $(0, 0)$ changes into an unstable focus and a stable periodic orbit of radius $\rho = \mu^{1/2}$ appears, figure 7.7. Note that $\dot{\rho} = 0$, $\dot{\theta} = b$ (b constant) describes a closed periodic orbit in polar coordinates.

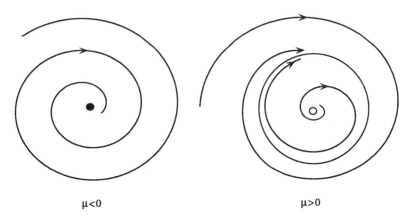

μ<0   μ>0

Figure 7.7. Hopf bifurcation.

Exercise 7.14: Identify the Hopf bifurcations of the Lorenz flow of chapter 3.

7.6.4   General features of fixed point bifurcations

Let us summarize the intuitive ideas present in all the bifurcation examples discussed up to now. What happens at a bifurcation point?

- There is a qualitative change in the flow.
- The bifurcation does not occur for general values of the parameter. A special (degeneracy) condition must be satisfied so that the continuation of solutions cannot be pursued further (failure in fulfilling the conditions of the implicit function theorem is the most common condition).
- The phase space can be split into 'singular' and 'regular' subspaces according to whether the dynamics in these subspaces presents a qualitative change or not, as a result of the change of parameters. Attention can be focused into the singular subspace where all the important changes take place.
- Additional, non-degeneracy, requirements imposed on the family of flows are necessary to establish the actual characteristics of the transition.
- The 'bifurcation' can be *unfolded*, i.e., by addition of the general terms that vanish at the bifurcation value one can find the *unfolding of the bifurcation*, i.e., all types of flow that may exist for parameter values in the vicinity of the bifurcation point.

We summarize the findings of this section in figure 7.8 and figure 7.9.

7.7   Bifurcations of maps and periodic orbits

The study of bifurcations of fixed points of maps (or periodic orbits of flows) parallels to a large extent the discussion for fixed points of flows. We present in the coming subsections the simplest cases in the same spirit as the previous section.

7.7.1   Saddle-node bifurcation

The saddle-node bifurcation for maps occurs when the linearized map at the fixed point presents one (and only one) eigenvalue equal to one. In most respects, the saddle-node bifurcation for maps (or periodic orbits) does not differ from the saddle-node in flows.

The truncated and scaled normal form reads

$$x_{n+1} = \mu + x_n \pm x_n^2 \tag{7.69}$$

and we can find fixed points satisfying

$$x = \mu + x \pm x^2, \tag{7.70}$$

i.e.,

$$0 = \mu \pm x^2 \tag{7.71}$$

Type	Bifurcation diagram	Remarks
Saddle-node $x' = \mu - x^2$ $y' = -y$	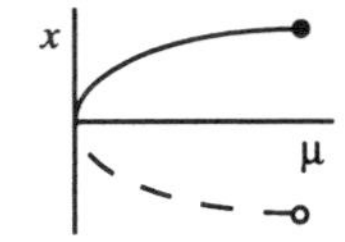	One zero eigenvalue at $\mu = 0$
Pitchfork $x' = x(\mu - x^2)$ $y' = -y$	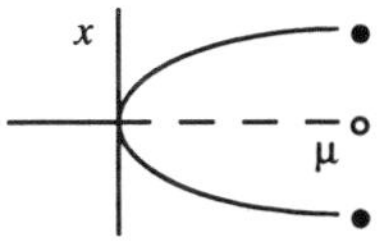	One zero eigenvalue at $\mu = 0$ Only odd powers of x are present
Transcritical $x' = x(\mu - x)$ $y' = -y$	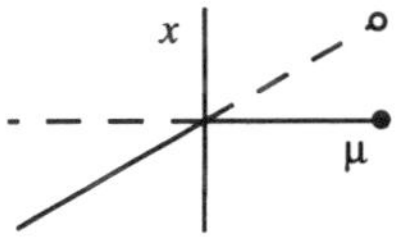	One zero eigenvalue at $\mu = 0$ No independent term (μ enters in the linear term)
Hopf $x' = \mu x + y - (x^2 + y^2)x$ $y' = -x + \mu y - (x^2 + y^2)y$	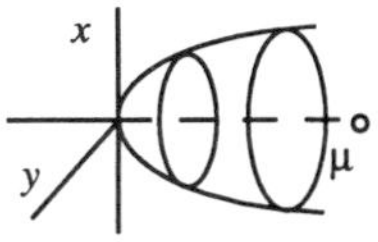	A pair of purely imaginary eigenvalues at $\mu = 0$

Figure 7.8. Normal form for the co-dimension-one bifurcations of flows.

with the same features as in the flow case.

The reader can also verify that the pitchfork and transcritical bifurcations can be treated in exact analogy with the flow case, the ultimate reason resting on the close relation between maps and flows.

Consider as an example a three-dimensional[†] flow having a periodic orbit (which undergoes a bifurcation). Let the centre manifold be two dimensional. Once the translation (and perhaps twisting) along the orbit has been accounted for, it is possible to view the centre manifold of the flow (locally) as a product of an interval times a periodic orbit. The dynamics in this product space is roughly decoupled, i.e., it can be approximately described as the product of the dynamics on the interval and a whole revolution round the orbit.

Consider a local transverse section cutting the orbit at one point. The centre manifold for the return map on the transverse section arises from the intersection of the flow with the section in the vicinity of the periodic orbit. It can be thought of as the (possibly slightly deformed) time-T image of the interval mentioned above by the flow where T is the revolution period of the

---

[†] Note that the dimension of the flow is less important (provided it is at least two) since we are interested in the 'slow' dynamics occurring at the centre manifold.

Type	Phase-space (flow) diagram		
	$\mu < 0$	$\mu = 0$	$\mu > 0$
Saddle-node			
Pitchfork			
Transcritical			
Hopf			

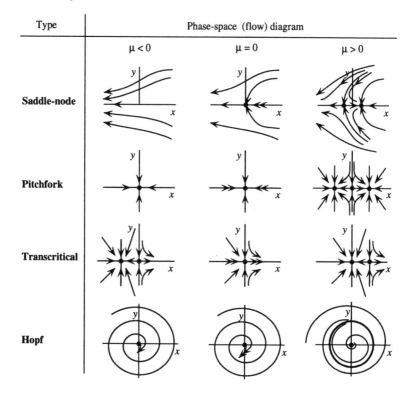

Figure 7.9. Sketch of the phase-space plots in co-dimension-one bifurcations of flows. Note that all sketches can be transported to the corresponding cases of bifurcations of maps and periodic orbits.

orbit (recall the discussion in chapter 5). If the periodic orbit undergoes e.g. a pitchfork bifurcation, the dynamics in the interval will undergo the same bifurcation. The pitchfork bifurcation for the return map can be seen as the time-T image of the flow bifurcation. We will come back to this idea in chapter 14 where we will have acquired the necessary tools to translate the idea into formulæ.

7.7.2 Flip bifurcation

The simplest type of bifurcation characteristic of maps which cannot be encountered in fixed points of flows is the *period doubling* or *flip* bifurcation. This bifurcation occurs at an eigenvalue -1 of the linearization of the map and has associated a one-dimensional centre manifold. The normal form reads

$$x_{n+1} = -(1+\mu)x_n \pm x_n^3; \tag{7.72}$$

the fixed point $x = 0$ (corresponding to the periodic orbit) does not disappear at the bifurcation. It changes stability when μ passes through zero. The negative side of the centre manifold maps onto the positive side and vice versa, at least close enough to zero, so there are no other possible period-one orbits near zero.

We may search for period-two orbits instead, regarding them as the fixed points of the second iteration of the map. Hence, we have

$$x_{n+2} = (1 + 2\mu)x_n \pm (-2)x_n^3 + \text{HOT} \qquad (7.73)$$

where we have dropped the higher-order terms in x as well as in μ. The result is an equation identical to the one corresponding to the pitchfork bifurcation in maps.

In the flip bifurcation a periodic orbit of period two is born out of a periodic orbit of period one that changes stability at a -1 eigenvalue. Each of the two points of the period-two orbit is represented by a fixed point in the second iteration of the map.

Why is it that this bifurcation cannot occur for fixed points of flows? Simply because in a one-dimensional space it is not possible to flow continuously from $x < x_0$ to $x > x_0$ without stopping at the fixed point x_0. Note also that the map associated with the flip bifurcation (restricted to the centre manifold) reverses directions, i.e., as such it cannot be a Poincaré map (although it may arise as a centre manifold restriction).

However, we may think of the flip bifurcation as occurring on a periodic orbit of a flow. This periodic orbit has to lie in a (at least) two-dimensional centre manifold of a (at least) three-dimensional flow (see the previous subsection). The linearized return map near the periodic orbit will be described by the monodromy matrix. Since the flip map reverses directions, the monodromy matrix of the flow must have at least two negative eigenvalues (it has positive determinant and one negative eigenvalue arising from the flip). Thus, the centre manifold of the periodic orbit results in a Möbius band (plus possibly a $2\pi n$ rotation around the orbit). The two negative eigenvalues of the monodromy matrix reflect the fact that the Möbius band *twists* around the periodic orbit. Hence, the return map reverses directions *both* in the centre manifold axis and in the perpendicular direction (see figure 7.10).

7.7.3 Hopf bifurcation (maps)

The so-called 'Hopf' bifurcation for maps[†] is also related to the Hopf bifurcation for flows, except for a few resonant cases.

The bifurcation occurs when a pair of non-real complex conjugated eigenvalues associated with the linearization of the map near a fixed point crosses the unit circle. The centre manifold is bi-dimensional and the linear operator L,

---

[†] Wiggins [wigg90, pp 374] remarks that this bifurcation theorem was first proved by Naimark in 1959 and independently by Sacker in 1965.

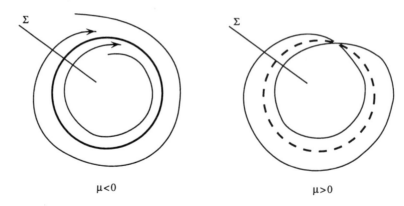

Figure 7.10. Flip bifurcation of a periodic orbit.

equation (7.57), associated with the normal form reads (in complex notation)

$$L(h(z, z^*)) = h(\lambda z, \lambda^* z^*) - \lambda h(z, z^*) \qquad (7.74)$$

with a corresponding equation for z^*. The resonant condition for monomials $z^n (z^*)^m$ reads now $\lambda^n (\lambda^*)^m = \lambda$ (recall that $|\lambda| = 1$). Such a situation is always satisfied when $n - m = 1$. This is the general case, which is valid regardless of additional properties of λ. In addition, if $\lambda = \exp(i 2\pi p/q)$, the resonant condition is fulfilled when

$$(n - m - 1)p/q = r \qquad (7.75)$$

for p, q and r integers.

Notice that there is a remarkable difference between the Hopf bifurcation for maps and that for flows. When we are dealing with flows, the bifurcation occurs in a family of dynamical systems where the real part of the eigenvalues associated to the fixed point cross zero, while keeping a non-zero imaginary part. This can be parameterized with a one-dimensional curve in parameter space, which is translated into bifurcation language as a *co-dimension-one bifurcation*. In the case of maps, what we have is a family of dynamical systems whose eigenvalues cross the unit circle, instead. Now we are in the presence of a two-dimensional problem, since the unit circle may be crossed at any possible angle. We refer to this process as a *co-dimension-two bifurcation*.

Let us consider the lowest-order resonant cases, i.e., when $2 \leq m + n \leq 3$ in the resonant monomials. For $n + m = 2$ the possible cases are:

$n = 2, m = 0, q = p = 1$ and $n = 1, m = 1, q = p = 1$: Saddle-node ($\lambda = 1$).

$n = 0, m = 2, q = 3, p = 1, 2$: Resonance at $2\pi/3$ and $4\pi/3$.

Finally, for $n + m = 3$ we have:

$n = 3$, $m = 0$, and $n = 1$ $m = 2$ ($q = 2$, $p = 1$ in both cases), and also
$n = 0$, $m = 3$, $q = 4$, $p = 2$: Flip bifurcation ($\lambda = -1$).
$n = 0$, $m = 3$, $q = 4$, $p = 1, 3$: Resonances at $\pm\pi/2$.

The general case of the Hopf bifurcation results in a normal form

$$z_{n+1} = \lambda z_n + a|z_n|^2 z_n \tag{7.76}$$

while the two *strong resonant* cases have normal forms

$$z_{n+1} = \lambda z_n + a|z_n|^2 z_n + (z_n^*)^2 \tag{7.77}$$

for the resonance at $2\pi/3$ and $4\pi/3$ and

$$z_{n+1} = \lambda z_n + a|z_n|^2 z_n + (z_n^*)^3 \tag{7.78}$$

for the cases at $\pm\pi/2$.

Note that the strong resonances for $\lambda = 1$ and $\lambda = -1$ are not the generic case for an eigenvalue 1 or -1 with multiplicity two.

7.7.3.1 Non-resonant Hopf bifurcation

The change of the dynamics for the non-resonant Hopf bifurcation can be split into the dynamics of the phase and that of $|z|$, the latter being decoupled from the former

$$|z_{n+1}| = |z_n| \ |\lambda + a(|z_n|)^2| \approx |z_n||\lambda| \ (|1 + (a/\lambda)|z_n|^2) \tag{7.79}$$

obtaining once again the equation for the pitchfork bifurcation in maps. Both λ and a will depend on the bifurcation parameter μ. Expanding to the lowest relevant order in μ we obtain

$$|z_{n+1}| \approx |z_n|(1 + a'\mu + b'|z_n|^2). \tag{7.80}$$

When λ crosses the unit circle a new constant solution is born or dies, depending on the sign of $\text{Re}(a\lambda)$ (equivalently, the solution will exist when $-\mu a'/b' > 0$). The solution can be seen as a fixed point of the $|z|$ equation. The characteristics of this new solution will be quite different from the flow case. To see this, let us consider the dynamics for the phase, ϕ:

$$\phi_{n+1} \approx \phi_n + \phi(\lambda) + \text{Im}(a/\lambda)|z_n|^2 \tag{7.81}$$

or

$$\phi_{n+1} \approx \phi_n + c' + \mu d' + e'|z_n|^2 \tag{7.82}$$

where the first row of the equation is obtained after approximating the phase of $1 + a|z_n|^2/\lambda$ by $\text{Im}(\ln(1 + a|z_n|^2/\lambda)) \approx \text{Im}(a/\lambda)|z_n|^2$ (note that this holds in general: $\text{phase}(x) = \text{Im}(\ln x)$). Standing on the *invariant circle* $|z| = \sqrt{-\mu a'/b'}$, the (truncated) dynamics for ϕ can be seen as a shift in the phase by a constant term. The detailed motion on the invariant circle will be either a collection of periodic orbits or a single quasiperiodic orbit, depending on the value of the shift.

7.7.4 Hopf bifurcation at a weak resonance

The normal form for the Hopf bifurcations at a weak resonance, i.e., when $p > 4$, does not differ from the non-resonant case up to order three. There are, however, some higher-order terms satisfying the resonant condition equation (7.75). For a resonance at an eigenvalue $\lambda = \exp(i2\pi q/p)$ the lowest-order resonant term in the normal form is proportional to $(z^*)^{p-1}$. After a proper scaling and keeping terms up to order $p-1$ the normal form reads

$$z_{n+1} = \lambda z_n + f(|z_n|^2)z_n + (z_n^*)^{p-1}/p \qquad (7.83)$$

where $f(w)$ is a function of the form $aw/p +$ HOT.

In order to analyse the latter equation it is convenient to take the p-iterate of the map obtaining (after approximating λ^p by unity in all but the linear term):

$$z_{n+p} = \lambda^p z_n + f'(|z_n|^2)z_n + (z_n^*)^{p-1} \qquad (7.84)$$

again with $f'(w) = aw +$ HOT.

We are now ready to look for periodic orbits of period p, i.e., fixed points of the p-iteration of the map. The equation to be solved reads

$$0 = (\lambda^p - 1)z + f'(|z|^2)z + (z^*)^{p-1}. \qquad (7.85)$$

Let us call $\epsilon = \lambda^p - 1$. Then we can scale z with ϵ defining $z = |\epsilon|^{1/2}y$ and solve equation (7.85), to the lowest meaningful order in ϵ. Separating in modulus and phase we get for the modulus

$$-\epsilon y = a|\epsilon||y|^2 y \quad \text{or} \quad |y| \approx (1/|a|)^{1/2} \qquad (7.86)$$

while for the phase we rewrite equation (7.85) as

$$z = \lambda^p z(1 + f'(|z|^2)/(1+\epsilon) + (z^*)^{(p-1)}/z(1+\epsilon)).$$

After dropping $1 + \epsilon = \lambda^p$ factors and approximating $\text{Im}(\ln 1 + f'(x) + g) \sim \text{Im}(f''(x)) + \text{Im}(g)$ (where f'' is a proper Taylor expansion of f'), we obtain

$$0 = \text{Im}(\ln \lambda^p) + \text{Im}(f''(|y|^2|\epsilon|)) + |\epsilon|^{(p-2)/2}|y|^{p-2}\sin(p\phi). \qquad (7.87)$$

Defining ω as the phase of the eigenvalue λ, $\lambda = |\lambda|e^{i\omega}$, then there are orbits of period p when the condition

$$|p*\omega + \text{Im}(f''(|y|^2|\epsilon|)) - 2\pi q| \leq |\epsilon|^{(p-2)/2}|y|^{p-2} \qquad (7.88)$$

is satisfied.

The condition equation (7.88) determines a narrow tongue with its tip at $\omega = 2\pi q/p$ in the parameter space $(|\epsilon|, \omega)$ where periodic orbits of period p can be found. At the edges of the tongue there is one (degenerate) periodic orbit while inside the tongue there is a pair of periodic orbits, one being stable in the angular direction while the other is unstable. The borders of the tongue correspond to saddle-node bifurcations of periodic orbits of period p and it is interesting to notice that the nodes have different points as saddle partners at both edges (see figure 7.11).

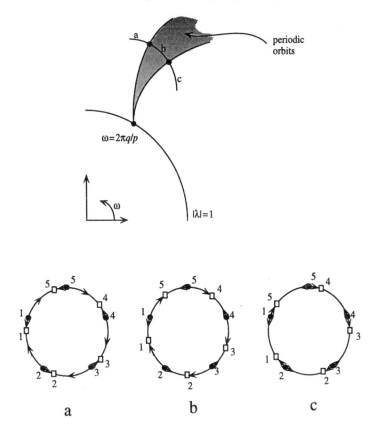

Figure 7.11. Bifurcation of periodic orbits in a Hopf bifurcation at a weak resonance.

7.7.4.1 *Hopf bifurcation at a 1/3 resonance*

For the 1/3 and 1/4 resonances we are only going to discuss the simplest details. These bifurcations are of a more complex nature than the previous one and some tools that are convenient for the discussion have not yet been introduced.

Consider the normal form in the 1/3 resonance after proper scaling it reads

$$z_{n+1} = \lambda z_n + a|z_n|^2 z_n/3 + (z_n^*)^2/3 \tag{7.89}$$

where λ is close to resonance, i.e., $\lambda \approx e^{i2\pi/3}$.

In contrast to what happens in the non-resonant case, the dynamics of $|z|$ is not decoupled from the dynamics of the phase.

We restrict our attention to the periodic orbits of period three which we expect to exist. Iterating three times equation equation (7.89), taking for λ its value at the bifurcation and disregarding terms of order higher than three we obtain (recall the previous section)

$$z_{n+3} = \lambda^3 z_n + a|z_n|^2 z_n + (z_n^*)^2 \qquad (7.90)$$

which has fixed points satisfying the condition

$$0 = (\lambda^3 - 1)z + a|z|^2 z + (z^*)^2. \qquad (7.91)$$

Clearly, if z is a solution of equation (7.91) also $z\ e^{i2\pi/3}$ and $z\ e^{i4\pi/3}$ are solutions. Letting $z = r\ e^{i\phi}$ the condition for fixed points reads

$$(\lambda^3 - 1)/r + ar = -e^{-i3\phi} \qquad (7.92)$$

which can be thought of as the intersection of a circle of radius one with a path in the complex plane obtained by changing r. Multiplying the previous equation by r and taking its square modulus we find that the points of intersection (i.e. the condition for the existence of fixed points) are characterized by:

$$r^2 = \frac{1 - 2\,\mathrm{Re}(a^*\epsilon)}{2|a|^2} \pm \frac{((2\,\mathrm{Re}(a^*\epsilon) - 1)^2 - 4|\epsilon|^2|a|^2)^{1/2}}{2|a|^2} \qquad (7.93)$$

where ϵ is the departure of the eigenvalue from the resonant condition $\epsilon = \lambda^3 - 1$.

We can ignore the $+$ sign in equation (7.93) since in that case $|z|$ does not scale with ϵ and such a solution is not in a small neighbourhood of 0. The remaining solution (choosing the $-$ sign in equation (7.93)) exists for all values of ϵ.

We conclude that there is a single period-three orbit that approaches zero for $\epsilon = 0$ but it persists for all parameter values close to the bifurcation point.

7.7.4.2 *Hopf bifurcation at a 1/4 resonance*

The normal form at the 1/4 resonance is

$$z_{n+1} = \lambda z_n + \frac{a}{4}|z_n|^2 z_n + (z_n^*)^3. \qquad (7.94)$$

Following the discussion of the 1/3 resonance we look for the periodic orbits of period four. After iterating four times the equation equation (7.94) we obtain the condition for the existence of periodic orbits in the form

$$0 = (\lambda^4 - 1)z + a|z|^2 z + (z^*)^3 \qquad (7.95)$$

which after the substitution $z = r\ e^{i\phi}$ reads

$$-\epsilon/r^2 = a + e^{-i4\phi}, \qquad (7.96)$$

with $\epsilon = 1 - \lambda$. In this case we have a circle of radius one centred in a and a straight line starting from the origin instead. If $|a| < 1$ there is only one intersection independent of the value of ϵ since the origin is within the circle; for $|a| > 2$ there are two intersections or no intersection depending on the value of ϵ. Hence, for $|a| < 1$ there is always a period-four orbit while for $|a| > 1$ there are saddle-node bifurcations generating pairs of period four-orbits. For $|a| = 1$ the situation is structurally unstable and higher-order terms are necessary in the normal form.

Exercise 7.15: Look for periodic orbits of period three and four in the 1/3 and 1/4 resonances using as a first approximation the solution

$$z_n = y_n \, e^{i2n\pi/p}.$$

$p = 3$ or $p = 4$ respectively. Is this approximation better or worse than the one we have been using?

We summarize the findings of this section in figure 7.12 and figure 7.13. In this last figure we only sketch the flip bifurcation since the other cases of figure 7.12 display phase space diagrams similar to those in figure 7.9.

7.8 Remarks

How to truncate.   The reader might have received the impression that the problem of truncating the normal form expansion consists simply in disregarding all terms but the non-zero lowest-order ones. This is certainly the case for the Hopf and saddle-node bifurcations studied but it is not the general case. The problem of the proper truncation has been addressed with some detail in [arno88a, guck86] and we refer the reader to these works for the discussion of this advanced topic.

Physical smallness.   The ideas of smallness for physicists and mathematicians might be different. For the physicist (or for any other natural scientist), there is often a limit for the detail that she/he can resolve. The resolution of the experimental set-up, the precision of the numerical calculation are physical limits which cannot be ignored (although they are improved—within certain bounds—from time to time).

As physicists we conceive smallness as 'small but yet meaningful', e.g., small but sufficiently larger than the noise level. The mathematician usually conceives small as 'small as needed' without any lower bound. To interpret mathematical smallness as physical smallness sometimes leads to serious misunderstandings.

Working with normal forms, the physicist will do well only if he/she performs those changes of coordinates having denominators bounded away from zero by a constant factor, i.e., the terms satisfying $|L(h)| \le \epsilon$ are kept in the normal form, for a given ϵ of physical interest. The normal form thus obtained might not be optimal from the mathematical point of view, but will be useful

Type	Bifurcation diagram	Remarks
Saddle-node $x_{n+1} = \mu + x_n - x_n^2$ $y_{n+1} = \gamma y_n$	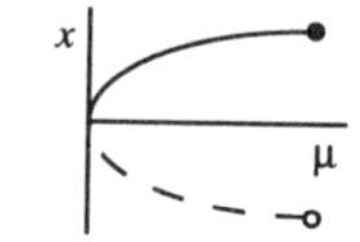	One eigenvalue 1 at $\mu = 0$
Pitchfork $x_{n+1} = x_n(1 + \mu - x_n^2)$ $y_{n+1} = \gamma y_n$	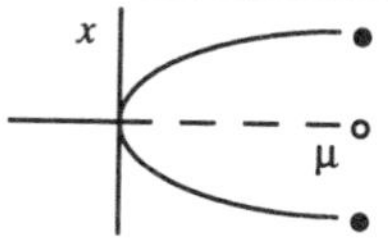	One eigenvalue 1 at $\mu = 0$ Only odd powers of x are present
Transcritical $x_{n+1} = x_n(1 + \mu - x_n)$ $y_{n+1} = \gamma y_n$	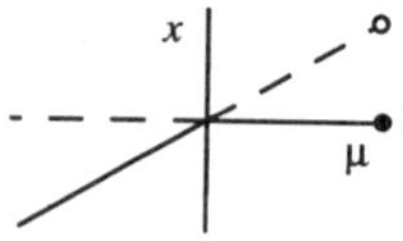	One eigenvalue 1 at $\mu = 0$ No independent term (μ enters in the linear term)
Flip $x_{n+1} = x_n(-1 - \mu - x_n^2)$ $y_{n+1} = -\gamma y_n$	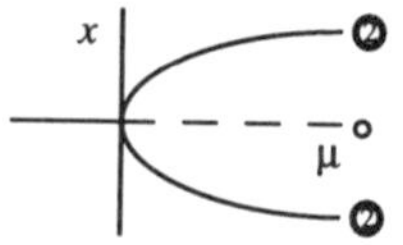	One eigenvalue -1 at $\mu = 0$ $x_{n+2} = g(x_n)$ undergoes a pitchfork, i.e., a period-two cycle appears. $\gamma > 0$ so the map is orientation preserving
Hopf $z_{n+1} = \lambda z_n + |z_n|^2 z_n$ (with $z = x + iy$)	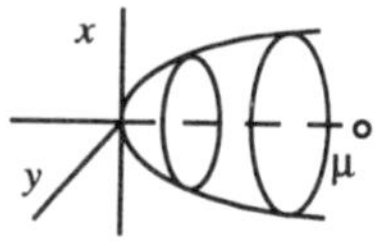	A complex eigenvalue crosses the unit circle, i.e., $|\lambda| = 1$ (note that λ^* crosses as well) Non-resonant case (see text)

Figure 7.12. Normal form for the co-dimension-one bifurcations of maps and periodic orbits.

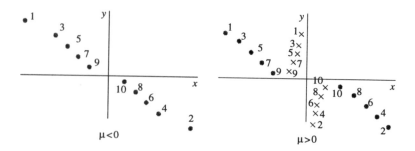

Figure 7.13. Sketch of a phase-space plot in a flip bifurcation. The numbers label the subsequent iterates. For $\mu > 0$ two initial conditions are shown.

in a larger region of parameter and phase space, which is, after all, what one wants.

Higher-co-dimension bifurcations. Higher-co-dimension bifurcations appear less often in natural sciences, but often enough to deserve attention. Usually, when more than one control parameter is used, there are points in parameter space where the bifurcation sets intersect and two (or more) bifurcations occur simultaneously. If in addition both bifurcations are mutually related (for example, both of them refer to the same fixed point) one is in the presence of a bifurcation of higher co-dimension, as announced by the need to use several parameters in the unfolding of the bifurcation.

In the same spirit as in the discussion about physical smallness, there are times when, even if one parameter is being used, considering a higher-co-dimension bifurcation helps in the understanding of the problem. A one-parameter sweep in a two-parameter space might then be viewed as a 'cut' in the parameter space of the bifurcation of higher co-dimension and a sequence of seemingly unrelated co-dimension-one bifurcations might acquire a unified meaning.

There are no magic recipes for the analysis of problems and systems; *the researcher will have to find the method that best suits the problem under study* [†].

7.9   Summary

The crucial ideas of this chapter have been:

- The identification of singular regions in parameter space in a family of dynamical systems (bifurcation points).
- The identification of the region in phase space where the long-term dynamics takes place, i.e., where the 'slow' variables dwell.
- The extension of the idea of coordinate changes in a systematic way, in order to simplify the long-term dynamics, retaining only the most relevant part.
- The analysis of the simplest cases of bifurcations of fixed points and periodic orbits.

Note that the whole discussion of this chapter has been of local character. These tools will be sufficient to study a large number of problems in nonlinear dynamics, but are as well far from being complete. We defer the discussion of global problems to a later chapter.

7.10   Additional exercises

Exercise 7.16: Consider the system of exercise 3.5.

[†] Please, note that the method must fit the problem and not necessarily the researcher!

(a) Show that in the neighbourhood of the fixed point at $x = y$, $z = 0$ and for parameter values close to $\mu = \pm 5b\cos(\theta)$, $\alpha^{-1} = -\tan(2\theta)$, the system can be approximated as

$$x' = y \tag{7.97}$$
$$y' = ax + by \mp x^3 - x^2 y. \tag{7.98}$$

Hint: Perform the following change of variables ($B^2 = x^2 + y^2$, $u = x^2 - y^2$). Check that the dynamics for parameter values close to the ones indicated above can be projected into the $B =$ constant manifold.

(b) Interpret the different solutions of the later system in terms of the original variables.

(c) Show that the $\mp$ sign in the reduced equations depends on the sign of $C = 2 - \alpha\tan(\theta) - 3\alpha\tan^3(\theta)$.

Exercise 7.17: Let

$$x'' + (\beta - cx)x' + \alpha x - bx^2 = 0 \tag{7.99}$$

with $b > 0$, $c < 0$. Show that there are 10 regions in the (α, β) plane in which the flow behaves in qualitatively different ways. Sketch the flow in each region. Study the bifurcations that take place in the separatrixes of the regions.

Exercise 7.18: A frequent phenomenon in stellar dynamics is star pulsation. The competition between gravity and pressure for the outer layers of the stars might end up either in an equilibrium or in a dynamical process that consists in a periodic expansion and contraction of the star's atmosphere. A simple model consists in assuming a hard core of mass M, and an infinitesimally thin layer of mass m at a position r. Then, the dynamics of this layer will be ruled by

$$mr'' = 4\pi r^2 P - GMm/r^2. \tag{7.100}$$

The interesting feature of this model is that the pressure P might be related to the size and temperature of the star in a very complex way. Assume that

$$P/P_0 = k(r/r_0)^\gamma, \tag{7.101}$$

where r_0, P_0 denote the equilibrium values of r and P respectively. Which are the different dynamical states that might be found as γ is varied?

Exercise 7.19: Let us study the truncated system of equations ruling the behaviour of the modes in term of which v was expanded in exercise 3.6.

(a) Choose a pair of parameters (a_0, l_0) in which two of the curves of marginal stability intersect. For parameter values close to (a_0, l_0), project the equations into the centre manifold and find the normal form of the vector field.

(b) Analyse the following statement: 'No Hopf bifurcations might arise for this system for any values of the parameters'.

Hint: Check that the system admits a potential form.

(c) Modify the original partial differential equation slightly so that Hopf bifurcations might occur.

Chapter 8

Numerical experiments

The process of abstraction, generalization and deduction started in chapter 3 was completed in chapter 7 and with it we were able to cover most of the classical subjects in nonlinear dynamics.

In this chapter we would like to initiate a new tour presenting another round of selected observations with the intention of proceeding immediately to their study in the coming chapters.

The exploration will be performed with the aid of computers to integrate a set of model equations numerically and display the outcome in a suitable form. This process is sometimes called a *numerical experiment*, a name with an unsuspected potential for generating arguments.

The polemic seems to be born in the different meanings (and affections) attached to the word 'experiment' by different scientists.

For the natural scientist, i.e., the scientist primarily concerned in understanding the laws of nature as they are, a numerical simulation only explores the proposed model and cannot be equated to an actual experiment that is directly ruled by the laws of nature. Numerical experiments are not experiments at all from this point of view.

In defence of the name we can look at the position that these experiments play in the scientific inquiry for a good number of theoreticians. Numerical simulations allow the researcher to observe that, for example, the appearance of irregular 'chaotic' behaviour in her/his models is always associated with the presence of homoclinic crossings between stable and unstable manifolds of periodic orbits. Following the inductive step of natural sciences she/he might conjecture that homoclinic crossings 'generate' irregular motions. This process of induction is identical to the process of induction leading to the discovery of natural laws, where from the outcome of a few and selected experiments a new law is proposed. On this ground the name 'numerical experiments' appears to us as fair.

It should be noted however that the validation methods for the conclusions induced from traditional experiments and from numerical experiments are different. Natural laws are confronted with larger sets of experiments whose results are predicted directly or indirectly (mediated by logical deduction and combination with other laws) by the law in question; whereas the laws conjectured from numerical experiments are to be confronted with direct (mathematical) deductions.

The value of these conjectures should not be taken lightly. Induction is a creative, 'open-ended' process, requiring a particular 'feeling' to realize *where* one can find interesting or useful properties, while deduction, at least when the number of rules and principles involved is limited enough, is a task that can in principle be automatized (there exist computer codes for proving theorems).

In the rest of this chapter we will present some of these numerical explorations focusing the attention not only in the observed changes of the attractors but also in the non-observable changes that accompany and prelude them.

8.1   Period-doubling cascades

When the sweeping of a parameter, within a finite parameter range, produces an infinite sequence of period doubling (or flip) bifurcations of an attractor (or any other periodic invariant set) we speak of *period-doubling cascades*. This cascades are rather common in systems that are low dimensional (i.e., their dynamics is dominated by few, often not more than three, variables) and specially when there is contraction of areas in phase space due to dissipation. For this reason period-doubling cascades have been observed in numerous experiments and even more numerous models.

We will discuss the case of the period-doubling cascades in the CO_2 laser model which reproduces qualitatively the experimental observations presented in chapter 1.

8.1.1   Period doubling in the CO_2 laser

We recall that a simple phenomenological model for the CO_2 laser with modulated losses consists of (3.10, 3.11)

$$dI/dt = I(W - R\cos(\omega t)) \tag{8.1}$$
$$dW/dt = 1 - \epsilon_1 W - (1 + \epsilon_2 W)I. \tag{8.2}$$

A candidate for a bifurcation diagram can be obtained from the numerical integration of these equations by, e.g., plotting some characteristic of the solution (for example the maximum intensity) versus the control parameter. In this example, the control parameter is the depth R of the losses, also called the *forcing term* in analogy with a forced oscillator. A typical bifurcation diagram for the laser system looks like figure 8.1.

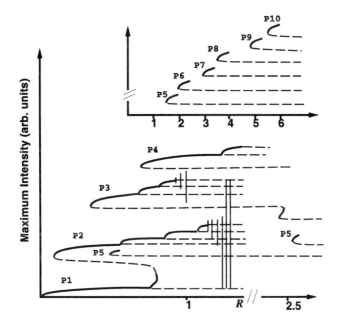

Figure 8.1. Bifurcation diagram, $\approx I_{max}$ versus R ($\epsilon_1 = 0.03$, $\epsilon_2 = 0.009$, $\omega = 1.5$) for the laser with modulated losses. Solid lines correspond to stable orbits; dashed lines correspond to unstable orbits. Vertical lines indicate chaos.

Here we have only one control parameter. In general, it is common to perform a numerical experiment by moving on a one-dimensional curve in parameter space. The description of the observed phenomena is usually made by borrowing (or abusing) the language of time evolution i.e., some behaviour may appear *before* or *simultaneously* (in parameter space) with some other.

The diagram can be described in words as follows. For $R = 0$, i.e., no external modulation of losses, the attractor consists of a period-one orbit which is the product of the 'circle' S^1 corresponding to the external modulation and the fixed point of the bi-dimensional unperturbed flow. This orbit continues to exist for small (and not so small) values of the external forcing, although its shape will change due to the effects of forcing ($R \neq 0$). Changing parameters by small steps we reach a critical value, $R \approx 0.9$, where the attractor loses stability in a flip bifurcation and a period-two attractor appears (the period measured in units of the period $\tau = 2\pi/\omega$ of the forcing term). For slightly higher values of the forcing term, the period-two orbit loses stability and a period-four attractor appears which does not visit the same regions of phase space as the previous attractor. This new bifurcation occurs at an eigenvalue $+1$ of the monodromy matrix and corresponds to a saddle-node bifurcation. The unstable period-two orbit involved was born at lower parameter value in another saddle-node bifurcation between periodic orbits.

This behaviour illustrates that nonlinear systems may have more than one coexisting attractor. Which particular attractor is reached for a given situation will depend on the initial conditions, i.e., on which basin of attraction the system is moving. When the old period-two attractor has disappeared, its basin of attraction has gone with it, leaving us within the basin of the period-four attractor. Note that the language of (reversible) time evolution to describe parameter excursions should not be taken too far.

The 'history' of the period-four attractor can be retraced by diminishing the value of R. The period-four attractor comes from a period doubling of a (different) period-two attractor. This last orbit is precisely the companion of the period-two saddle orbit that 'annihilated' our original period-two orbit. A fresh glance at figure 8.1 may be of help at this point.

Following the 'branch' of the period-four attractor for larger values of R, we observe more period-doubling bifurcations to attractors of periods eight, 16, 32, etc. Each one of these period-doubling bifurcations leaves behind an unstable periodic orbit of period 2^{n-1}. The values of these bifurcation points are increasingly close. In the laser experiment one can observe only up to period 16 while in the numerical exploration one can go (with considerable computational effort) to periods up to $2^9 = 512$. At the end of this period-doubling cascade we can observe an attractor that presents a 'noisy' period-2^n periodicity (n may take values from ∞ to some n_0). This *noisy periodic orbit* is seen in the Poincaré section in the form of a recurrence among 2^n spots in an orderly way but without ever closing into a periodic orbit or repeating a point (at least to the degree to which the experiment and/or the simulation can be trusted). Figure 8.2 shows a noisy periodic attractor in the CO_2 laser model.

Other branches of solutions develop as we increase the parameter R. These solutions coexist as we have already pointed out. They are characterized by being initially born at a saddle-node bifurcation of period $p = 2, 3, \ldots$ (the upper limit is in principle infinity and large periods can be actually observed in theory and experiment). These orbits are easy to identify because they consist of one strong laser pulse (the longer the period the stronger the pulse) followed by $n - 1$ periods of almost zero intensity. Each one of these branches of periodic solutions evolves (as the parameter R is increased) through period-doubling cascades, reaching an accumulation point for the corresponding critical values of the control parameter, R_c, presenting noisy periodic attractors and finally becoming globally unstable, not longer visiting the region of phase space they used to. This phenomenon is usually called a *crisis* and we will refer to it later. The change is easy to notice since the maximum values of the intensity drop abruptly after R exceeds the parameter value for the crisis.

Period-doubling cascades proceed approximately as described for all systems. A cascade might be interrupted at some level (like the cascade initiated at the period-one orbit in the example above). However, the cascades that reach the accumulation point proceed in a standard form, after some period $2^n p$, with

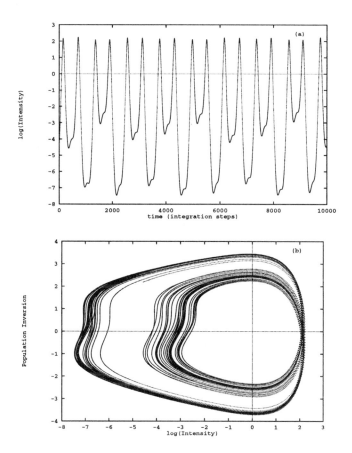

Figure 8.2. A chaotic attractor after the period-doubling bifurcations in the laser model. Parameters are $\Omega = 1.5$, $R = 1.018$, $\epsilon_1 = 0.03$ and $\epsilon_2 = 0.009$. (a) Time-trace of the intensity. (b) Orbits projected in the (I, W) plane.

p the original period. In other words, it is not the beginning of the cascade that is similar in all the period-doubling cascades but the way they reach the accumulation point, i.e., some small neighbourhood in parameter space of the critical parameter value.

8.1.1.1 Manifold organization

Let us take a deeper look into the process of period doubling. Consider a one-parameter family of dynamical systems, having a period-doubling cascade initiated at period one and reaching the accumulation point. At each parameter value R_n the period of the attractor doubles, becoming 2^n and $R_n \rightarrow R_c$ as

$n \rightarrow \infty$. Let us further assume that we are dealing with an attractor of a three-dimensional flow or two-dimensional map. None of these assumptions is essential (recall however the previous paragraph) but they make the description simpler.

Each period-doubling bifurcation proceeds in the same way from the local point of view. At R_n the orbit becomes unstable in one direction with an associated eigenvalue -1 of the monodromy matrix. For parameter values slightly beyond R_n an attractor of double period has been born. In addition, the unstable manifold of the mother orbit is *feeding* the daughter[†]. The new orbit is stable with an eigenvalue very close to $1 = (-1)^2$. Further increasing the parameter this eigenvalue 1 changes until it acquires a non-zero imaginary part (simultaneously with a second eigenvalue). The flow around the daughter orbit acquires a rotation in addition to the twisting that already carries from the mother. The pair of complex conjugate eigenvalues migrate through the complex plane and rejoin on the negative axis where they split. One of the two negative eigenvalues crosses the unit circle at -1 for $R = R_{n+1}$, producing the next flip bifurcation of the cascade, see figure 8.3.

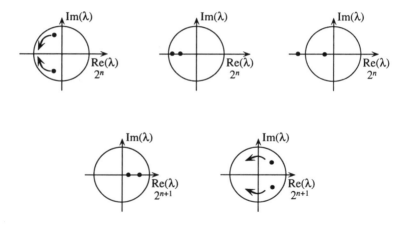

Figure 8.3. Evolution of eigenvalues during a period doubling.

Along the process of modification of the local flow around the daughter orbit, the unstable manifold of the mother has continued to feed the attractor. For $R = R_{n+1}$ it is wrapping around the region of the new bifurcation. As a result of this process, when the grand-daughter is born the unstable manifold of the mother accumulates onto the unstable manifold of the daughter and feeds with the daughter the newborn orbit. At this point we can be more specific and say that orbit A *feeds* orbit B precisely when there exists a heteroclinic trajectory from A to B.

[†] We will call the newborn attractor the *daughter* of the former one.

The result of this process is a hierarchy $\ldots \geq O_{n-1} \geq O_n \geq O_{n+1} \geq \ldots$ where $A \geq B$ means that there is a heteroclinic orbit from A to B. Note that it can very well be that $A \geq B$ and $B \geq A$ simultaneously: such a situation can and will appear for example when a heteroclinic cycle involving both A and B is formed[†], see figure 8.4.

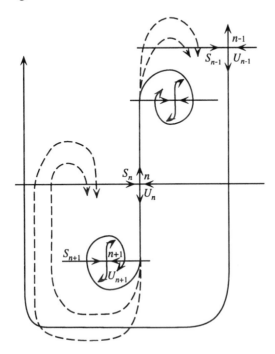

Figure 8.4. Evolution of the manifold organization during a period doubling.

We turn now to the idea of torsion. This concept is easy to present for three-dimensional flows although it can be extended to any higher dimension [uezu82]. Consider a coordinate frame consisting of the stable and unstable directions of a saddle orbit and the velocity along the orbit. If we follow this frame along the periodic trajectory it will evolve continuously in time having rotated $\tau 2\pi$ times around the direction of the local velocity when reaching back to the initial condition. The number τ is the *torsion* of the orbit and it is necessarily a half-integer. Consider now the flow for a slightly different parameter value. If no bifurcation is involved, there will be a continuous path in parameter space so that the flow deforms continuously from the initial to the final point. The torsion will therefore 'vary' continuously as well, remaining always a half-integer! We realize immediately that this number is *invariant* in the sense that it cannot be

---

[†] To our knowledge, the first presentation of this phenomenon was made by Chen, Györgyi and Schmidt [chen87].

modified when parameters are changed provided the saddle continues to be a saddle along the change of parameters [beie83].

When an orbit undergoes a period-doubling bifurcation it modifies the torsion inherited from its mother adding (or subtracting) half a turn (π). As a result of this process, there is a relation between the torsion of mother and daughter at the parameter value of their respective period-doubling bifurcations: $\tau_{n+1} = 2 * \tau_n \pm 1$.

A more restrictive relation has been observed [beie83], so far without exception. The relation between the torsion of mother, daughter and grand-daughter follows the empirical rule $\tau_{n+1} = 2 * \tau_{n-1} + \tau_n$. This relation is global in character and has not yet been theoretically explained although some of its implications have been discussed in the literature [gonz89].

This section has been structured having in mind three-dimensional problems. We have the strong feeling that, under adequate conditions, the results will in some sense be transportable to higher dimensions[†]. Our hint is based on the fact that for the case of strongly contracting flows admitting a Poincaré section, when the contraction rate is arbitrarily strong, the structure of periodic orbits of the Poincaré map 'dressed' with their stable manifolds is equivalent to the structure on the two-dimensional reduction to the centre manifold [nati94]. Current practices in physics rely strongly on the hope that this result will hold even under weaker conditions, but it is unclear to what extent this result may hold for systems with a weaker contraction rate. Our coming 3-d flows/2-d maps picture may in the best of cases be regarded as the underlying structure of *some* higher-dimensional problems.

8.1.2 Scaling ratios

Period-doubling cascades belonging to different systems have features that do not depend on the details of the system but rather on a few aspects of it. This should not sound surprising for the reader who verified in chapter 7 that, for example, all the co-dimension-one bifurcations proceed in the same way and are equivalent to one general normal form for each case. However, the presence of regularities across systems in a certainly more complex situation made a remarkable impact in the scientific community, especially in the physics community during the late seventies and early eighties.

The simplest class of dynamical systems presenting sequences of period-doubling bifurcations is the class of *unimodal maps of the interval* $f : [0, 1] \to [0, 1]$. Roughly speaking, a (continuous) map is unimodal when it has a single maximum in the interval $[0, 1]$ (see [coll86] for a discussion at length of maps of the interval).

The period-doubling cascade for unimodal maps was studied from the topological point of view by Metropolis, Stein and Stein [metr73] who

---

[†] How about lower dimensions? Go back a couple of chapters if you are not sure of the answer.

established the qualitative regularities of this process, first comparing the period-doubling cascade for the systems

$$x' = \lambda x(1 - x) \qquad \text{for } 3 < \lambda < 4 \tag{8.3}$$

$$x' = \lambda \sin(\pi x) \qquad \text{for } 0.715 \simeq \lambda < 1 \tag{8.4}$$

$$x' = 3\lambda y(1 - y + y^2), \quad y \equiv 3x(1 - x) \qquad \text{for } 0.875 \simeq \lambda < 64/63 \tag{8.5}$$

and then generalizing these observations. Equation (8.3) describes the *logistic map* already discussed in chapter 4.

Suppose now for a moment that, for a given system, you want to determine the values λ_n for which a period-doubling bifurcation occurs within a given cascade. Since the parameter values accumulate at λ_c the distance $\lambda_{n+1} - \lambda_n$ will decrease towards zero requiring at each step more and more precision in the determination of the bifurcation values. Searching for bifurcations with constant increments for the parameter values is bound to fail close enough to the accumulation point. How should one decrease these increments with increasing periodicity? The solution to this practical problem was found by Feigenbaum [feig78], who observed that for a number of unimodal maps the ratio

$$\delta = (\lambda_{n+1} - \lambda_n)/(\lambda_{n+2} - \lambda_{n+1}) \tag{8.6}$$

has a limit

$$\delta \approx 4.669\ 201\ 609\ 1029\ldots.$$

Furthermore, he observed a second regularity. The 2^n cycle of f can be seen as a period 2^q orbit for the map $F_{n-q} = f^{2^{n-q}} = f(f(\ldots))\ 2^{n-q}$ times. The structure of periodic orbits for F_{n+1-q} and $\lambda_n \le \lambda \le \lambda_{n+1}$ ($q \ll n$) clustered around a fixed point of F_{n+1-q} is a scaled down (and reversed in orientation) version of the cluster of periodic orbits of F_{n-q} for the parameter interval $\lambda_{n-1} \le \lambda \le \lambda_n$. The scale of this reduction has also a constant limit

$$\alpha = 2.502\ 907\ 875\ 0957\ldots.$$

These two observations are the key to a renormalization procedure and the understanding of many properties of the period-doubling cascade and its accumulation point [feig79]. If the previous conjectures are correct (i.e. that δ and α are the same for a (large?) class of maps irrespective of the choice of f within the class, we can expect the following limit to exist (under 'appropriate' choices of the metric in the space of maps and x restricted to a suitable interval!):

$$g_q(x; \lambda) \equiv \lim_{n \to \infty} (-\alpha)^n F_{n-q}(x/(-\alpha)^n; \lambda_n + \lambda) \tag{8.7}$$

$$0 \le \lambda \le (\lambda_{n+1} - \lambda_n). \tag{8.8}$$

We also expect the rescaling property to hold for the set of functions $\{g_q\}$, namely

$$g_q(x; \lambda) = -\alpha g_{q+1}(-x/\alpha; \lambda/\delta).$$

Then, if the limit $g(x) = \lim_{n \to \infty} g_n(x; \lambda)$ exists and is continuous with respect to x (and somehow independent of λ since $0 \le \lambda \le (\lambda_c - \lambda_c) = 0$) the function g should have the renormalization property

$$g(x) = -\alpha g(g(-x/\alpha)). \tag{8.9}$$

It is usual to rescale and shift the coordinate origin so that the interval becomes $[-1, 1]$ and the unimodal maximum is at $x = 0$. Consequently, we will have $g'(0) = 0$ and we can standardize our maps so that $g(0) = 1$.

Exercise 8.1: Give a geometrical meaning to the renormalization conjecture drawing the expected position of the periodic orbits of F_n and F_{n+1} for properly chosen parameter values (give a meaning to the choice).

Equation (8.9) can be regarded as a fixed point equation on some functional space [arno88b]. This fixed point happens to be of saddle type where δ is the eigenvalue associated to the unstable direction. The equation has not one but *many* solutions. In fact, the logistic map studied by Feigenbaum can be rewritten as $x_{n+1} = 1 - \mu x_n^2$ and can hence be regarded as the first element of the set of unimodal maps of type $x_{n+1} = 1 - \mu x_n^{2p}$, $p = 1, 2, \ldots$. Each p value has an associated limit function g_p with corresponding scaling rates δ_p and α_p [zeng84]. The case $p = 1$ is in some sense the most natural, since any small perturbation of, e.g., unimodal maps of type $x_{n+1} = 1 - \mu x_n^4$ which *does not* alter unimodality is likely to include quadratic terms.

These scaling properties are often called *universal* in the literature.

8.1.3   Period halving: the inverse (noisy) cascade

What happens when the parameter is moved beyond the accumulation point of the period-doubling cascade?

The answer to this question depends on the degree of area-contraction of the system. The situation for extremely dissipative systems (such that they generate Poincaré maps that are essentially one dimensional) is well understood and from this we can understand which changes have to be made in order to adapt this limit case to more general cases.

For simplicity we will restrict our discussion to the case of attracting sets.

Beyond the accumulation point of the period-doubling cascade, the phase-space region where the cascade took place does not change stability (i.e., it continues to be attractive). The attractor undergoes a series of metamorphosis at specific parameter values. In general the attractor will take the form of 2^n balls that are visited periodically, each ball 'centred' in one of the points belonging to the period 2^n orbit. Although the visit to each ball is periodic, the way in which a ball is revisited after 2^n steps is aperiodic ('chaotic'). The system displays periodicity up to a certain resolution but is 'chaotic' when observed with sufficient magnification. This situation is known as 'noisy or chaotic periodicity' [lore80, hao 82]. We will not discuss further at this point what one exactly means by 'chaotic'.

When a critical value, in parameter space, is reached, half of the islands merge pairwise with the other half producing an attractor with 2^{n-1} islands. This process is therefore known as a *period-halving cascade*. We show an example for the case $n = 1$ in figure 8.5. The upper left part of figure 8.5(a) corresponds to the odd iterations of the two-parameter *Hénon map* (a definition of this map is presented in chapter 11), making one of the two islands. The lower right part of the same figure corresponds to the even iterations, yielding the second island. In figure 8.5(b) the two islands have merged into one after a slight parameter change. Both odd and even iterations extend over the whole island. The situation of the stable and unstable manifolds of the period-one (flip) orbit is displayed in figure 8.6. The merging of the chaotic islands is an example of a crisis, as we will soon see.

The singularity of one-dimensional maps consists in that there is a strict order (in parameter space) between the local bifurcations and the global bifurcations responsible for the period-doubling cascade and the period-halving cascade. In the general (higher-dimensional) case these two phenomena evolve with a certain independence. As a consequence of this independence the period-doubling cascade may be *bi-stable* with a 2^n noisy periodic attractor (i.e., both attractors coexist although they have different basins of attraction) and, consequently, the period-halving cascade will not present its steps in a sequential order.

8.2 Torus break-up

8.2.1 Curry and Yorke's experiment

J H Curry and J A Yorke [curr78] studied the transition to chaos in a map, starting from an invariant circle. Their model was motivated by two observations: first, that many physical systems contract areas in phase space and, furthermore, present a (compact) trapping region from which trajectories cannot leave; and second, that quite often the fixed point inside the trapping region (its existence being a consequence of Banach's fixed point theorem) loses stability in a Hopf bifurcation.

One of the simplest (in an algebraic sense) models with these properties can be written as a composition of two maps. The first map is a rotation by a fixed amount plus a contraction in the radial direction, while the second map presents a quadratic nonlinearity. The Curry–Yorke model reads

$$\psi(x, y) = \psi_2(\psi_1(x, y))$$
$$\psi_1(\rho, \theta) = (\epsilon \ln(1 + \rho), \theta + \theta_0) \qquad (8.10)$$
$$\psi_2(x, y) = (x, y + x^2)$$

with $(x, y) = \rho(\cos(\theta), \sin(\theta))$.

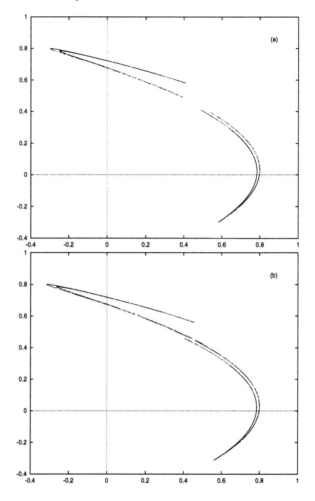

Figure 8.5. Noisy periodic attractor and period-halving crisis for the Hénon map. (a) The noisy period-two attractor for $\mu = 2.01$, $\epsilon = 0.3$ (see text). (b) Same as (a) for $\mu = 2.03$. The two islands have merged.

The map is invertible and presents a fixed point at the origin. The fixed point changes stability at $\epsilon = 1$ going from stable to unstable through a Hopf bifurcation when ϵ is increased.

In the numerical study θ_0 was kept at a constant value of two. Consequently, the Hopf bifurcation is non-resonant and the emerging attractor is an invariant circle, at least within numerical accuracy.

The attracting circle grows as ϵ is increased. For $1.28 \le \epsilon \le 1.3953$ the attractor is an orbit of period three. The period-three orbit is marginally stable for $\epsilon \approx 1.3953$ where it presents an eigenvalue unity (i.e., it is about to

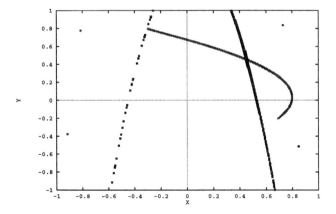

Figure 8.6. Stable and unstable manifolds for the period-one (flip) orbit in the Hénon map just before the crisis $\mu = 2.01$.

undergo a saddle-node bifurcation). For ϵ beyond 1.4 the attractor is again a (highly deformed) circle but this circle appears to be folded onto itself. The attractor is actually not a topological circle but a more complex structure. The structure can be better observed for even larger values of ϵ. For $\epsilon = 1.52$ the circle displays very clearly several wrinkles. The extremes of the wrinkles are successive images of a point in the 'smooth' part of the circle as displayed in figure 8.7. These extremes are therefore as smooth as the pre-image. It can also be observed in the figure that the extension of the wrinkles increases with the interaction and that the wrinkle becomes finally indistinguishable from the circle, i.e., the circle is not smooth but built from infinitely thin layers.

Increasing ϵ beyond 1.63 produces a new attractor which displays a noisy periodic recurrence among four islands. At $\epsilon = 1.7$ the attractor is again a new strange circle.

8.2.2 Analysis of the map

The Curry–Yorke map has been studied as a one-parameter family in [curr78] displaying phenomena that can be better understood in terms of a two-parameter family of maps. The fixed point at the origin loses stability in a Hopf bifurcation (maps) that is a co-dimension-two problem (see chapter 7). The frequency of the Hopf bifurcation is two which, although corresponding to an irrational factor of 2π, is not too far from $2\pi/3 \approx 2.1$.

In the same line of thinking we observe that in the Curry–Yorke study the attractor changes from a period-three attractor to a noisy period-four attractor when the value of ϵ is increased.

Since the Hopf bifurcation for maps is in general a co-dimension-two

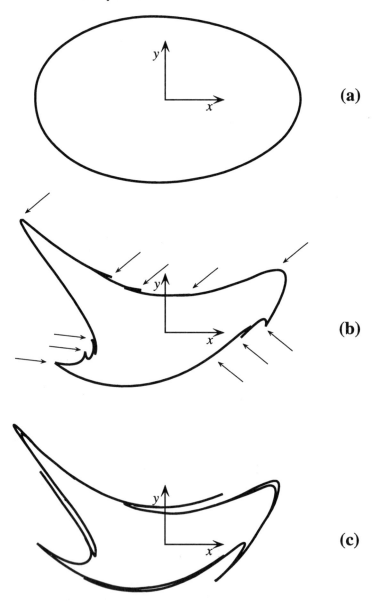

Figure 8.7. Evolution of the Curry–Yorke attractor. (a) $\epsilon = 1.01$; (b) $\epsilon = 1.45$; (c) $\epsilon = 1.52$.

problem (see chapter 7), and since there is evidence that two different resonant Arnold tongues are crossed (the 1/3 and either the 1/2 or the 3/4 tongues), it is sensible to perform a two-parameter study of the map.

Such a study was performed [aron82] for a different map presenting similar behaviour. We reproduce here a small part of their discussion.

A two-dimensional study of the map (8.10) shows how the selected one-dimensional path cuts across several resonant tongues, figure 8.8. Each tongue is delimited in parameter space by two lines of saddle-node bifurcations to the corresponding periodic orbit.

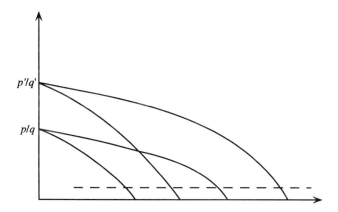

Figure 8.8. A one-dimensional path in a two-dimensional unfolding of the map (schematic).

The attracting set inside a resonant tongue is, at the tip of the tongue, a topological circle composed by a periodic node, its saddle partner at the saddle-node bifurcations and the unstable manifolds of the saddle that connect the saddles to the sinks. When the tongue is traversed from one saddle-node line to the other, the sinks exchange bifurcation partner.

The saddle-node bifurcations corresponding to different 'levels' along the tongue present different global organizations. Far enough from the tip of the tongue the invariant set is no longer a circle (see figure 8.7(a),(c)) but a folded structure still composed (at least) of the saddle, the node and the unstable manifold of the saddle (see figure 8.9). When the one-dimensional path leaves the tongue the attractor is a wrinkled circle.

This description should be taken with some caution. Actually, there is an infinite number of tongues and they can even overlap in parameter space (their associated attractors coexist). In any path cutting across different tongues we expect periodic and aperiodic attractors (quasiperiodic or 'chaotic') nested at all scales and possibly coexisting. The associated tongues will be very thin, so that they may fit in a finite region of parameter space. When scanning parameter space in numerical experiments, it is very likely that one will miss these tongues if the numerical search is performed carelessly.

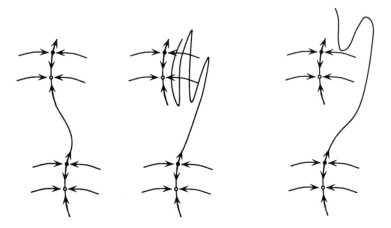

Figure 8.9. Saddle-nodes with different global connections.

8.3 Homoclinic explosions in the Lorenz system

The dynamics of the Lorenz system also presents chaotic attractors. They can be detected for $r \approx 24.06$, and they are easily found after $r = 470/19 \approx 24.74$. For $r = 28$ the attractor is the (famous) Lorenz butterfly (see figure 8.10). At this point the reader might wonder about the concept of *chaotic attractors*. For the time being, it is enough to recall that we have described as chaotic some observed orbits in numerical experiments (on, e.g., the Lorenz system) which are neither periodic nor quasi-periodic, although they move somehow recurrently within a bounded region of phase space.

How was the unstable periodic orbit involved in the Hopf bifurcation created? What is the origin of the butterfly?

It is possible to follow the periodic orbit through parameter space using computer programs designed to the effect. The orbit exists also for $r \leq 470/19$. Its period increases continuously, apparently diverging for $r \approx 13.926$ [spar82] where the periodic orbit passes very close to the symmetric periodic point. Certainly, the same is true for the periodic orbit related by symmetry to the one studied.

At $r \approx 13.926$ there is a second important observation to make. The (one-dimensional) unstable manifold of the fixed point at $(0, 0, 0)$ undergoes a qualitative change for this parameter value. From $r = 1$ to $r < 13.926$ the two branches of the unstable manifold of the origin end up at the respective stable fixed points on the 'same side' (measured with respect to the plane formed by the local stable manifold of the origin) suffering a smooth deformation as the parameter is changed. For $r > 13.926$ the manifolds 'exchange partners': the left branch now feeds the right fixed point and the right branch feeds the left fixed point (see figure 8.11).

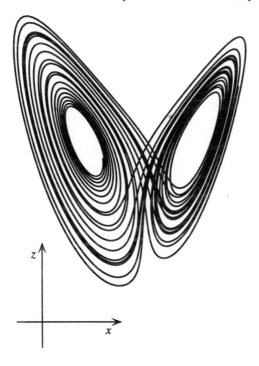

Figure 8.10. Lorenz butterfly. The attractor for the Lorenz system for $b = 8/3$, $\sigma = 10$ and $r = 28$.

Since for $r < 13.926$ the basins of attraction of both attractors were divided by the stable manifold of the origin we can speculate that at $r \approx 13.926$ the unstable manifold becomes *part of* the stable manifold (both manifolds referring to the fixed point at the origin). In this situation, a point on the unstable manifold will tend to the origin both forwards and backwards in time. We recall here that orbits bi-asymptotic to an invariant set are called *homoclinic* orbits (see chapter 6).

There are numerical techniques to confirm the existence of the homoclinic orbit for $r \approx 13.926$ [spar82]. One plausibility argument would be to perform a numerical integration starting 'on' the unstable manifold of the origin (within numerical accuracy) and integrating for a fixed period of time τ for different values of r. This finite segment of orbit will spiral around one fixed point for $r < 13.926$ while for $r > 13.926$ it will approximate the other fixed point. Regarded as a function of r, this change of behaviour will appear to be continuous (as long as τ is finite, but this is always the case in numerical (or other) experiments!), so there is a range of parameters where the segment of orbit will not have a definite bend towards either of the fixed points. This picture will persist after increasing the time τ except that the 'uncertainty interval' around

$r \approx 13.926$ will become narrower. We may conjecture that for infinite time and critical r the trajectory will stop at the origin.

In the parameter range $13.926 < r < 24.06$ there exist chaotic transients, i.e., trajectories that look 'chaotic' for a long time but eventually settle down to a more regular behaviour (see next section): these chaotic transients resemble the chaotic attractor found for $r > 24.06$.

The numerical and analytic studies by Sparrow show that a *strange attractor* (i.e., the Lorenz 'butterfly') evolves from a chaotic unstable set created at the same time as the homoclinic orbit at $r \approx 13.926$ of which the periodic orbits involved in Hopf bifurcations at $r \approx 470/19$ are part.

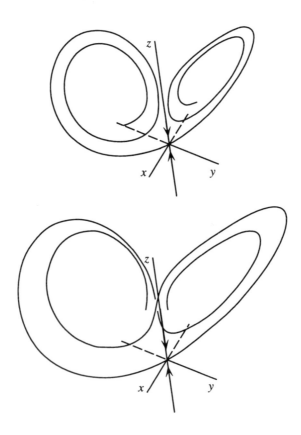

Figure 8.11. Changes in the behaviour of the unstable manifold of the symmetric fixed point for $r \approx 13.926$. (a) $r < 13.926$; (b) $r > 13.926$.

8.4 Chaos and other phenomena

We have seen that one of the landmarks of the systems so far studied in this chapter is the existence of dynamical situations in which an attractor exists which is not organized in as orderly a fashion as fixed points, periodic orbits or quasi-periodic orbits. We have named these attractors 'chaotic'. They correspond to recurrent situations where the nature of the recurrence is less perfect than in periodic and quasi-periodic orbits. Let us mention their more striking characteristic: 'the divergence of nearby trajectories', which corresponds to the fact that two initial conditions that are very close to each other will evolve with time following trajectories that (in general) depart from each other at an exponential rate as long as they remain very close. The time when they separate beyond a distance ϵ can be made as long as we want if we select sufficiently close initial conditions (i.e., within a ball of adequate radius δ). According to the divergence of nearby trajectories we need a relation

$$\delta = \exp(-\lambda t)\epsilon.$$

Since nearby trajectories diverge, two different points in a chaotic attractor, no matter how close they are, will end up far apart (i.e., farther than a given distance ϵ) after sufficiently long time. This phenomenon is called 'sensitivity to the initial conditions'. Together with its immediate consequence, 'the rapid deterioration, with time, of our ability to predict the future', they are the most important and persistent elements in the phenomenology of chaotic motion.

It is perhaps already clear for the reader that experiments in the form of numerical integration of a dynamical system for different initial conditions, although giving strong *hints* about the existence of e.g., a chaotic attractor, have to be imbedded in further theoretical and numerical work. The finite orbits in a numerical experiment (which in addition are registered with finite precision) could as well be part of a periodic orbit of comparatively large period (this holds also for laboratory experiments). These experiments are in any case relevant not only because they contribute to a better understanding of the dynamics but also because they serve as inspiration and motivation for further work.

8.4.1 Crisis

We have already met the crisis phenomenon in the discussion of the merging of chaotic bands after the period-doubling cascade is completed. In general, the name *crisis* has been given to the phenomenon of a 'sudden' expansion of a chaotic attractor or its sudden disappearance. The word *sudden* is meant as something that happens during an exploration in parameter space, somehow 'unexpectedly' for a given value of the parameter [lefr94].

Since the phenomenon is characteristic of attractors we have to discuss what happens in such a situation with the basin of attraction of the attractor.

We will discuss two slightly different cases first analysed by Greboggi and Ott [greb83].

The first case consists of the sudden expansion of an attractor and can be very well represented by the situation in the period-halving cascade. Assume for the argument that we are in the presence of a period-two noisy cascade about to undergo a period halving. If we observe the system before the merging, we will only see one of the chaotic islands every two iterations. At the merging of chaotic bands we will observe a sudden expansion of the attractor corresponding to visiting the regions previously frequented by the other merging island.

The basin of attraction of the two different islands (before the sudden expansion) was divided by the stable manifold of the (saddle) period-one orbit, while the different sides of the unstable manifold were feeding one attractor each (see figure 8.12). We think here as usual of a three-dimensional problem.

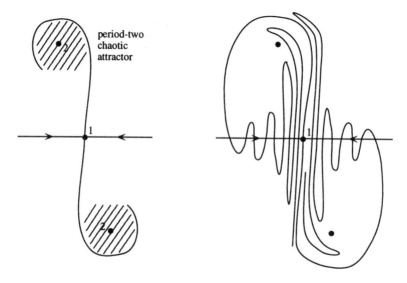

period-two
chaotic
attractor

Figure 8.12. The sudden expansion of an attractor.

At the crisis value the stable and unstable manifolds of the saddle become tangent and after the crisis they intersect each other transversely. In this last situation the stable manifold of the period one orbit fails to divide the phase space in two invariant regions and the trajectories are no longer confined to one or the other side of the stable manifold, hence the sudden expansion of the attractor due to the merging of both basins of attraction.

Certainly, the situation described is not peculiar to the period-halving cascade. The fact that both sides of the unstable manifold become tangent to the stable manifold at the same parameter value, however, is a structurally unstable situation (a small deformation of the flow may destroy this property) unless some restriction applies to the map (e.g., being the second iterate of another map as in this case) so that both sides of the manifold are one the image of the other. Such a restriction can be found in cases where symmetries are present [chos88].

In the general case of a crisis producing a sudden expansion of an attractor we can expect that one of the sides of the unstable manifold of the orbit in the basin boundary will be intersecting the stable manifold even before the crisis. The second branch of the unstable manifold will be feeding the chaotic attractor and at the crisis value it will undergo a tangency with the stable manifold (see figure 8.13). The 'net' result is that the attractor will visit previously forbidden regions, closely following the unstable manifold of the orbit in the basin boundary and eventually reentering the region of the former attractor. Figures 8.14 and 8.15 illustrate the effects of a crisis in the laser model.

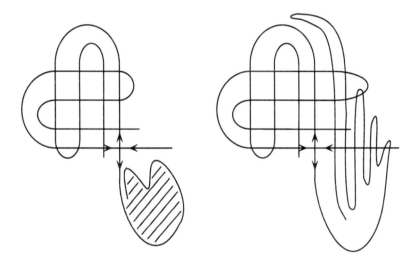

Figure 8.13. Sudden expansion of an attractor at a crisis: the general case.

The case of the sudden disappearance of the attractor is a variation of the other case in which there is no pre-existent (to the crisis) homoclinic crossing. One branch of the stable manifold of the periodic orbit in the basin boundary is feeding the attractor about to undergo the crisis while the other is feeding another attracting region. After the crisis, the trajectories in the region of the former chaotic attractor will wander following the unstable manifold of the saddle, eventually reaching the region of the remaining attractor. In this crisis, the basin of attraction of the chaotic attractor shrinks to the attractor itself which in this way loses the attracting character. The situation is depicted in figure 8.16.

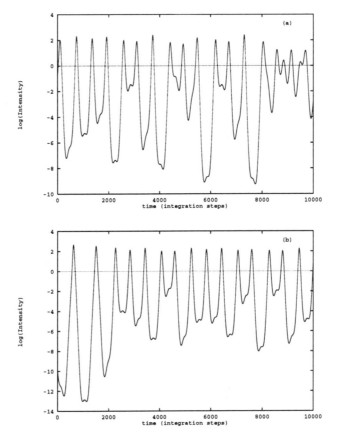

Figure 8.14. A chaotic attractor for the laser model, before and after a crisis. The parameters are: $\Omega = 1.5$, $\epsilon_1 = 0.03$, $\epsilon_2 = 0.009$ and $R = 1.05$, 1.08. (a) Time trace of the intensity, $R = 1.05$. (b) Time trace of the intensity, $R = 1.08$.

8.4.2   Chaotic basin boundaries and chaotic transients

What happens when an attractor disappears in a crisis? Is there any residual left?

The answer is yes! In the extinction of an attractor at a crisis it is the *attractive character* (with the shrinking of the basin of attraction) that disappears. An unstable invariant set (most likely of saddle type) is left as a residual of the chaotic attractor.

This invariant set will manifest itself in at least two different forms. First, the region of phase space formerly occupied by the strange attractor will present long chaotic transients, i.e., there will be orbits that follow the strange set for a

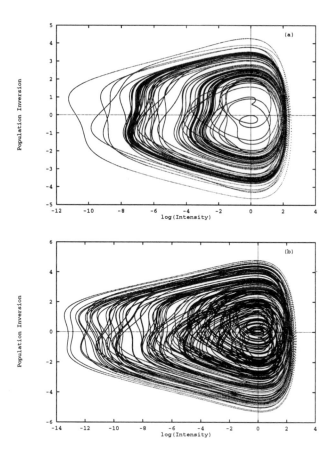

Figure 8.15. A chaotic attractor for the laser model, before and after a crisis. (a) Projection of the attractor in the (I, W) plane, $R = 1.05$. (b) Projection of the attractor in the (I, W) plane, $R = 1.08$.

long time resembling the former attractor. We can certainly say that the strange set belongs to the boundary of the basin of attraction of the new attractor (the one reached after the crisis). Note that we use the expression *strange set* since we can no longer call this invariant set a strange *attractor*.

The second manifestation of the residual invariant set will become apparent if two different attractors can be reached after the crisis from the region of the former attractor. In this case the strange set will belong to the common boundary of the basins of attraction [esch89]. The basins of attraction will be intertwined at every scale, a situation that is often referred as 'fractal basin boundaries' [mcdo85], see figure 8.17.

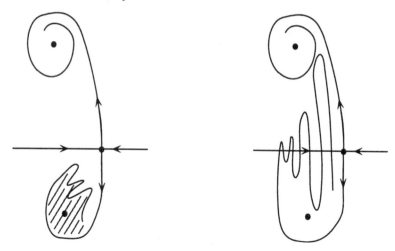

Figure 8.16. Crisis of an attractor and its disappearance.

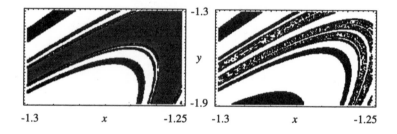

Figure 8.17. Fractal basin boundary between three attractors.

8.4.3 Chaotic scattering

Transient chaotic behaviour can also be found in scattering processes. The most commonly studied situation [rod 73, bleh90] consists of a bi-dimensional potential (i.e., a Hamiltonian system with two degrees of freedom, or equivalently, a four-dimensional phase space) which presents three repulsive hills (see figure 8.18), although the situation is not exclusive of repulsive potentials [fend92].

In this sort of situation an unstable strange set is associated with the unstable periodic orbits that are 'bouncing' among the hills. As a result of the existence of the unstable invariant set, 'particles' reaching the region from infinity can be trapped in the proximity of the strange set for long times. The stable manifolds of the periodic orbits that extend towards infinity are associated with singularities in the scattering matrix since they correspond to initial conditions that are asymptotic to bound states.

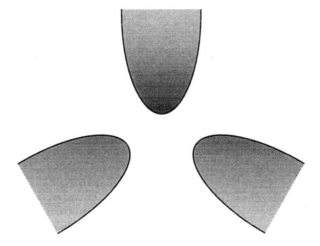

Figure 8.18. Chaotic scattering in a monkey-saddle.

8.5 Summary

The aim of this promenade through the realm of numerical experiments was to present a number of phenomena that are characteristic of chaotic motion. Nonlinearity has more exciting and surprising phenomena than the local bifurcations of fixed points or periodic orbits.

We have illustrated that:

(i)   Limit sets may be more complex than just periodic orbits.
(ii) The basins of attraction of certain invariant sets may shrink, merge, grow or even disappear completely (with or without the invariant set).
(iii) Manifolds may experience 'jumps' at certain parameter values.
(iv) Stable and unstable manifolds of a limit set may approach and eventually cross each other. This may have far-reaching consequences for the motion in the neighbourhood of this set.

The numerical experiments turned the scientists' attention to these phenomena, but part of the work is still to come. Taking for example the case of the Lorenz equations, we know that there exists a trapping region. We know also from the mentioned work of Sparrow [spar82] about the existence of a homoclinic point with a transversal intersection of the invariant manifolds and the numerical evidence for the existence of a chaotic set is overwhelming.

Chapter 9

Global bifurcations: I

In chapter 8 we presented evidence regarding the existence of irregular (chaotic) motion. We have emphasized the association of this dynamical situation with the existence or creation of homoclinic orbits to a periodic orbit (meaning by this in most examples just a fixed point).

In this chapter we will study the dynamics in an 'extended neighbourhood' of the periodic orbit with associated homoclinic trajectories to gain further insight into the mechanisms of chaos.

We will first analyse the situation of a transverse homoclinic orbit showing that it is the limit of a family of periodic orbits. Next, we will consider how these periodic orbits are created at a homoclinic tangency; and finally, we will show how Smale's horseshoes are associated with homoclinic orbits. The analysis of the Smale horseshoe is left for the next chapter because of its fundamental role in our understanding of chaos.

Trying to economize efforts, this chapter will assume the simplest case of a bi-dimensional map.

9.1 Transverse homoclinic orbits

We recall that a homoclinic orbit to a periodic point is an orbit bi-asymptotic to the periodic point. In other words, a point Q belongs to a homoclinic orbit associated with the periodic orbit P with respect to the map F iff Q belongs to $W_s(P) \cap W_u(P)$ ($W_s(P)$ and $W_u(P)$ are the stable and unstable manifolds of P). In the next subsection we will discuss the case of P being a hyperbolic fixed point of a map.

A homoclinic orbit is said to be *transverse* if $W_s(P)$ intersects transversally $W_u(P)$, i.e., the tangents to $W_s(p)$ and $W_u(p)$ at Q span the full space. A homoclinic tangle showing transverse intersections at the points Q_i is presented in figure 9.1. By homoclinic tangle we mean the intricate organization of the stable and unstable manifolds of P.

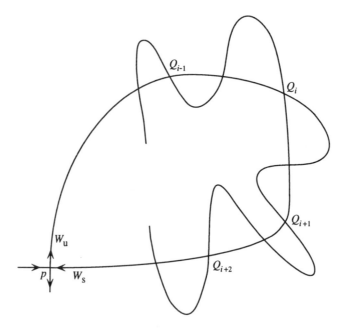

Figure 9.1. Homoclinic tangle associated with the periodic orbit P. The stable and unstable manifolds intersect transversely at Q_i.

9.1.1 Linear approximation

We would like to regard the homoclinic orbit as the limit of periodic orbits. We will concentrate on the hyperbolic case to begin with.

Let P be a hyperbolic fixed point of a map F whose stable and unstable manifolds cross transversely at Q. We can divide phase space in two regions, a ball B around P where the equivalence with the linearized map holds (Hartman–Grobman theorem) and the complement of this region, i.e. the rest of phase space. Within the ball B we will choose local coordinates so that the map is effectively linear, hence taking the form

$$
\begin{aligned}
x_{n+1} &= \lambda_- x_n \\
y_{n+1} &= \lambda_+ y_n
\end{aligned}
\tag{9.1}
$$

where $\lambda_- < 1 < \lambda_+$. The x coordinate describes locally the stable manifold of P while y describes in the same way the unstable manifold.

Within the region B we can hence find an infinite number of points belonging to the homoclinic orbit since P is the accumulation point of sequences of both backward and forward iterations of the map F. Pick such a point $Y = (0, Y) \in B$ and consider $X = (X, 0) \in B$ (whether we speak of the point or the coordinate of the point should be clear by context) such that $F^n(Y) = X$

for some n (see equation (9.1)). The reader can verify that such a pair of points always exists for adequate n (we will come back to this point in the following sections). The map $F^n(x, y)$ when restricted to a small neighbourhood, $U \subset B$, of Y will take the (approximate) form

$$\begin{aligned} x' &= X + ax + b(y - Y) \\ y' &= cx + d(y - Y) \end{aligned} \tag{9.2}$$

with a, b, c, d constants, $ad - cb \neq 0$ since the map is considered invertible.

The condition of a *transversal intersection* at X is now written as

$$d \neq 0. \tag{9.3}$$

Once a trajectory reenters the ball B we can apply the linear expression of the map (9.1) to estimate higher iterates of the map. Thus, if we are looking for a periodic point of period $k = n + m$ lying inside B and close to Y we have to look for fixed points of

$$\begin{aligned} x' &= \lambda_-^m(X + ax + b(y - Y)) \\ y' &= \lambda_+^m(cx + d(y - Y)). \end{aligned} \tag{9.4}$$

Taking $\delta = y - Y$, the fixed points will be given by

$$\begin{pmatrix} -\lambda_-^m X \\ Y \end{pmatrix} = \begin{pmatrix} \lambda_-^m a - 1 & \lambda_-^m b \\ \lambda_+^m c \cdot & \lambda_+^m d - 1 \end{pmatrix} \begin{pmatrix} x \\ \delta \end{pmatrix} \tag{9.5}$$

or, dividing the second line by λ_+^m

$$\begin{pmatrix} -\lambda_-^m X \\ \lambda_+^{-m} Y \end{pmatrix} = \begin{pmatrix} \lambda_-^m a - 1 & \lambda_-^m b \\ c & d - \lambda_+^{-m} \end{pmatrix} \begin{pmatrix} x \\ \delta \end{pmatrix}. \tag{9.6}$$

The equation (9.6) will have solutions at least for large k if $d \neq 0$ since λ_-^m and λ_+^{-m} go to zero with m going to infinity and the determinant of the matrix goes to $-d$ which is non-zero by the transverse condition. The right hand side of (9.6) also goes to zero with m and correspondingly with k.

It follows that, at least in a first approximation, whenever there is a transverse homoclinic orbit, there are periodic orbits of period k greater than a certain k_0 that converge towards the points of the homoclinic orbit when $k \to \infty$ (see figure 9.2).

We can also guess that the smaller d, the larger k_0, since our approximation is only valid in the small neighbourhood U where the linearization of the map around Y can be considered plausible.

When d is zero the stable manifold intersects the unstable manifold at X but they are tangent, i.e., the intersection is no longer transverse.

We finally note that transverse homoclinic crossings are structurally stable features (in any reasonably defined space of maps) since any small deformation of the map will produce only a small deformation of the stable and unstable manifolds going from P to Q being these pieces of manifolds the union of images of the local manifolds, $F^{m_-}(W_s^{loc})$ and $F^{m_+}(W_u^{loc})$, for m_- and m_+ smaller than n.

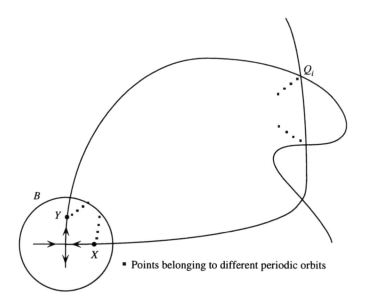

Figure 9.2. Periodic orbits converging to the homoclinic point.

9.2   Homoclinic tangencies

What happens at a *homoclinic tangency*, i.e., when $d = 0$ in (9.3)?

The homoclinic tangency is a structurally unstable situation since nearby maps (again, in a reasonably defined space of maps) might have no manifold crossing at all or (in general) two transverse crossings. The problem should be studied as a family of maps that contain the homoclinic tangency in a structurally stable way; i.e., we are at a bifurcation point and would like to unfold the situation.

What happens to the periodic orbits associated with the homoclinic orbit? Will they disappear also? Or, exactly the same, are periodic orbits created at a homoclinic tangency?

We will answer these questions using a nonlinear version of the analysis of the previous section.

We modify the map (9.2) to

$$
\begin{aligned}
x' &= X + ax + b(y - Y) \\
y' &= \mu + cx + e(y - Y)^2.
\end{aligned}
\tag{9.7}
$$

Such a map presents the desired feature of a homoclinic tangency for $\mu = 0$ which in general will be quadratic.

Composing (9.7) with an appropriate iteration of the local map we obtain the equation for the periodic orbits of period $k = n + m$ in the form

$$
\begin{aligned}
x &= \lambda_-^m (X + ax + b(y - Y)) \\
y &= \lambda_+^m (\mu + cx + e(y - Y)^2)
\end{aligned}
\tag{9.8}
$$

with solutions

$$
x = \lambda_-^m \frac{X + b(y - Y)}{1 - \lambda_-^m a}
$$

$$
2e(y - Y) = \left(\frac{1}{\lambda_+^m} - \frac{\lambda_-^m bc}{1 - \lambda_-^m a} \right)
\tag{9.9}
$$

$$
\pm \left((\frac{1}{\lambda_+^m} - \frac{\lambda_-^m bc}{1 - \lambda_-^m a})^2 - 4e(\mu + \frac{c\lambda_-^m X}{1 - \lambda_-^m a} - \frac{Y}{\lambda_+^m}) \right)^{1/2}.
$$

Clearly, two periodic orbits of period $k = n + m$, for k large enough, are born or destroyed when

$$
\mu_k = \left(\frac{1}{\lambda_+^m} - \frac{\lambda_-^m bc}{(1 - \lambda_-^m a)} \right)^2 /4e - \left(\frac{c\lambda_-^m X}{1 - \lambda_-^m a} - \frac{Y}{\lambda_+^m} \right).
\tag{9.10}
$$

It is now easy to verify that $\mu_k \to 0$ when $k \to \infty$. The homoclinic tangency is a point of accumulation for saddle-node bifurcations.

9.3 Homoclinic tangles and horseshoes

The dynamics close to the homoclinic orbits present many more interesting features than the periodic orbits already discussed.

Let $U_{0,0}$ be the piece of the unstable manifold of the periodic orbit P that goes from the homoclinic point Q_0 to the homoclinic point Q_0' (see figure 9.3). Since the extremes of the segment $U_{0,0}$ accumulate at P under forward iterations of the map (they lie on the *stable* manifold) we will have that (at least in the region, B, dominated by the local dynamics) the forward iterates of this piece of manifold will become narrow and long fingers that accumulate onto the unstable manifold. The same can be said of a piece of the stable manifold $S_{0,0}$ that under the action of backward iterations will accumulate onto the stable manifold stretching along the direction of W_s and being squeezed in the direction of W_u. The combined result of the two actions is that more intersections of W_s and W_u which are not part of the original homoclinic orbit will appear, hence more periodic orbits and more homoclinic orbits and so on (cf the discussion in the previous sections).

In order to put some order into this proliferation of periodic and homoclinic orbits we will explore a geometrical–topological construction: the Smale horseshoe [smal67].

Consider again a homoclinic tangle like the one in figure 9.4. Consider a small square $abcd$ in the region B nearby the saddle P. The square is chosen so

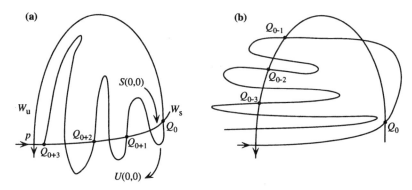

Figure 9.3. (a) Forward iterates of a piece of the unstable manifold accumulate onto the stable manifold. (b) Backward iterates of the stable manifold.

that there are four secondary homoclinic points corresponding to the intersection of some backward, say m-, iterate of a piece S of the stable manifold (see figure 9.4), with some forward n-iterate of a piece U, of the unstable manifold. It is clear that we can always find n and m so that U_n intersects S_{-m} in the proposed way (the reasons have been stated in the previous paragraph).

We now take the backward n-image of the box to produce a rectangle $a_{-n}b_{-n}c_{-n}d_{-n}$ with the sides a_nb_n and c_nd_n parallel to S. In the same way the forward m-image of the box $abcd$ will have the same general shape as U with the sides b_mc_m and a_md_m parallel to U. The box $a_mb_mc_md_m$ has a *horseshoe* shape. Hence, the $(n+m)$- image of the rectangle $a_{-n}b_{-n}c_{-n}d_{-n}$ is a horseshoe $a_mb_mc_md_m$ with non-empty intersection with this rectangle. We will call the map $H \equiv F^{n+m}$ restricted to the region $a_{-n}b_{-n}c_{-n}d_{-n}$ a *horseshoe map*.

The relevance of the horseshoe map will become clear in the coming sections. We would like to point out that since the conditions imposed on the secondary homoclinic orbits are fulfilled for $k > k_0$ whenever they are satisfied for $k = k_0$ (k_0 plays the role of $n+m$ above), actually there are horseshoe maps associated with the k-iteration of the map F for all $k \geq k_0$. This fact should be intuitively clear from the figures.

In the case of the *horseshoe tangle* illustrated in figure 9.5, the Smale horseshoe can be realized as the first iteration of a map. The borders of the 'square' $abcd$ and its image, the horseshoe, $a'b'c'd'$ are pieces of the stable and unstable manifold.

9.4 Heteroclinic tangles

Cyclic transverse heteroclinic connections are also able to introduce complex behaviour in the dynamics. To fix ideas, let us consider a cycle consisting of only two hyperbolic periodic orbits P_1 and P_2 of the same periodicity. Figure

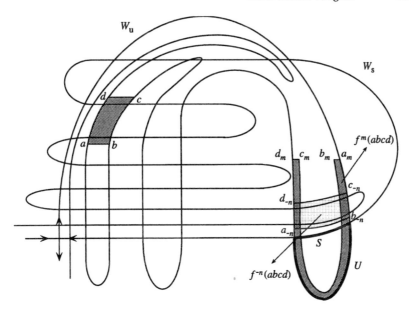

Figure 9.4. From the homoclinic tangle to the horseshoe.

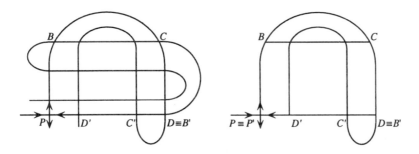

Figure 9.5. A homoclinic tangle completely developed into a period-one horseshoe.

9.6 describes the dynamics of an associated Poincaré map. We consider the case when $W_s(P_1)$ intersects transversely $W_u(P_2)$ and $W_s(P_2)$ intersects transversely $W_u(P_1)$. These transversal intersections imply that the stable (unstable) manifold of P_1 folds and accumulates into the stable (unstable) manifold of P_2 and an identical statement can be made exchanging P_1 with P_2.

The transverse heteroclinic crossings with the folding and accumulation processes produce secondary crossings between the stable and unstable manifolds of each periodic orbit hence producing (secondary) homoclinic orbits with all the properties described in the previous section including the existence of horseshoe maps for certain regions of the phase space.

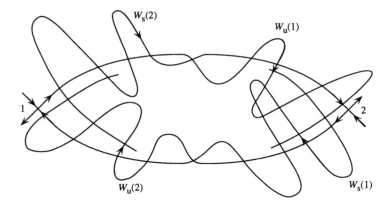

Figure 9.6. Transverse heteroclinic cycle.

Although these homoclinic orbits would be enough to justify the claim of complexity in the dynamics we would like to sketch how to address the dynamics nearby the (primary) heteroclinic orbits. In the same way as in the previous section, we can look for small squares S_1 (S_2), close enough to P_1 (P_2) that the full map is equivalent to the linearized map restricted to this region. Under backward iteration these squares will stretch and move away from the periodic points (orbits) following the stable manifold. Under forward iterations the squares will stretch, fold and move away from the periodic points following the unstable manifolds. The squares can be tailored so that when, after n iterations, the forward image of S_1 intersects S_2, $F^n(S_1)$ has the form of a horseshoe. The same can be said exchanging 1 with 2 for some forward iterate (say m) of F. The situation is illustrated in figure 9.7. The dynamical structures that will emerge from this map will have the property of sequentially visiting the neighbourhoods of the primary periodic orbits P_1 and P_2 and the neighbourhoods of some points belonging to the heteroclinic orbits.

Exercise 9.1: Following the ideas of the linear approximation to homoclinic orbits find the approximate location of the periodic orbits that accumulate at the transverse heteroclinic points.

Exercise 9.2: (Challenge!) *Crisis.* Let P be a saddle of a certain map F with associated negative Floquet multipliers (we call it a *flip* saddle. Think for example of a periodic orbit that has undergone a flip bifurcation). First note that both branches of the stable (unstable) manifold are images of each other (Hint: Consider F^2). Suppose that the manifolds have no intersections for $\mu < 0$ and that for $\mu = 0$ the stable and unstable manifolds become tangent. Show that for

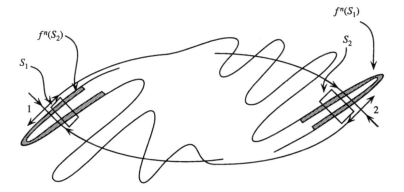

Figure 9.7. Maps associated with the heteroclinic cycle between two periodic orbits.

$\mu > 0$ there are bifurcations to periodic orbits of F^2 that visit the neighbourhood of both branches of the stable and unstable manifolds (see figure 9.8).

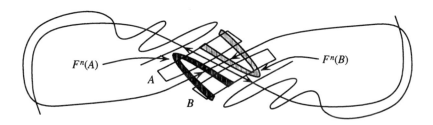

Figure 9.8. A flip saddle with a transverse manifold intersection ($\mu > 0$).

9.5   Summary

In this chapter we have started the analysis of the features present when the stable and unstable manifolds of a periodic orbit cross. We have considered the case of hyperbolic fixed points of two-dimensional maps, showing that

(i)   There is an infinite number of periodic orbits associated with a transverse homoclinic crossing of the stable and unstable manifolds.
(ii) There is an infinite number of saddle-node bifurcations associated with a homoclinic tangency.
(iii) Transverse homoclinic crossings have an associated *horseshoe map* (the features of this map will be the main topic of the following chapter).
(iv) Heteroclinic points present an orbit structure having similar features to that of homoclinic points.

Chapter 10

Horseshoes

We hope that the discussion of chapter 8 (section 8.3) was sufficient to suggest to the reader the important role that homoclinic orbits play in generating complex dynamics. Further, in chapter 9, some of the implications of having transverse homoclinic crossings in two-dimensional maps were discussed, namely the existence of orbits of any period higher than a given one, and the existence of a *Smale horseshoe*. With the word horseshoe we described the fact that a portion of phase-space in the vicinity of the *homoclinic point* was mapped back onto approximately the same region by a (possibly high) iteration of the map, folding and bending the original region (in the shape of a horseshoe).

The complex dynamics that arises in the presence of these facts is so rich and interesting that it deserves some formalization and systematic study. Hence, in this chapter we will study Smale's horseshoe maps, concentrating on the general properties of horseshoes (in other words, concentrating on general dynamical features associated with the existence of homoclinic orbits in maps) without reference to the specific facts of the dynamics where it may appear.

The word *horseshoe* will be used a little loosely to indicate either the map (that bends its domain in the form of a horseshoe) or the invariant set Λ of this map (which we will describe in the coming sections), or both.

10.1 The invariant set

Let us begin analysing the horseshoe from a geometrical–topological point of view. The horseshoe map, F, takes a (topological) square $S = ABDC$ into a horseshoe shaped figure $F(S)$ in such a form that the intersection between the original square and its image are two strips (see figure 10.1). The map can be decomposed in three successive steps.

- Deform the square S into a narrow strip contracting the sides AB and CD.
- Stretch the rectangle elongating the sides AC and BD.
- Fold into an upside-down U shape.

Note that the image of the sides AB and CD fall outside (under) the original square while there is a middle strip going from the side AC to the side BD that is mapped above the square.

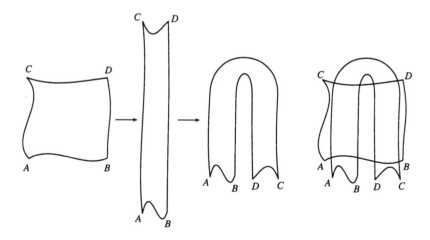

Figure 10.1. Topological horseshoe.

From the discussion in chapter 9, we notice that the horseshoe map can be seen as the restriction of a more general map H to a particular region of phase space in the presence of a transverse homoclinic crossing. We can therefore safely assume that F is defined in a larger region than just the square S, and that it has an equally well defined inverse[†]. In any case, the more interesting features of F will appear within S.

The inverse of the map F, F^{-1}, can be obtained reversing the steps. Such a map will have to

- Unfold the upside-down U shape into a rectangle.
- Compress the rectangle along its longest direction.
- Stretch the rectangle along the shortest direction to obtain a square.

In the same way as F can be assumed to be defined in a region larger than just S, we can assume F^{-1} to exist in a region larger than just $F(S)$. The natural procedure is to follow the steps taken in chapter 9 to define the horseshoe, in reverse order. In figure 10.2 we show this procedure and the deformation produced by F^{-1} on the original square. Note that $F^{-1}(S)$ is a sidewise U that intersects S in two horizontal strips.

We are interested in determining the (maximal) invariant set Λ contained in the square S. A point in S belongs to Λ if and only if its forward and backward

---

[†] If one insists in having F defined *only* on S some (solvable) difficulties arise when computing powers of F. For instance $F^{-1}(S)$ would not make sense.

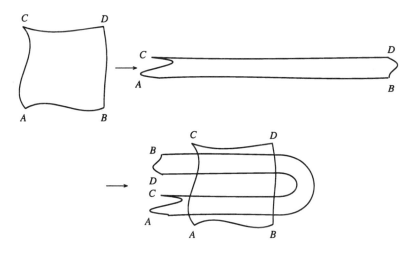

Figure 10.2. Inverse of the topological horseshoe.

orbits belong to S. Hence,

$$\Lambda = \left\{ x \in S : F^n(x) \in S \text{ for } n = 0 \ldots \infty \right\}$$
$$\cap \left\{ x \in S : F^n(x) \in S \text{ for } n = 0 \ldots - \infty \right\}. \qquad (10.1)$$

It is convenient to consider Λ as an intersection of sets rather than as a union of orbits. In order to characterize Λ we consider a family of sets, starting by $\Lambda_0^f = S$. The f refers to the *forward* iterates of F. The next member of the family is the set $\Lambda_1^f = F(\Lambda_0^f) \cap S \equiv F(S) \cap S$, which consists of the two vertical strips V_0 and V_1 displayed in figure 10.3. The following step is to consider $\Lambda_2^f = F(\Lambda_1^f) \cap S \equiv F(S \cap F(S)) \cap S \equiv F^2(S) \cap F(S) \cap S$, which can be constructed 'by hand' from the figure. F will map the strips V_i onto narrower U shaped strips. Portions of these strips fall back onto $F(S) \cap S$. Taking the intersection with S we obtain four thin strips placed pairwise within V_0 and V_1. We may call these strips V_{00}, V_{01}, V_{10} and V_{11}. The procedure is illustrated in figure 10.3.

This procedure can be iterated in a straightforward way. After n iterations of F, the set $\Lambda_n^f = F(\Lambda_{n-1}^f) \cap S \equiv F^n(S) \cap F^{n-1}(S) \cap \ldots \cap S$ will have 2^n vertical strips, pairwise contained within the 2^{n-1} strips of the previous iteration.

Let us consider now negative n values. Which portion of S lands on V_0, V_1 by the action of F? In other words, what is $F^{-1}(V_i)$? We can convince ourselves that $F^{-1}(S \cap F(S)) \equiv F^{-1}(S) \cap S \equiv F^{-1}(\Lambda_1^f)$ consists of a pair of horizontal strips H_i. Let us call this set Λ_1^b. Correspondingly, the b refers to *backward* iterations of F.

We can proceed in a similar way to construct the negative nth iteration. The question to answer now is which portion of S will land on one of the

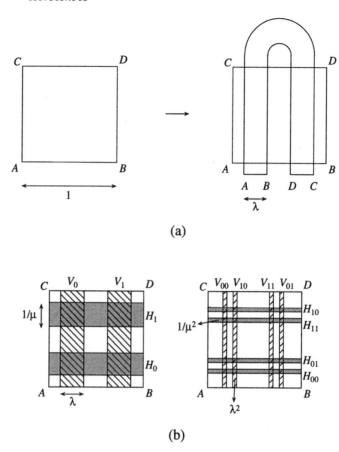

Figure 10.3. (a) The Smale horseshoe map. (b) First steps in the construction of a horseshoe.

2^n vertical strips after the action of F^n? In other words, we look for the set $\Lambda_n^b = F^{-n}(\Lambda_n^f) \equiv F^{-n}(S) \cap F^{-n+1}(S) \cap \ldots \cap S$, which consists of 2^n horizontal strips pairwise placed within the 2^{n-1} strips of the previous step. It is clear that we have constructed these strips so that after n iterations of F, they will form the vertical nth-order strips.

The intersection of both backward and forward iterations results in a set

$$\Lambda_n = \Lambda_n^f \cap \Lambda_{n-1}^f \cap \ldots \cap \Lambda_1^f \cap \Lambda_0^f \cap \Lambda_1^b \cap \ldots \cap \Lambda_{n-1}^b \cap \Lambda_n^b \quad (10.2)$$

$$= F^n(S) \cap F^{n-1}(S) \cap \ldots \cap S \cap \ldots \cap F^{-n+1}(S) \cap F^{-n}(S) \quad (10.3)$$

consisting of $(2^n)^2$ small squares fourfold contained in the squares of the $(n-1)$th step. The reader may have guessed that in order to obtain the invariant set Λ we let $n \to \infty$.

We can now rephrase our original definition of Λ using the horizontal and vertical strips constructed above. Hence, the set of points whose *forward orbit* belongs to S is the intersection of the horizontal strips, i.e.

$$\Lambda_+ = \cap_{n=0}^{\infty} F^{-n}(S) \tag{10.4}$$

since those points will remain within V_0 and V_1 after the action of F, within V_{00}, V_{01}, V_{10} and V_{11} after F^2 and so on. In a similar way, the set of points whose *backward orbit* belongs to S is the intersection of the vertical strips:

$$\Lambda_- = \cap_{n=0}^{\infty} F^{n}(S) \tag{10.5}$$

and the invariant set is the intersection

$$\Lambda = \Lambda_+ \cap \Lambda_-. \tag{10.6}$$

10.1.1 A horseshoe prototype

Let us assume we have made a clever choice of coordinates so that the topological square becomes the unit square (we use however the same name, S, for both squares). Let us further represent the horseshoe bending by an invertible map [guck86] $F : \mathbb{R}^2 \to \mathbb{R}^2$ which is a combination of two maps: a linear compression and expansion F_1 composed with a bending F_2. Choosing Cartesian coordinates x, y, the map F_1 restricted to S reads

$$\begin{pmatrix} x \\ y \end{pmatrix} \mapsto \begin{pmatrix} \lambda & 0 \\ 0 & \mu \end{pmatrix} \begin{pmatrix} x \\ y \end{pmatrix} \tag{10.7}$$

where $0 < \lambda < 1/2$ is the contraction rate and $\mu > 2$ is the expansion rate. These constraints assure that the strip will be thin and long enough to place largely within the unit square after the action of F_2 (see below). The actual values of λ and μ are not very important, provided that $1/\mu, \lambda < 1/2$.

The exact details of how F_2 works are again less important. The reader may be convinced by now that a larger or smaller deformation may be accounted for with a suitable coordinate change.

A possible candidate for the portion of F_2 that maps back onto S could be

$$\begin{aligned} F_2(w) &= w & w = (x, y) \in [0, \lambda] \times [0, 1] \\ F_2(w) &= (1 - x, \mu - y) & w = (x, y) \in [0, \lambda] \times [\mu - 1, \mu]. \end{aligned} \tag{10.8}$$

In this model the width of the 2^n vertical strips in equation (10.2) is λ^n while the height of the 2^n horizontal strips is μ^{-n}. The model is sketched in figure 10.3.

The same conclusion can be achieved if the F_1 map is replaced by e.g.,

$$\begin{pmatrix} x \\ y \end{pmatrix} \mapsto \begin{pmatrix} \lambda(y) & 0 \\ 0 & \mu(x) \end{pmatrix} \begin{pmatrix} x \\ y \end{pmatrix} \tag{10.9}$$

provided that $\lambda(y) < 1/2$ and $\mu(x) > 2$. Note that if these conditions were not satisfied, it would be possible (in principle) for some of the squares to remain with finite size in the limit $n \rightarrow \infty$. This would make Λ a patchy subset of S. In what follows we assume that such a possibility has been ruled out.

The crucial point, then, is that the rates of contraction λ^n and μ^{-n} guarantee (1) that the horseshoe can be constructed and (2) that the size of each of the intersection squares formed in Λ_n shrinks towards zero. Hence, we are allowed to expect that in the limit each square will determine a unique point.

10.2   Cantor sets

Before trying to establish the properties of Λ we will attempt a characterization of the iterative procedure we have developed.

The underlying process producing the sequence of horizontal (or vertical) strips in the previous section consists of extracting (open) segments from an interval (to fix ideas consider the side AC of the square S). This construction is known as a *Cantor set* [kolm57]. Some properties of Cantor sets can also be found in [barn88]. Let us look at this set in detail.

Consider the unit interval on the line. Subtract from it the open central interval $(1/3, 2/3)$. Subtract then the open central interval of size 3^{-2} from each remaining (closed) segment. The remaining set after n steps will consist of 2^n disjoint closed intervals of length 3^{-n} each. We illustrate the beginning of the procedure in figure 10.4.

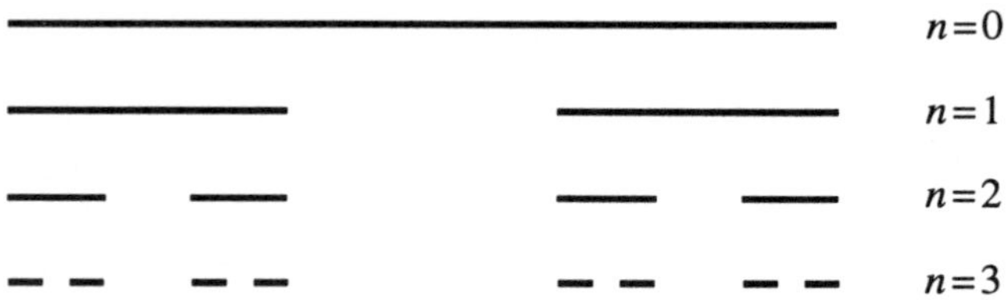

Figure 10.4. First steps in the construction of a Cantor set.

Since all intervals at each step are created in pairs from a parent interval from the previous step, we can use a two-symbol labelling, say 0 for the leftmost interval of a pair and 1 for the other. At step n each interval can be labeled by a string of n 0s or 1s indicating the label of itself and all its parent intervals all the way up to step 1. Taking the limit $n \rightarrow \infty$ we obtain a Cantor set $\mathcal{P}$. We will learn about some properties of this set through a sequence of exercises.

Exercise 10.1: Show that the Cantor set has as many points as the whole interval [0, 1].
Hint: Use the binary representation of numbers in [0, 1].

Exercise 10.2: You may have noticed that the complement of $\mathcal{P}$ in $[0, 1]$ is a countable union of disjoint open intervals. Compute the measure of this complement set by summing the length of the component intervals (this way of computing measures is one of the main ingredients of the *Lebesgue measure* [roma75]). What would you say then is the measure of the Cantor set?

The construction of a Cantor set scales with the iteration number n, i.e., at each step one 'extracts' a central open interval (for the horseshoe, the distance between two horizontal strips) whose size is a fraction of the available size. In the first iteration we extract an interval of size $\alpha < 1$, leaving two closed intervals of size, say, $(1 - \alpha)/2$. In the second iteration we extract two intervals of size $\alpha(1 - \alpha)/2$, and so on. The reader may convince herself/himself that the measure of the extracted intervals is independent of $\alpha < 1$, i.e., the same measure as the one obtained in the exercise above!

Exercise 10.3: Construct a set $\mathcal{Q}_\epsilon$ by extracting an open central interval of length $(1/3 - \epsilon)^n$ at each step (to fix ideas, let $\epsilon = 1/12$). Repeat the previous exercises for this new construction and note similarities and differences.

Exercise 10.4: Show that $\mathcal{P}$ does *not* contain any open interval. Hint: Ternary numbers! Write each number in $[0, 1]$ in base 3 (using the symbols 0, 1 and 2. The number $2/3$ is written 0.2) and notice that the elements of $\mathcal{P}$ are those numbers that do not have a 1 in their ternary representation. Show then that *any* non-empty open interval (a, b) contains numbers having 1s in its ternary representation[†].

The argument in the last exercise can be adapted to show that all points in $\mathcal{P}$ are accumulation points, i.e., limit points of some sequence in $\mathcal{P}$. Take a point x in the set and an arbitrary ball around it, of radius ϵ. Notice that for any positive radius there is always another point of $\mathcal{P}$ within the ball, just by checking that in the open intervals $(x, x + \epsilon)$ and $(x - \epsilon, x)$ there are points that do not have a 1 in their ternary representation.

10.3   Symbolic dynamics

We can regard Λ as consisting of the intersection of two Cantor-like sets: $\Lambda = \Lambda_h \cap \Lambda_v$, where Λ_h is the Cartesian product of the unit interval in the horizontal direction times a Cantor set in the vertical direction. The actual

---

[†] The ternary representation (or the decimal for what it matters) has a minor ambiguity. All numbers having ternary representation with the tail $1000\ldots$ can be also represented using the tail $0222\ldots$ instead (cf decimal representations $0.1000\ldots$ and $0.0999\ldots$). We adopt the convention of rejecting the first representation and choosing consistently the second one. Hence, $1/3$ (which belongs to the Cantor set) is correctly (and uniquely) represented in ternary language by $0.0222\ldots$.

width of the substracted interval depends on $1/\mu$ and on the specific choice for the folding. We can describe Λ_v in a similar way. Each point in Λ can then be characterized by two infinite strings of 0s and 1s, indicating its position on the horizontal and vertical Cantor sets. We will presently define and use an alternative characterization. But first:

Exercise 10.5: Show that Λ is compact, does not contain any open set, has uncountably many points and all points are accumulation points.

The map F is injective. This is a crucial property, since 'real-life' maps F representing e.g., physical systems are described by injective mappings (or by a Poincaré mapping of a dynamical system near a homoclinic orbit) because of the unicity of solutions of differential equations. As a consequence, each point in Λ has a *unique* trajectory for all past and future iterations of the map. We will then characterize every point $x \in \Lambda$ by a property of its trajectory, $\{F^i(x)\}$, $-\infty \leq i \leq \infty$, namely by the bi-infinite sequence of 0s and 1s indicating on which of the two initial horizontal strips H_0 and H_1 the iteration i falls.

Exercise 10.6: Show that two different points $x, y \in \Lambda$ have different sequences.

We can formalize these ideas by saying that there is a one-to-one mapping ϕ between Λ and Σ, the set of all bi-infinite sequences of the form $a = \ldots a_{-2} a_{-1} \cdot a_0 a_1 a_2 \ldots$, or in short, $a = \{a_i\}_{i=-\infty}^{\infty}$, where $F^i(x) \in H_{a_i}$. When Σ is equipped with the metric defined by

$$d(a, b) = \sum_{i=-\infty}^{\infty} \frac{|a_i - b_i|}{2^{|i|}}, \tag{10.10}$$

the mapping is continuous in both directions, i.e., a homeomorphism.

The action of F restricted to Λ induces a map on Σ, which can be regarded as a symbolic representation $\sigma : \Sigma \to \Sigma$ of F or the *symbolic dynamics* associated with F:

$$\sigma(a) = \phi(F(\phi^{-1}(a))). \tag{10.11}$$

So, if $b = \sigma(a)$, we have $F^i(F(x)) \in H_{b_i}$ and hence $F^{i+1}(x) \in H_{b_i}$ which amounts to say $b_i = a_{i+1}$, i.e. b is obtained by shifting forward the dot in the sequence of a. σ is usually called a *shift automorphism*. Understanding the orbit structure of the invariant set Λ will prove to be much easier using the symbolic dynamics.

10.3.1   Restriction of the map to the invariant set

The action of F on S can yield one of two results. Either all the iterates remain on S, i.e., we are concerned with a point of Λ, or a higher power F^k of the map will eventually 'kick' the point out of S. In a 'real-life' problem this last possibility would mean that the horseshoe-map approximation is no longer valid and nothing more can be said about the future of the system without a closer study of the problem. On the contrary, in the first case it is possible to draw conclusions of a general character, i.e., valid for *any* system that holds a horseshoe map and consequently a symbolic dynamics σ. Let us then further consider $F|_\Lambda$ or equivalently σ.

Notice first that the symbolic name a, b of points $x, y \in \Lambda$ defines (as well as the points themselves) the (past and future) *itinerary* that the point in question covers along the dynamics, described in terms of H_0 and H_1.

10.3.2 Periodic points of F^n

The map σ has clearly two fixed points, namely $a = \ldots000.000\ldots$ and $b = \ldots111.111\ldots$. In the same way, periodic sequences of period n, i.e., when $a_i = a_{i+n}$, describe the periodic orbits of σ.

We can compute $\sigma^n(a)$ as an n-element shift in the sequence a_i. An n-periodic orbit of σ is a fixed point of σ^n. There are as many fixed points of σ^n as there are symbol sequences of length n, i.e., 2^n. Note that with this number we are counting all the periodic orbits of σ of period n or a divisor of n (for example, two consecutive cycles of a period-two orbit will form one of the possible symbol combinations of $n = 4$). What are the stability properties of these periodic points?

We see then that σ has periodic points (orbits) of *any* period $n \geq 1$: it has in fact infinitely many periodic points. Notice also that there are infinitely many more *non*-periodic orbits, i.e., *uncountably* many. Just consider points having the symbol sequences of irrational numbers written in binary form, on either side of the central dot.

Not only the periodic orbits but also a subset of their (local) stable and unstable manifolds belong to Λ. The points with symbolic sequences of the form $x.p^\infty$, where x is any infinite sequence and p^∞ is the sequence of the periodic orbit p, represent a horizontal line that contains the point p^∞. Every point of the form $x.p^\infty$ has $p^\infty.p^\infty$ as its t-infinite limit, i.e., $x.p^\infty$ represents a point belonging to the local stable manifold. By the same argument, $p^\infty.x$ represents a point belonging to the local unstable manifold of p.

Exercise 10.7: Show that for every pair of hyperbolic periodic points (orbits) p, q in Λ, $W_s(q) \cap W_u(p) \neq \emptyset$.

10.3.3 Orbit dense in Λ

The distance $d(a, b)$ defined above ensures that nearby points in Λ have nearby sequences in Σ and vice versa. This fact will help us to build an orbit (represented by a sequence in Σ) passing arbitrarily close to any point in Λ. This orbit is called *dense* since the orbit will have points within any non-empty open neighbourhood 'in' Λ (an open subset of Λ is defined to be the intersection of Λ with an open set of S) of any point $x \in \Lambda$.

Notice first that two points $a, b \in \Sigma$ which are closer than $\epsilon = 2^{-k}$ (for positive k) have symbolic sequences that coincide in their central parts up to index $\pm(k + 1)$. We can now build a sequence γ by starting in the central dot and writing on both sides of the dot all finite sequences of 0s and 1s of length $2k + 1$, for $k = 1, 2, 3, \ldots$. The orbit we thus get will pass (both in future and past times) closer than any given positive ϵ to any point $y \in \Lambda$. It is a matter of iterating forward the string γ so that its central part coincides with the central part of y up to index $|k| = [1 - \ln_2(\epsilon)]$ or higher.

Exercise 10.8: Use the existence of a dense orbit to prove that no periodic point of F in Λ can be a hyperbolic sink (i.e., all periodic points are saddles). In any neighbourhood of a periodic point p, find the sequence of a point which does not approach p. Note: Because of this property, Λ is called *indecomposable*.

10.3.4   Stable and unstable foliations

The stable and unstable manifolds for periodic orbits can be generalized for any other orbit. The stable set of a point, x, will be the set of points that approaches $o(x)$ (the orbit of x) when time (or the number of iterations) goes to infinity. The local part of this set will include the point x. Thus, if the symbolic sequence associated with $o(x)$ is $o^-.o^+$, the points with sequences $a.o^+$ (a standing for any sequence) make a subset of the local stable set of x. We leave as an exercise to verify with the horseshoe distance that $a.o^+$ converges uniformly towards $o^-.o^+$ (certainly, the reciprocal is also true!).

The same sort of construction can be performed with the unstable set. The stable and unstable sets are horizontal and vertical lines defined by the equivalence relation just described. We can say that they 'foliate' the invariant set Λ and that the stable and unstable foliations intersect transversely.

We will come back to the stable and unstable foliation in the subsection dedicated to hyperbolicity.

10.3.5   Sensitivity to initial conditions

The counterpart of the periodic, non-periodic and dense orbits we found in the previous subsections is the fact that two different points $x, y \in \Lambda$, no matter how close they are to each other, will have sequences with different central parts after sufficiently high forward or backward iteration of F. Just take a point x and another point y such that $a_n(x) \neq a_n(y)$. Hence $a_0(\sigma^n(x)) \neq a_0(\sigma^n(y))$. In other words, no matter how close to each other x and y are, their iterates will sooner or later, in the past or future, land in different horizontal rectangles a finite distance apart.

If we consider for $x \in \Lambda$ a neighbourhood of points in the invariant set $U = \{y : |x - y| < \epsilon\} \cap \Lambda$ we will have that for a sufficiently large (and positive) n, $F^n(U)$ will consist of a collection of vertical strips filling the square. Our initial uncertainty grows to the point of making any prediction of the future impossible. This is a 'practical' example of *sensitivity to initial conditions*, carried to its utmost expression. The size[†] of U necessary to keep the size of $F^n(U)$ smaller than a given positive δ decreases like μ^{-n} for fixed λ and some $\mu^* > 1$. Note that in the prototype horseshoe, equation (10.7), $\mu^* = \mu$.

10.4 Horseshoes and attractors

Now that we have discussed the peculiar properties of the dynamics of a horseshoe, it would be interesting to establish the relation between horseshoes and attractors.

We observe in our prototype equation (10.7) that the leaves of the unstable foliation consist of vertical lines. Only a piece of these lines (a Cantor set to be precise) belongs to Λ. We conclude that the invariant set Λ has an associated unstable set that is larger than Λ. A similar reasoning can be followed with the stable set. Hence, Λ is a set of saddle type and *cannot* be an attractor. This problem is not a particularity of our prototype but rather a general feature for bi-dimensional systems. No subset of a bi-dimensional horseshoe can be an attractor for identical reasons. Chaotic attractors of bi-dimensional systems are more subtle sets than 'simply' horseshoes. Their relation with horseshoes has yet to be fully elucidated. The following theorem due to Katok [kato80] shows that there is in fact an important relation between chaotic attractors and horseshoe maps.

Theorem (Katok): If F is a C^n ($n > 1$) diffeomorphism of a compact two-dimensional manifold and its *topological entropy*[‡], $h(F) > 0$, then F has a hyperbolic periodic point with a transversal homoclinic point and consequently there is an F-invariant subset Λ where F restricted to Λ is a horseshoe map.

Roughly speaking, the theorem says that if one has a mapping F with a chaotic attractor, then the mapping contains *also* a horseshoe. But the attractor (whatever it is) has to be more complex than just a horseshoe, since the horseshoe is not attractive. The last part of Katok's theorem is in fact part of Smale's theorem discussed in the next section.

The situation is different for three-dimensional maps. Smale [smal67] gives a rather simple example of a strange attractor. The map takes a solid torus into a new torus that revolves twice around the centre, see figure 10.5. A cut of the

---

[†] The size of $F^n(U)$ is the diameter of the smallest disc that completely covers $F^n(U)$.
[‡] The reader can safely read F is chaotic. The exact concept of topological entropy will be discussed in the coming chapters.

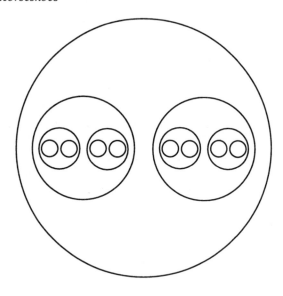

Figure 10.5. Cut of the Smale attractor.

attractor for a given angle will show a Cantor set structure as displayed in figure 10.5.

10.4.1 Smale's theorem

What are the conditions on a diffeomorphism F in order to assure the existence of a horseshoe? Roughly speaking, we have seen that we need a map F that takes part of its domain (in the shape of something like a deformed disc), stretches it and bends it onto itself so that (at least) two strips land back on the original domain. To specifically answer our question, we will have to concentrate on matters such as how big should the expansion (contraction) be, how 'vertical' should the vertical strips be, and so on.

On the other hand, from the dynamical systems point of view, one would like to know which *dynamical* property of F assures the existence of a horseshoe.

Both problems have been addressed thoroughly [smal67, guck86, wigg90]. The first question looks more 'academic' to the physicist while the other seems more 'practical'. They are however highly related to each other.

Concerning the first question we can state the following:

Theorem (Moser) [guck86]: If F is a two-dimensional homeomorphism for which (i) there exist $N \geq 2$ disjoint horizontal strips H_i and disjoint vertical strips V_i such that $F(H_i) = V_i$, sending horizontal (vertical) boundaries of H_i onto horizontal (vertical) boundaries of V_i, (ii) F contracts the vertical strips uniformly by a factor $v < 1$ and (iii) F^{-1} contracts the horizontal strips uniformly also by some factor $v' < 1$, then there exists an invariant set Λ where

F is topologically equivalent to a shift σ. The bi-infinite sequences have in this case N symbols.

The relation between orbits homoclinic to a (saddle) periodic orbit and horseshoe maps has been discussed from an intuitive point of view in chapters 8 and 9. This relation is spelled out by Smale's theorem [smal67]:

Theorem (Smale): Suppose x is a transversal homoclinic point of a diffeomorphism $F : M \rightarrow M$ (M a manifold). Then there is a Cantor set $\Lambda \subset M$, $x \in \Lambda$, and m a positive integer such that $F^m(\Lambda) = \Lambda$ and F^m restricted to Λ is topologically a shift automorphism.

A sketch of the proof of the Smale theorem for the case $M \subset \mathbb{R}^2$ was given in chapter 9 when the Smale horseshoe was presented as one of the features that are present in transverse homoclinic crossings. The general proof is a formalization of the same argument, i.e., checking that the candidate horseshoe map F (e.g., that of chapter 9) fulfills the hypotheses of the previous theorem.

10.5 Hyperbolicity

Some of the properties discussed so far rely heavily on the fact that the invariant set Λ is *hyperbolic*, i.e. that (i) at each point $z \in \Lambda$ one can define two directions E_z^u and E_z^s for which $Df(z)E_z^u = E_{f(z)}^u$ and $Df(z)E_z^s = E_{f(z)}^s$, (ii) that $Df(z)$ contracts vectors along E_z^s and also $Df^{-1}(z)$ contracts along E_z^u *both* at a rate smaller than some $\lambda < 1$ and (iii) E_z^u and E_z^s vary continuously with $z \in \Lambda$.

The main idea is to follow the same procedure as used to find out that the stable and unstable sets of a fixed point or periodic orbit were actually manifolds, but now applied to the stable and unstable foliation.

We need to find the analogous of the linearization of the field in a neighbourhood of a fixed point for more complex orbits. Let us first recall that the stability matrix for a periodic orbit, $o(x)$, of period p was given by $M = Df^p(x)$ and the stable and unstable subspaces tangent at x to the stable and unstable manifolds were spanned by the eigenvectors of M with eigenvalues smaller or larger than one respectively. We can try to generalize this idea as follows:

$$v \in E_x^u \text{ iff } |Df^{-n}(x)v| \le C\lambda^n|v|$$
$$v \in E_x^s \text{ iff } |Df^n(x)v| \le C\lambda^n|v|$$

for all positive n with $\lambda < 1$ and C a constant independent of n.

Exercise 10.9: Formally verify that with the above definition we have $Df(z)E_z^u = E_{f(z)}^u$ and $Df(z)E_z^s = E_{f(z)}^s$.

A discussion of this hyperbolicity property can be found in [wigg90]. Apart from its use in some of the (not shown) proofs, we used explicitly a consequence of this in the exercises, namely the hyperbolicity of the periodic orbits in Λ.

To establish that a given system does have an invariant set with a hyperbolic structure may in general be a difficult task.

Hyperbolic invariant sets have a property that is crucial to understand the meaning of numerical experiments done on a horseshoe. This is called the *shadowing lemma*. Let us consider it more closely.

Let us assume we perform numerical experiments on a map f having a hyperbolic horseshoe Λ as (part of an) attracting set. What would be the outcome of a large number of iterations started somewhere in the basin of attraction of this attracting set? After some 'transient' the iterates would approach the set exponentially (if we had infinite precision available), so our iterates will jump to and fro in the surroundings of points of Λ.

One thing we can be 'sure' of: that the iterates *do not* follow the orbit of a point $z \in \Lambda$. There are two reasons to support this statement. First, since we are mainly interested in invertible maps, we can be sure that if we start anywhere *outside* Λ, we will not reach the invariant set in any finite number of steps. We will just *approach* it. The second observation slightly contradicts the previous one. Numerical experiments are done with finite precision. Hence, the iterations do not carry x_n to $F(x_n)$ but rather to some point x_{n+1} in the surroundings (x_{n+1} is in fact the finite-precision truncation of $F(x_n)$). So except for rare coincidences, our numbers will be the iterates of a *different* map jumping somewhere *outside* (occasionally inside) but very close to Λ.

This strange wandering is called an α-*pseudo-orbit*. It is illustrated in figure 10.6. α is the size of a ball around x_{n+1} that is big enough to contain $F(x_n)$, uniformly for all iterates n in a finite segment $a \le n \le b$. For the case in discussion, α is of the order of the precision of the numerical experiment.

The way out of the conflict is the following:
Theorem (shadowing lemma) [bowe70]: Let Λ be a hyperbolic invariant set. Then for every $\beta > 0$ there is an $\alpha > 0$ such that every α-pseudo-orbit is closer than β to the iterates of some point $y \in \Lambda$ (we say that y β-*shadows* the sequence $\{x_i\}$).

10.6   Structural stability

The problem of structural stability of horseshoes can be restated in the following way: are transverse homoclinic crossings of the manifolds $W^s(p)$ and $W^u(p)$ of a fixed point p structurally stable? The keyword of the problem is *transverse*. A sufficiently small perturbation of a map f cannot destroy a transverse crossing. Hence, all sufficiently 'close' maps $f + \epsilon g$ will have a horseshoe together with f (for a sharper definition of 'close', recall the discussion in chapter 5). Moreover, the location of the ϵ-horseshoe in phase space will be ϵ-close to the original one, just because the maps are only slightly different, and hence the regions which are mapped onto themselves will only differ slightly.

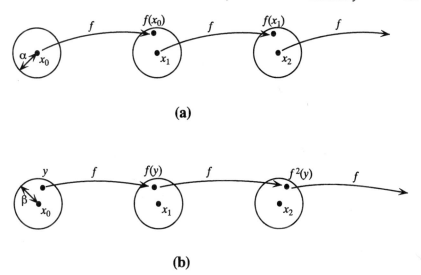

(a)

(b)

Figure 10.6. (a) A pseudo-orbit. (b) A true orbit shadowing the pseudo-orbit.

10.7 Summary

This chapter elaborates further on the observations of chapter 9, namely that a transverse crossing of the (stable and unstable) manifolds of a hyperbolic saddle point in a two-dimensional map leads to the fact that a small region of phase space is folded and mapped onto itself. The central topological property is the folding–shrinking process, which is explicitly realized by our prototype example (with contraction rates λ and $1/\mu$).

The main properties of the invariant set Λ of a horseshoe are:

- Λ has infinitely many periodic orbits of all periods.
- Λ has uncountably many non-periodic orbits.
- Λ has a dense orbit (passing arbitrarily close to any point in Λ).
- Sensitivity to initial conditions: any two different points x, y in Λ land a finite distance apart after a sufficiently high (positive or negative) iteration of F.

All these properties are persistent under small perturbations of the map F.

10.8 Additional exercises

Exercise 10.10: Let us consider a horseshoe map $F : S \rightarrow S$, with S the unit square.

(a) Label the sixteen components of $F^2(S) \cap F(S) \cap S \cap F^{-1}(S) \cap F^{-2}(S)$.

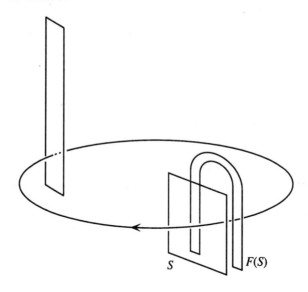

Figure 10.7. A horseshoe map in the Poincaré section of a 3-d flow.

(b) Locate in S the orbit associated with the periodic bi-infinite sequence
    ...01110111.01110111....
(c) Locate in S the orbit associated with the periodic bi-infinite sequence
    ...011011.011011....

Exercise 10.11: Let us assume that the horseshoe map $F : S \rightarrow S$, with S the unit square, is obtained through a Poincare section of a flow that lives in a three-dimensional phase space (see figure 10.7).

(i) Draw the orbit associated with the periodic sequence ...0111.0111... as a closed curve in the three-dimensional phase space.
(ii) In the same drawing, locate the curve corresponding to the orbit associated with the periodic sequence ...011.011....

Chapter 11

One-dimensional maps

We have two main reasons to study one-dimensional maps: first, there are many experiments that display first return maps that are almost one dimensional (one dimensional within experimental resolution); second, one-dimensional maps are among the simplest non-trivial dynamical systems. Correspondingly, we know a good amount about them.

One-dimensional maps appear as models of nonlinear problems in two different forms: as we mentioned in chapter 4 some processes that evolve from generation to generation (for example some insect populations) are naturally described by 'logistic maps' like $x_{n+1} = Ax_n(1 - x_n)$ with a linear term representing exponential (Malthusian) growth for small populations and a quadratic term that models limitations in population growth due to, for example, a finite *carrying capacity* of the ecosystem (the carrying capacity describes the fact that ecosystems, being finite in size and resources, only admit a finite population). The second form in which one-dimensional models appear is as extreme (singular) approximations to highly dissipative physical problems (high contraction of areas in phase space).

In figure 11.1 a peak-to-peak map of the Belousov–Zhabotinskii reaction is shown [rich87]. The map was obtained from a sampling of the concentration of bromide at equidistant intervals of times and plotting the concentration at a maximum (peak) against the value of the concentration at the next peak. A similar map for the intensity of a CO_2 laser with saturable absorber [papo91] is shown in figure 11.2. Although both systems are modelled by differential equations that produce n-dimensional ($n > 1$) Poincaré maps we can verify that up to the precision of the measurements the effective map is one dimensional.

This chapter will focus mainly on the 'combinatorial aspects' of one-dimensional maps. A thorough examination of one-dimensional maps can be found in Collet–Eckmann's book [coll86].

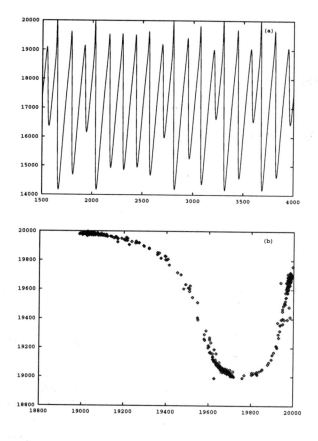

Figure 11.1. Experimental data from the Belousov–Zhabotinskii reaction. (a) Time series. (b) Peak-to-peak map.

11.1   Unimodal maps of the interval

Let us first introduce one of the (algebraically) simplest maps, the Hénon map [heno76]

$$(y_{n+1}, x_{n+1}) = F(y_n, x_n) = (x_n, -\epsilon y_n + \mu - (x_n)^2). \qquad (11.1)$$

The equation (11.1) defines an automorphism in $\mathbb{R}^2$ that contracts infinitesimal areas at a rate

$$\det(DF) = \epsilon. \qquad (11.2)$$

In the limit of infinite dissipation, $\epsilon \to 0$, the map becomes singular (it no longer has a unique inverse) and the information carried by the variable y turns out to be useless for the prediction of the value of x, i.e., the relevant dynamical information is carried by the variable x and the dynamics is given by the one-

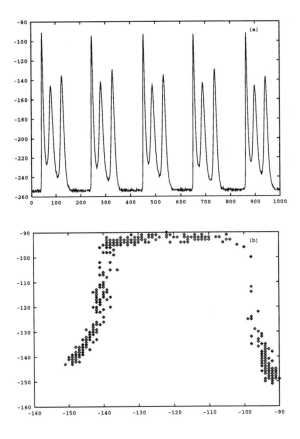

Figure 11.2. Experimental data from a CO_2 laser with saturable absorber. (a) Time series. (b) Peak-to-peak map.

dimensional map

$$x_{n+1} = f(x, \mu) = \mu - (x_n)^2 \tag{11.3}$$

which can be carried into the logistic equation with an affine change of coordinates.

The singularity of the map equation (11.3) is manifest in that the derivative for $x = 0$ is zero. The point $x = 0$ separates two branches of the map: $f(x, \mu)$ is a monotonically increasing function for $x < 0$ while for $x > 0$ it is monotonically decreasing. The two branches are called left (L) and right (R), while the central point $x = 0$ will be called C. We will often refer to the L branch as the orientation-preserving branch since given two points $x_1 > x_2$ ($x_1, x_2 \in [-1, 0)$) we have that $f(x_1) > f(x_2)$; in the same form we call R the orientation-reversing branch since for $x_1 > x_2$ ($x_1, x_2 \in (0, 1]$) we have that $f(x_1) < f(x_2)$.

The role of the orbit of C is most relevant in understanding one-dimensional maps that are qualitatively similar to equation (11.3). We will call a periodic orbit *superstable* if $x = 0$ belongs to it. The reason is that the eigenvalues associated to such an orbit (the eigenvalues of the linearization of f^p, where p is the period) will be identically zero. Hence, such a periodic orbit will have the highest possible stability (note that we are assuming that the map is differentiable; some peculiarities appear when the map is continuous but not differentiable).

We will call a map, $f : [-1, 1] \to [-1, 1]$ a *unimodal map of the interval* (see figure 11.3) if it satisfies that:

- f is continuous in $[-1, 1]$
- $f(0) = 1$
- f is strictly decreasing in $(0, 1]$ and strictly increasing in $[-1, 0)$.

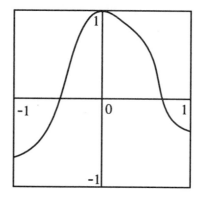

Figure 11.3. Unimodal function of the interval.

11.1.1 Itineraries

Metropolis, Stein and Stein [metr73] established in their study of unimodal maps, already mentioned in chapter 8, that the order in parameter space in which the periodic attractors appeared was the same independently of the maps considered in their studies. Since it is hard to conceive that the maps considered by them have identical behaviour just by pure chance, we are inclined to seek the reasons for the above-mentioned regularity. A key element in their study consisted in labelling the points belonging to periodic orbits (and its forward iterates) according to their position to the left (L) or to the right (R) of zero. We intend to show that despite the differences among unimodal maps their behaviour can be summarized by the concept of *itinerary*.

Consider the points in the forward orbit of x, $f^i(x)$, $i = 0, \ldots, \infty$, for $x \in [-1, 1]$. We will write an L (R or C) when the $f^i(x)$ lies in $[-1, 0)$ $((0, 1]$

or is 0). The itinerary of x, $I(x)$ is the (infinite length) letter-string constructed in this fashion. Letter-strings are also called *words*. For example, the word $RLLL\ldots$ identifies an orbit where $x > 0$, $f(x) < 0$, $f^2(x) < 0$, $f^3(x) < 0,\ldots$. In principle, it is possible for several orbits to have the same itinerary. One exception is those itineraries that begin with a C: since the C identifies a unique point, there is only one point whose orbit begins with a C. Note that unlike the symbolic sequences of chapter 10, we are not going to define a distance in the space of itineraries.

Exercise 11.1: Show that if there is an itinerary with more than one C, then this itinerary is composed by a (finite) first word followed by a second word which is repeated infinitely many times, i.e., $I(x) = W_1(W_2)^\infty$. We can take W_2 beginning with a C. Show that in this case the orbit of $x = 0$ is periodic.

Exercise 11.2: Can you find any relation between this construction and the symbolic sequences described for the horseshoe?

Note that the itinerary of $f(x)$, $I(f(x))$ is the itinerary of x where the first letter has been dropped. We will abuse notation and let f act on itineraries. Dropping the first letter of $I(x)$ is an operation that we can perform without actually knowing the value of x. Hence, f induces an automorphism in the 'space of itineraries' that we should denote by something like $\hat{f}$ with $\hat{f}(I(x)) = I(f(x))$, but we abuse notation and simply write $f(I)$.

11.1.1.1 Order on the itineraries

Which itineraries are possible for a given unimodal map? Which information is contained in the itineraries?

To answer these questions let us imagine the following situation: consider the orbits of x and y and their respective itineraries $I(x)$ and $I(y)$. Let us denote by $I_n(x)$ the first $n \geq 1$ symbols (from 0 to $n-1$) in the itinerary of x. Assume $I_n(x) = I_n(y)$ but $I_{n+1}(x) \neq I_{n+1}(y)$. For definiteness let $I_1(f^n(x)) = R$ and $I_1(f^n(y)) = L$ (we encourage the reader to verify that the symbol appearing in the last position of $I_{n+1}(y)$ is exactly $I_1(f^n(y))$). Then, we know that x and y lie on the same side of the interval and the same can be said for the $n-1$ images by f of x and y. Moreover we know that

- if $x < y$ and $f^i(x)$, $f^i(y)$ have visited an even number of times the orientation-reversing branch of the map (i.e., $I_n(x) = I_n(y)$ has an even number of R), then $f^n(x) < f^n(y)$;
- if $x < y$ and $f^i(x)$, $f^i(y)$ have visited an odd number of times the orientation-reversing branch of the map (i.e., $I_n(x)$ has an odd number of R), then $f^n(x) > f^n(y)$.

Now, since we know that $f^n(x) > f^n(y)$ and we can count the number of times that the orientation-reversing branch has been visited, we can decide with this information whether $x > y$ (if $I_n(x)$ contains an even number of R) or $x < y$ (if $I_n(x)$ contains an odd number of R).

We usually refer to I_n as an *odd sequence* (*even sequence*) when I_n contains no C and an odd (even) number of R respectively.

We conclude that comparing the itineraries of two points, x and y, we can decide which is their relative order in the interval of x and y.

Exercise 11.3: Show that if $I(x) = RLL\ldots$ and $I(y) = RLR\ldots$, then $x > y$. Order in the interval the points corresponding to the periodic sequences $R^\infty, (RL)^\infty, (RLLL)^\infty, (LR)^\infty, (LLLR)^\infty, (LLRL)^\infty, (LRLL)^\infty$.

We can formalize this information introducing an order of the sequences so that it coincides (as much as possible) with the order in the interval. This order has the following expression:

(i) $L < C < R$
(ii) $WL < WC < WR$ if the word W contains an even number of R
(iii) $WL > WC > WR$ if the word W contains an odd number of R
(iv) let W and W' be finite length sequences, and $W > W'$ ($W < W'$), then
      $W\ldots > W'\ldots$ ($W\ldots < W'\ldots$)

Exercise 11.4: Show that if $RA > RB$ then $A < B$, where A and B are two arbitrary, non-empty words.

It is now immediate to show that:

• if $I(x) > I(y)$ then $x > y$;
• if $x > y$ then $I(x) \geq I(y)$.

After all, we constructed the order of the sequences so that the above statements would be true!

11.1.1.2 Reasonable itineraries

We can now turn our attention to the first question posed: for a given unimodal map, which itineraries are possible? There are at least two restrictions: (1) no point in $[-1, 1]$ might have an itinerary such that $I(x) < I(-1)$ since -1 is the left extreme of the interval, and (2) no point in $[-1, 1]$ might have an itinerary such that $I(x) > I(1)$ since 1 is the right extreme of the interval and $I(x) > I(1)$ would imply that $x > 1$. Moreover, no forward image of x lies outside the $[-1, 1]$ interval. Hence, any *reasonable sequence*, I, will satisfy $f^n(I) \leq I(1)$. In particular this requirement applies to the sequence of the maximum,

$$f^n(I(1)) \leq I(1). \qquad (11.4)$$

We will call a sequence, J, *reasonable* if

$$I(-1) \leq J \leq I(1)$$
$$f^n(J) \leq I(1) \tag{11.5}$$
if J contains a C in the position n then $J = J_n I(1)$.

Exercise 11.5: Show that $f^n(I(1)) \geq I(-1)$ and $f^n(I(-1)) \leq I(1)$.

Exercise 11.6: Note that we do not require explicitly in the definition of reasonable sequences that $I(-1) \leq f^n(J)$. Show that this property is a consequence of the definition. Hint: use induction and consider separately the cases when $f^n(J)$ begins with L, C and R.

Because of the condition in equation (11.4), not every sequence can correspond to $I(1)$. The possible sequences for $I(1)$, called *maximal sequences*, may have one of the forms $RL\ldots$, R^∞, L^∞, C^∞ or $(RC)^\infty$, according to the condition equation (11.4). The sequences starting with L or C cannot belong to an unimodal map (although they fulfill the condition), since for such maps $x = 1$ always belongs to the R branch.

Exercise 11.7: Prove that the possible maximal sequences of a map of the interval (unimodal or not) can only have the expressions stated above.

Moreover, depending on the sequence $I(1)$ the reasonable sequences might be particularly simple.

(i) If $I(1) = (RC)^\infty$ the interval $[0, 1]$ maps into itself, hence once an R appears in a reasonable sequence, it only can be followed by $(CR)^\infty$ or by R^∞. In terms of sequences, since $RJ \leq (RC)^\infty$, $J \geq CI(1)$ and either $J = (CR)^\infty$ or $J = R\ldots$, hence $J = RJ'$ and any R has to be followed by a new R or a C. Thus, the reasonable sequences are of the form $I(-1) \leq L^n R^\infty$ or $I(-1) \leq L^n R^k (CR)^\infty$, for any choice of non-negative integers n and k.

(ii) If $I(1) = R^\infty$ the only reasonable sequences are of the form $I(-1) \leq L^n R^\infty$ and $L^n C R^\infty$ by a similar argument as in the previous case.

The most interesting cases will appear when $I(1) = RL\ldots$. Note also that the largest maximal sequence is of the form RL^∞ and that if $I(1) = RL^\infty$ any infinite collection of R, L, C is a reasonable itinerary.

11.1.1.3 Reasonable itineraries that are represented by some point

We can now ask whether there are more laws that itineraries $I(x)$ have to obey, or which is the same, whether there is at least one point $x \in [-1, 1]$ for each reasonable itinerary.

There are two key observations to make:

- Sequences of infinite length represent some sort of limit in the space of sequences. We have no natural notion of distance in that space, hence we have to be aware of possible problems related to limit points.
- We have defined unimodal maps as continuous functions having two monotonic branches. These two facts will play a most relevant role in understanding the properties of unimodal functions.

Let us explore the consequences of continuity of the unimodal functions on the relation among points and their itineraries. In figure 11.4 we illustrate most of the following statements.

Consider the sequences of length $k > 0$ such that

$$I_k(-1) \leq J_k \leq I_k(1)$$
$$f^n(J_k) \leq I_k(1) \text{ for } n < k \qquad (11.6)$$
$$\text{if } J_k \text{ contains a } C \text{ in position } m \text{ then } J_k = J_m I_{k-m}(1).$$

We have the following facts:

(i) Let $I_k(x)$ be the itinerary[†] of a point and assume that $I_k(x)$ has no C. Then there is an open interval containing x, ($x \in (x_1, x_2)$), such that every point in (x_1, x_2) has the same itinerary (up to element k) as x i.e., $I_k(y) = I_k(x)$ $\forall y \in (x_1, x_2)$. The reason is simply that the continuity of the map assures that the image by f^n of a point y can be kept as close to $f^n(x)$ as we want simply by selecting y close enough to x. Hence, by restricting the interval (x_1, x_2) to points close enough to x we can assure that all of their images by f^n, $n \leq k$, lie on the same side of zero as $f^n(x)$ and thus have the same itinerary I_k as x.

(ii) The set of points defined as $\{x \in [-1, 1] : I_k(x) = A_k\}$ is connected. We can simply check that if this statement were not true we would have at least two points $x < y$ with the same itinerary, $I_k(x) = I_k(y)$, and at least a third point z lying between x and y, $x < z < y$, that has a different itinerary $(I_k(z) \neq I_k(x) = I_k(y))$. Since the order of the itineraries is consistent with the order of the points in the interval, we have $I_k(x) \leq I_k(z) \leq I_k(y)$, hence $I_k(x) = I_k(z) = I_k(y)$, which is a contradiction. Hence, the set of points x having $I_k(x) = A_k$ is connected.

(iii) We now consider the itineraries of the points close to 0. We take $x_1 > 0$ sufficiently small and consider the itineraries of $x \in (-x_1, 0) \cup (0, x_1)$. The image of all these points lies in the open interval $E \equiv (\min(f(x_1), f(-x_1)), 1)$. We can make x_1 small enough so that $I_{k-1}(y)$ is the same for all $y \in E$. This fact follows from the continuity argument

---

[†] We will sometimes call the finite string $I_k(x)$ also an *itinerary*.

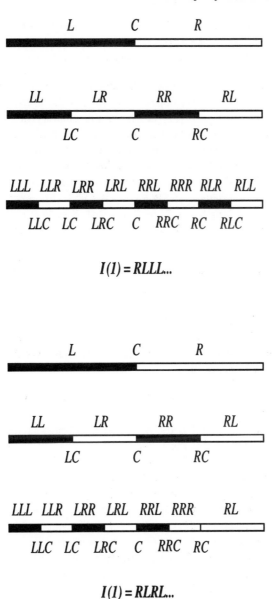

Figure 11.4. Points having different itineraries considering sequences of one, two and three symbols for two different maximal sequences.

again, since if x_1 is small, then $f(\pm x_1)$ gets very close to unity. Letting $B = I_{k-1}(y)$, $y \in E$, we notice that the points $x \in (0, x_1)$ have itineraries $I_k(x) = RB$ while the points $x \in (-x_1, 0)$ have itineraries $I_k(x) = LB$.

(iv) What is the relation between the itineraries of the points close to 0 with the itinerary of 0? The answer depends on whether the itinerary of 1, $I_{k-1}(1)$, contains C or not. If $I_{k-1}(1)$ has no C the itinerary of 1 is the same as the itinerary of the nearby points, hence $I_k(0) = CB$ (with the definitions of the previous statement). However, if $I_{k-1}(1)$ has a C we have that $I_{k-1}(1) > B$. In order to see the relation between B and $I_k(1)$, let $I(1)$ be $I(1) = (WC)^\infty$. If W is odd (even) then the sequence B is made up of the first $k - 1$ letters of $(WR)^\infty$ $((WL)^\infty)$. For example, taking for simplicity W of length $k - 2$, we have that B is WR or WL, depending on the parity of W. For both possibilities we have that B is even and that $I_{k-1}(1) = WC > B$.

(v) Let us finally consider the case when $I_k(x)$ has a C, i.e., $I_k(x) = I_q(x)CI_{k-q-1}(1)$. This case is simply the composition of the previous cases. The itineraries of the points close to x will share with x the first q letters of the itinerary. If $I_{k-q-1}(1)$ has no C they will have itineraries $I_q(x)LI_{k-q-1}(1)$ or $I_q(x)RI_{k-q-1}(1)$. If $I_{k-q-1}(1)$ has a C then they will have itineraries of the form $I_q(x)LB$ or $I_q(x)RB$ with $B < I_{k-q-1}(1)$ as in the previous statement.

(vi) The reader might have noticed that sequences with a C are very special. C is different from L and R in that there is a whole interval of points associated to the symbol L (or R), while C belongs exclusively to $x = 0$. For example, notice that the sequence B defined above cannot have a C. How can we verify this? We observed above that $I_k(x) = LB$ for all points in the open interval $E_l = (-x_1, 0)$. Assume that B starts with a C. In such a case, for all $x \in E_l$ we have that $I(f(x)) = CG$, with G some other sequence. This means that the whole set $f(E_l)$ maps to zero by another iteration of f, i.e, $f(f(E_l)) = 0$. But this is impossible since $f(E_l)$ is itself some open interval. Part or all of this interval lies in one of the two monotonic branches of f, and a monotonically increasing (or decreasing) function cannot take the same value in a whole interval. So B does not start with a C. Repeated use of this argument will convince the reader that B has no C at all.

Exercise 11.8: Show a unimodal map f where there is a C in the itinerary of infinitely many points. Show another map where only one point has a C in its itinerary.

We turn now our attention to the question: given any unimodal map and a reasonable sequence A, can we find a point $x \in [-1, 1]$ such that its itinerary $I(x)$ is $I(x) = A$? We divide the discussion into two steps: first we consider finite sequences and later we will discuss infinite sequences.

We claim that all sequences of type equation (11.6) are realized for any unimodal map.

We can reason as follows. Let A be a reasonable sequence and $\hat{S}_{A_k} = \{x \in [-1, 1] : I_k(x) < A_k\}$, $\check{S}_{A_k} = \{x \in [-1, 1] : I_k(x) > A_k\}$ and $\dot{S}_{A_k} = \{x \in [-1, 1] : I_k(x) = A_k\}$, then $[-1, 1] = \hat{S}_{A_k} \cup \dot{S}_{A_k} \cup \check{S}_{A_k}$. What we have to show is that $\dot{S}_{A_k} \neq \emptyset$. To see this, we will work mainly with the sets $\check{S}_{A_k}$, $\hat{S}_{A_k}$, in order to show that some part of $[-1, 1]$ is left out of them.

Since 1 and -1 have different itineraries, they belong to different sets. Hence, if one of the sets $\check{S}_{A_k}$, $\hat{S}_{A_k}$ were empty, the proposition is proved. Also, if both those sets were closed, there would be some point in $[-1, 1]$ outside the union, since they are disjoint. The same would hold if both sets were open[†] because $[-1, 1]$ is connected.

The only possibility that is left for consideration is that e.g., $\hat{S}_{A_k} = [-1, y]$ and $\check{S}_{A_k} = (y, 1]$ for some $y \in (-1, 1)$, or the symmetric alternative with the open and closed ends in y interchanged. Consider the itinerary $I_k(y)$. Recall that this possibility requires that $I_k(y) < A_k$, since $y \in \hat{S}_{A_k}$ and also that $I_k(x) > A_k$ for $x \in \check{S}_{A_k}$. We want to show that under this assumption there is no reasonable itinerary between $I_k(y)$ and $I_k(x)$ thus contradicting the hypothesis that there is a sequence A_k such that $I_k(y) < A_k < I_k(x)$ used to define the intervals.

We reach the conclusion studying the itinerary of y. $I_k(y)$ must have a C since otherwise all the points near y would have the same itinerary, then $I_k(y) = I_q(y)CI_{k-q-1}(1)$. Assuming that $I_q(y)$ is even (an equivalent reasoning can be performed in the case where $I_q(y)$ is odd) and that $I_{k-q-1}(1)$ has no C, we have that for points $x > y$ and arbitrarily close to y, $I_k(x) = I_q(y)RI_{k-q-1}(1)$. The inequality being considered reads $I_q(y)CI_{k-q-1}(1) = I_k(y) < A_k < I_k(x) = I_q(y)RI_{k-q-1}(1)$. We see that there is no sequence A_k that can satisfy this relation since A_k must be of the form $I_q(y)B$ with $CI_{k-q-1}(1) < B < RI_{k-q-1}(1)$, B must begin with an R (otherwise $A_k = I_k(y)$) and in such a case $f(B) > I_{k-q-1}(1)$ (recall that $I_q(1)R$ is odd). Then A_k is not among the reasonable sequences and we have reached a contradiction.

The case when $I_{k-q-1}(1)$ contains a C can be treated along the same lines recalling that in such a case $I_{k-q-1}(1)$ are the first $k - q - 1$ letters of $(WC)^\infty$ and that points close to 1 have sequences formed by the first $k - q - 1$ letters in $(WR)^\infty$. We reach an identical conclusion since there is no reasonable sequence between $(WR)^\infty$ and $(WC)^\infty$. This concludes the proof.

[†] Intervals of the form $[-1, b)$ and $(b, 1]$ can be regarded as open *relative* to $[-1, 1]$ [roma75]. In general, when we consider a subspace Y (like $[-1, 1]$) of a larger space X (like $\mathbb{R}$) there is a natural way to make the smaller space inherit the properties of the parent. In this case, an open set relative to Y is defined as the intersection between Y and an open set of X. We used this fact 'spontaneously' in chapter 10, when considering open sets of the horseshoe regarded as a subspace of $\mathbb{R}^2$.

As a consequence of the above observation, any reasonable sequence containing a C will be realized by some $x \in [-1, 1]$ for all unimodal maps with given $I(-1)$ and $I(1)$ since the sequences with a C can be considered as sequences of finite length followed by the sequence $I(1)$ and every reasonable sequence of finite length belongs to the itinerary of a point $x \in [-1, 1]$.

Let us now consider infinite sequences. We need to consider only those that do not contain a C since we already know that those containing a C are realized. We can also skip the sequences $I(-1)$ and $I(1)$ since they are trivially represented by a point in $[-1, 1]$. We let A be our infinite sequence and consider those sequences containing a C that are just greater or smaller than A up to k symbols: $A_k^- < A_k < A_k^+$. The points whose sequences are A_k^- and A_k^+ are the limits of the interval of points with the sequence A_k (see figure 11.4).

Exercise 11.9: Show that between points with sequences $A_k \neq A_k'$, A_k and A_k' not containing a C, there is a sequence $\dot{A}_k$ containing a C.

Letting $k \to \infty$ and defining $x_A^- = \limsup_{k\to\infty}\{x : I_k(x) = A_k^-\}$ and $x_A^+ = \liminf_{k\to\infty}\{x : I_k(x) = A_k^+\}$, we know that $x_{A^-} \leq x_{A^+}$. We also have that $I(x_A^-) \leq A$ and $I(x_A^+) \geq A$. We can guarantee that those sequences such that $I(x_A^-) = A = I(x_A^+)$ will be realized by all unimodal maps having the proper itineraries $I(1)$ and $I(-1)$ (see figure 11.5).

11.1.1.4 Some examples

It is time for some relief. Let us show an example of a reasonable sequence that might or might not be realized. Consider the sequence $A = RI(1)$, we have that $RI(1) > CI(1)$ (we should think of a point very close to the right of the maximum). There is no sequence B such that $A = RI(1) > B > CI(1)$ since B would have to be of the form RB' with $B' > I(1)$ and hence B' would not be reasonable. We find also that $x_A^- = 0$ whose itinerary is $CI(1)$. According to the previous reasoning the sequence $RI(1)$ might not be realized since $I(x_A^-) \neq A$. Let us consider the problem further. Assume there is a point $y \neq 0$ whose itinerary is A. We have that all points in the interval $(0, y]$ must have the same itinerary (because of the ordering of itineraries and the fact that $x_A^- = 0$). Thus, either there is an infinity of points with the same itinerary A (if the assumption was true) or no point at all has the sequence $RI(1)$.

The last situation can be realized with the map $f = 1 - 2|x|$, sometimes called the *tent map*, where $I(1) = RL^\infty$, $I(0) = CRL^\infty = CI(1)$ and $I(-1) = L^\infty$. To see that there is no point in $[-1, 1]$ having the sequence $A = RI(1)$, the reader can prove that the points $x \in (0, 1/2^n)$ have partial sequences of the form $I_{n+1}(x) = RRL^{n-1}$. Letting $n \to \infty$ we note that the set of points with the sequence $RI(1)$ is empty. Note that the map expands distances by a factor of two, i.e., two different points cannot for ever lie on the same branch!

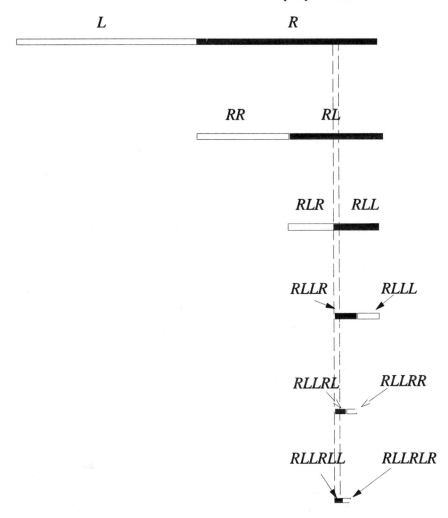

Figure 11.5. Successive approximations to the locus of the sequence *RLLRLLL....*

Correspondingly, an example of a map having infinitely many points with the sequence $A = RI(1)$ could be a unimodal map where 0 belongs to the basin of attraction of a stable fixed point $x > 0$. A simple example is displayed in figure 11.6, where $I(1) = RLR^\infty$, $I(0) = CI(1) = CRLR^\infty$ and $I(y > 0) = RI(1)$ for y sufficiently small.

Exercise 11.10: Show that there are either infinitely many points or no points with the sequence $LI(1)$.

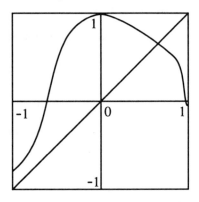

Figure 11.6. A unimodal map having itineraries of the form $A = RI(1)$.

11.1.1.5 The conclusion

Let us finally give some sufficient conditions for a sequence A not containing a C to be realized by all unimodal maps for which the itinerary is reasonable.

We would like to show that:

- If $I(1)$ does not contain a C and $f^n(A) < I(1)$ $(n \geq 0)$ then $I(x^+) = I(x^-) = A$.
- If $I(1)$ contains a C, i.e., $I(1) = (WC)^\infty$ and also

$$
\begin{aligned}
f^n(A) &< (WR)^\infty \text{ if } W \text{ is odd, or} \\
f^n(A) &< (WL)^\infty \text{ if } W \text{ is even,}
\end{aligned}
\tag{11.7}
$$

then $I(x^+) = I(x^-) = A$. In any case, there is at least a point x that has the sequence A.

In the first case, let $A^- = I(x^-)$ and $A^+ = I(x^+)$. Assume that A^+ has a C after position k and not before: $A^+ = A_k C I(1)$. Clearly, A_k coincides with the first k members of the sequence A (otherwise something would be wrong with A^+). If A_k is even (the case of odd A_k can be handled analogously), we have $A = A_k f^k(A) < A_k C I(1) = A^+$. Then $A \leq A_k LI(1) < A^+$. We can say a bit more about A. Since A does not have a C, $A = A_k f^{k+1}(A)$. Since $f^{k+1}(A) < I(1)$, we have $A < A_k LI(1) < A^+$. This means that there has to be a sequence containing a C between A and $A_k LI(1)$, which is a contradiction (our hypothesis was that A^+ was the infimum of the sequences greater than A containing a C). The contradiction is rooted in the assumption we made, i.e., that A^+ had a C. We have thus proved that A^+ contains no C and using once again the fact that between two different sequences containing no C there is at least one containing a C, we realize that since A^+ is the infimum of the sequences larger than A and it contains no C, then $A^+ = A$. We invite the reader to devise a similar argument to show that $A^- = A$.

The second case is analogous to the first case. We have only to modify the argument slightly since now $A_k LI(1)$ will again contain a C. This modification turns out to be rather simple, since the restrictions imposed above on A allow us to replace the infinite C in the sequence of $A_k LI(1)$ by L or R depending on the case. The rest of the proof follows the same lines as the previous proof and is left as an exercise for the reader.

The statement we have just proved corresponds to theorem II.3.8 in Collet–Eckmann [coll86] although our approximation is rather different. We warn the reader that our definition of itineraries corresponds to that of 'extended itineraries' in [coll86] while our definition of reasonable sequence is *a priori* (i.e., long before demonstrating that reasonable sequences $A < I(1)$ not having a C are always realized) rather than *a fortiori* as in [coll86, p 90] and has therefore a slightly different content. An immediate consequence of these differences is that the results presented here are stronger than the results in [coll86].

11.2   Elementary kneading theory

11.2.1   Kneading of periodic orbits

An important consequence of the analysis of itineraries is the existence of periodic orbits. The organization of periodic orbits is one of the most interesting subjects concerning unimodal maps. The key question motivating this research is: 'if one knows that a map f has a periodic orbit of period p, does this orbit force the existence of some other periodic orbit(s)?' In simpler words, if for some reason one happens to 'need' a period-seven orbit, is it enough to find a period-three in order to assert its existence? The theory of orbit organization is called *kneading theory*. The *kneading sequence* of an unimodal map $K(f)$ is just the itinerary of $x = 1$, i.e., $K(f) = I(1)$. The role of $I(1)$ in determining properties of f is even more important that what we have seen up to now. For instance, two unimodal maps which are conjugated[†] have the same kneading sequence [coll86] (although the converse is not true). Also, $I(1)$ determines the existence of periodic orbits, as we shall presently discuss. So, loosely speaking, no matter how wildly f and its iterates fold the interval $[-1, 1]$ into itself (as when one is kneading a dough), the kneading sequence describes some of the invariance properties of f.

Let us consider a periodic itinerary $A = P^\infty$ ($A_p = P$). Assume further that if $I(1)$ contains no C then $I(-1) \le A < I(1)$ and $f^k(A) < I(1)$ or alternatively the inequality equation (11.7) if $I(1)$ contains a C (note that we have to check only p sequences to verify the inequality!). We know that A is realized by at least one $x \in [-1, 1]$ as $I(x)$. Moreover, x lies within another (smaller) closed, non-empty interval, namely $[x_A^-, x_A^+]$ (recall that x_A^- (x_A^+)is the supremum (infimum) of all points having partial k-itineraries containing a C and

---

[†] f_1 is conjugated to f_2 if there exists a C^0, one-to-one function g—a change of coordinates—such that $g(f_2(x)) = f_1(g(x))$.

smaller (larger) than A, and that $I(x_A^-) = A = I(x_A^+)$). However, x does not need to be a periodic orbit of f. We would like to know under which conditions there is also a periodic orbit $y = f^P(y)$ and whether $I(y) = A$.

We can find a candidate in the following form: let $x \in [-1, 1]$ have the itinerary $I(x) = A$ and consider the accumulation points of the sequence $\{x_n\}$, with $x_n = f^{np}(x)$ (note that all elements of the sequence have the same itinerary). Such accumulation points exist since $[-1, 1]$ is compact and the whole sequence is contained within this interval. Moreover the sequence fits also within $[x_A^-, x_A^+]$. We can be sure that all points of this interval have the itinerary A, because of the definition of the extremal points. As a corollary, 0 does not belong to this interval. This last fact is crucial. All the iterates x_n belong to the same monotonic branch of f! Hence, if P is even $x_i < x_{i+1} < \dots$ and the sequence has a limit point $x^* = f^P(x^*)$. For the case of odd P, the sequence of f^2, namely $x_k = f^{2kp}(x)$ has even 'period' PP and the result holds. x^* has hence period p if P is even and period p or $2p$ if P is odd.

We can go even further and note that if P is odd and x^* has period $2p$ and itinerary P^∞ (this rules out the possibility of a C in P) then there is *another* point $z^* = f^P(x^*)$ which also has period $2p$ and the same itinerary as $I(x^*)$. To fix ideas, let $x^* < z^*$ (otherwise, just switch their names). All the points in $[x^*, z^*]$ have the same itinerary. The continuous real-valued function $d(x) = x - f^P(x)$ has a zero within the interval, since it is negative in one end and positive in the other end. Hence, there is a point w with itinerary P^∞ such that $d(w) = 0$, i.e. $w = f^P(w)$.

11.2.1.1 Orbit implication

We have now a striking possibility, that the existence of a periodic orbit implies the existence of infinitely many periodic orbits! A famous example is the 'period three implies chaos' theorem announced in [li-y75]. We discuss this possibility in the lines of our previous argumentation.

Assume that the orbit $(RLR)^\infty$ exists and that $f^k((RLR)^\infty) \leq I(1)$. Note that of all the sequences that can be built using the 'syllables' R and LR, i.e., $W = R^{n_1}(LR)^{n_2} \dots$, the greatest sequence is $(RLR)^\infty$. If $(RLR)^\infty$ is reasonable all the sequences constructed with the syllable R and LR are also reasonable. In particular we have periodic sequences of the form P^k with $P \in \{(RLR)^n(LR)^m, (RLR)^n R^m\}$ for example. We notice that all these sequences are reasonable and moreover $P^\infty < (RLR)^\infty \leq I(1)$ and thus P^∞ is realized as the itinerary of some periodic point.

In some sense, we can say that the periodic orbits that we have considered 'accumulate' on the period-three $(RLR)^\infty$.

Exercise 11.11: Prove the that of all the sequences that can be built using the 'syllable' R and LR, i.e., $W = R^{n_1}(LR)^{n_2} \dots$, the greatest sequence is $(RLR)^\infty$. Hint: Consider the cases of sequences that begin with $R^2 \dots, LR \dots$

and $RLR\ldots$.

To grasp a little bit more of this situation consider the case when the orbit $I(1) = (RLC)^\infty$ exists. We have that any unimodal map presenting this maximal sequence will map the interval $[f(1), 0]$ into $[0, 1]$ and $[0, 1]$ into $[f(1), 0] \cup [0, 1]$, hence any reasonable sequence not containing a C will follow the rules:

(i)   Each L is followed by an R.
(ii) An R can be followed by an L or R without further restrictions.

Any combination of R and LR makes a reasonable sequence and we can produce any periodic combination of them as we wish to generate the itinerary of a periodic orbit! We obtain infinitely many periodic orbits having all possible periods.

The two simple rules we have obtained can be represented by a matrix, M, displaying the allowed transitions.

$$M = \begin{pmatrix} 0 & 1 \\ 1 & 1 \end{pmatrix}. \tag{11.8}$$

Imagining a space with the two 'directions' L and R, the 0 in M represents the fact that L cannot be followed by another L, while all other three pairwise combinations are allowed.

The matrix M is useful to compute the number of reasonable periodic sequences for the map and also to estimate the rate of growth of the number of reasonable periodic orbits with size (length) n. Let k_L^n (k_R^n) be the number of reasonable sequences of length n with last letter L (R). From them we can construct sequences of length $n + 1$ tacking an R to sequences terminated in L and either an R or an L to the sequences terminated in R. For sequences of length $n + 1$ the number of reasonable sequences are $(k_R^n, k_L^n + k_R^n)$ or, in terms of vectors in our two-dimensional space,

$$\begin{pmatrix} k_L^{n+1} \\ k_R^{n+1} \end{pmatrix} = M \begin{pmatrix} k_L^n \\ k_R^n \end{pmatrix} \tag{11.9}$$

$$\begin{pmatrix} k_L^{n+1} \\ k_R^{n+1} \end{pmatrix} = M^n \begin{pmatrix} 1 \\ 1 \end{pmatrix}. \tag{11.10}$$

The number of reasonable sequences of length n will increase approximately with the largest eigenvalue of M raised to the nth power, i.e. $((1 + \sqrt{5})/2)^n$, producing a topological entropy[†] of $\ln((1 + \sqrt{5})/2) > 0$.

[†] The topological entropy can be defined as $h(f) = \lim_{n \to \infty}(1/n) \ln N_n$ in the present context, with N_n the number of monotonic pieces in f^n (this number coincides with the number of different sequences of length n, which one can compute with M^n). For a more general definition see chapter 13. The topological entropy is a measure of the complexity of the map.

The complete set of implications has been studied by A N Šarkovskii [sark64]. A nice discussion of Šarkovskii's theorem can be found in [bloc80] showing that periodic orbits of unimodal maps have an implication order $\succ$, such that if an orbit of period q is present in a given map, then periodic orbits of all other periods p such that $q \succ p$ will be present. The ordered sequence of periodic orbits reads $3 \succ 5 \succ 7 \succ \ldots \succ 3 \cdot 2 \succ 5 \cdot 2 \succ \ldots \succ 3 \cdot 2^k \succ 5 \cdot 2^k \succ \ldots \succ 2^n \succ \ldots \succ 4 \succ 2 \succ 1$.

We also refer the reader to the book by Collet–Eckman [coll86] for more results on unimodal maps of the interval, specially those involving differentiable maps, measure theory and Markov partitions.

11.3 Parametric families of unimodal maps

We turn now our attention to parametric families of unimodal maps that depend continuously of a real parameter (we equip the unimodal maps with the C^0 topology, otherwise the statement does not make sense).

Even without all the acquired knowledge on itineraries of unimodal maps we can make the following observation: if two hyperbolic periodic orbits exist for all members of a parametric family of unimodal maps the order in which the points belonging to the periodic orbits lie on the interval is the same for all members of the family. The reason is as follows: if the exchange of order were permitted, there would be a member of the family for which the two orbits would coincide point by point since the orbit of a point (for positive times) is uniquely determined by the map. In turn, the orbits can coincide if and only if they are undergoing a local bifurcation, but this is forbidden by the condition of hyperbolicity.

We also note that the condition of being hyperbolic can be dropped if the orbits are of different period or having the same period they have a different associated permutation (we can naturally associate the permutation $(f_1, \ldots, f_p)$ with an orbit: the numbers $1, \ldots, p$ label the points of the orbit according to their relative order in the interval while f_i is the order of the point $f(x_i)$). The reason is again the impossibility of exchanging the order of the points belonging to periodic orbits.

In order to avoid dealing with the p points pertaining to an orbit of period p we are going to name the full orbit with the itinerary of its last point, p (the one furthest to the right), i.e., with the maximal sequence associated with the orbit. In this way the orbits inherit an order from the order of itineraries. Is this order meaningful?

Assume that we have a parametric family of unimodal maps, $f(x; \lambda)$, for which the complexity is non-decreasing, in the sense that if $\lambda_1 < \lambda_2$ then $I_{\lambda_1}(1) \leq I_{\lambda_2}(1)$. Whenever we can produce periodic reasonable itineraries P such that $I_{\lambda_1}(1) < P < I_{\lambda_2}(1)$, there will be periodic orbits in $f(x; \lambda_2)$ which are absent from $f(x; \lambda_1)$. We can then ask: at which parameter values are created the periodic orbits present in $f(x; \lambda_2)$ and absent from $f(x; \lambda_1)$? We

know that the order of creation is unique since the existence of a given periodic
sequence implies the existence of all periodic orbits having sequences that are
smaller than the considered orbit. Thus, orbits are incorporated for increasing λ
according to the order of their itineraries.

Let us consider the simplest example among all, the period-doubling
sequences studied by Metropolis, Stein and Stein [metr73]: $R \prec RL \prec$
$RLRR \prec RLRRRLRL \prec \ldots$, starting at the rightmost end of the $\succ$-ordered
sequence.

To get some intuition, let us follow for a while the fate of the period-one
orbit of the unimodal family $f_\mu(x) = 1 - \mu x^2/2$, figure 11.7, starting with
$\mu = 1$. As all unimodal maps, the maps in this family have a period-one orbit
at some $x > 0$. The orbit is stable for $\mu = 1$ being $f'(x_1)$ negative at the fixed
point ($x_1 = \sqrt{3} - 1$). Its itinerary has the odd minimal-period word $W = R$.
Note that f^2 also has a fixed point at the same place. At $\mu = 3/2$ there is a
period-doubling bifurcation, i.e., at the fixed point ($x_{3/2} = 2/3$) we have now
$f'(x_{3/2}) = -1$ and f^2 undergoes a pitchfork (see figure 11.8). A pair of period-
two orbits is born, both having the even word RR. If the stable period-two
orbit has to bifurcate in another period-doubling bifurcation for higher μ, its
word will have to change from even to odd somewhere along the way (so that
the eigenvalue of the orbit becomes negative), and hence pass through C (the
eigenvalue passes through zero), so we have the transitions $RR \rightarrow RC \rightarrow RL$.
When the period-two orbit suffers the period doubling its (odd) word is RL.

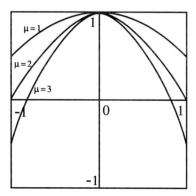

Figure 11.7. Members of a parametric family of unimodal maps.

In general, the order of the 'harmonics' can be summarized as follows:
From the odd sequence W of length p, make WW and change the last symbol
to R if it was an L and to L if it was an R to make W'. The idea behind this
construction is the following. A node is created either at saddle-node bifurcation
or a period-doubling bifurcation with an even sequence (period p, word W
and eigenvalue $\lambda = 1$) at some parameter value. Changing the parameter the

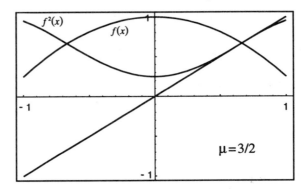

Figure 11.8. Flip bifurcation for f and pitchfork bifurcation for f^2.

node will eventually period double. At that parameter value, it has to have an eigenvalue -1. Its word, W' (at that parameter value), has to be odd. Since the eigenvalue of the node passes from positive to negative, it has passed through 0. Hence the initial word W has changed one letter from L through C to R (or vice versa) to make W'. Moreover, the letter changed is the last letter since we name orbits by their maximal sequence and when the name of the orbit has a C (i.e., it is superstable) its sequence becomes the maximal sequence $I(1) = I(f(0))$, hence the name of the orbit ends with a C. This process is what our law for the construction of harmonics resumes.

The harmonic sequence obtained is odd and has length $2p$. We now have to be able to show that there is no other periodic sequence between an orbit and its harmonic. We leave this as an exercise for the reader.

Exercise 11.12:

- Show that $W > W_{p-1}C > W'$.
- Show that if $W^\infty < B < (WW')^\infty$ then $B = WD$ with $W' < D < W$, and hence $D = W_{p-1}C$.
- Show that if W is reasonable, then $B = WDI(1)$ and $DI(1)$ are not reasonable.
- Show that there is no B such that W, B, WW' are reasonable and $W^\infty < B < (WW')^\infty$.

There are many other results that can be obtained in terms of the combinatorial analysis of unimodal maps (perhaps with additional restrictions) as well as extensions to maps of the interval having several branches, i.e., with sequences made up with more than three symbols (these maps could be named 'multimodal' maps) but we leave our discussion here and refer to the bibliography for further results.

11.4 Summary

We have learned how to classify the points of $[-1, 1]$ by their itinerary $I(x)$, learning which itineraries are always present for *any* unimodal map with given $I(-1)$, $I(1)$.

Further, we saw how itineraries help to classify periodic orbits of unimodal maps, and moreover that the existence of a periodic orbit forces the existence of many (possibly infinitely many) others, according to an ordering sequence which is the same for all unimodal maps. This helped us to infer the ordering in which e.g., period-doubling sequences appear in parametric families of maps with increasing complexity (larger kneading sequence $I(1)$).

We hope the reader has grasped the idea that there is a very rigid organization of unimodal maps which is independent of the details of the map and is basically forced upon them by the strong restriction that a one-dimensional space imposes.

Chapter 12

Topological structure of three-dimensional flows

12.1 Introduction

The existence of chaotic attracting sets has had an important impact in the natural sciences. As they have high sensitivity to initial conditions, an experimenter trying to repeat her/his experiment with all the knobs in the same positions will notice that repeated outputs differ from each other. For a discipline that has built its cultural power through history on reproducibility, this is quite a revolution. It is for this reason that we are going to study the structure of the solutions of three-dimensional flows: three is the minimal dimensionality in which chaotic phenomena occur (see chapter 6). In other words, the reader will not find other chapters dealing with 4-d flows, etc. Besides, the study of the structure of n-d flows with $n > 3$ is an open problem. Which of the tools used in this chapter to understand 3-d flows can be extended to higher dimensions might be an interesting question to keep in mind throughout these pages.

We shall begin by revisiting one of the examples presented in chapter 3 and chapter 8, a model for a CO_2 laser with modulated losses [sola87]. A variety of solutions were found as the parameters were changed. Among them, we found periodic solutions arising in saddle-node bifurcations, periodic solutions undergoing period doubling with accumulation points satisfying the usual scaling properties and chaos.

These phenomena resemble the ones observed in our study of the solutions of one-dimensional families f_μ of unimodal maps of the interval

$$f_\mu : [-1, 1] \rightarrow [-1, 1]. \tag{12.1}$$

In the case of the laser, we also had a family of dynamical systems depending on a parameter, i.e., the amplitude R of the modulation of the losses. Yet, deep differences exist between the solutions of the laser model and those of a one-dimensional map. In the laser case, the order in which different periodic solutions appear (while varying the parameter R) changes depending on the values of the damping rates. This means that the rigid order among periodic solutions that we described when we presented the kneading theory for one-dimensional unimodal maps seems to fall apart.

It can be argued that such differences should not surprise us, since the comparison is inappropriate. The flow of the laser model can be studied in terms of a *two-dimensional* (2-d) Poincaré map, while kneading theory was conceived for one-dimensional (1-d) maps. A motivation to pursue this comparison programme is the fact that in many physical systems having a natural description in terms of 2-d maps a control parameter (usually the dissipation) allows us to 'collapse' the 2-d maps into 1-d ones. Recall the discussion in chapter 11 about the (2-d) Hénon map.

Certainly, the search for an order organization in 2-d maps similar to the kneading theory in 1-d maps requires new tools. This chapter deals with the development of these tools. Finally, we should point out that throughout this chapter, we shall restrict ourselves to a subset of all possible 3-d flows, namely,

$$dx_1/dt = f_1(x_1, x_2, t) \tag{12.2}$$
$$dx_2/dt = f_2(x_1, x_2, t) \tag{12.3}$$

with f_1, f_2 smooth and T-periodic functions. That this is an autonomous 3-d flow can be readily seen rewriting it as

$$dx_1/dt = f_1(x_1, x_2, \theta) \tag{12.4}$$
$$dx_2/dt = f_2(x_1, x_2, \theta) \tag{12.5}$$
$$d\theta/dt = 1 \tag{12.6}$$

with $\theta \in \mathbb{S}^1$. Moreover, the planes with constant θ are intersected transversely by the flow (note that the vector field can never be parallel to such planes since the third component is always equal to one), and hence a global Poincaré section can be defined.

Many of the tools to be presented can be extended to more general autonomous three-dimensional dynamical systems that do not have a global two-dimensional Poincaré section. We shall indicate them as they are defined.

12.2 Forced oscillators and two-dimensional maps

Let us review the model for the CO_2 laser with modulated losses. The relevant variables were the intensity I and the population inversion W. The rate equations read

$$dI/dt = (W - R\cos(\Omega t))I \tag{12.7}$$
$$dW/dt = (1 - \epsilon_1 W) - (1 + \epsilon_2 W)I \tag{12.8}$$

where ϵ_1, ϵ_2, R and Ω are the damping parameters, the amplitude of the modulation (the forcing term) and its frequency, respectively. As mentioned above, the phase space for this dynamical system is $\mathbb{R}^2 \times \mathbb{S}^1$, where the circle $\mathbb{S}^1$ parameterizes the time direction. For this system we can define a global Poincaré map, by selecting a cross section $\Sigma = (I, W, \Omega t = \theta_0)$, and measure periodicity in terms of this 'clock'. In figure 12.1 we show the bifurcation diagrams as the parameter R is varied, for two choices of the damping parameters (smaller in figure 12.1(a)). Notice that although in both cases there is multistability, there is a shift in the relative location at which the branches of solutions come to existence. For example, in figure 12.1(a), the period-three branch is born while the period-two branch of stable solutions still exists. In figure 12.1(b), the period two has undergone two steps of the period-doubling process at the parameter value R at which the period-three orbit is born. If the damping is further increased, we would see the familiar scenario of the period-three orbit being born *after* the period-doubling cascade of the period-two branch is completed.

The fact that a parameterized system with a natural 2-d Poincaré map displays, for certain range of parameters, a dynamics similar to that of a 1-d map is not surprising. We saw in chapter 11 that 1-d maps might arise when a natural reduction of the dimensionality takes place, for example, through dissipation. The kind of question that concerns us now is *how different might the behaviours of a system be when the parameters that control the dissipation are varied*? For example, the period three, in the case of high values of the damping, is born after the period-doubling cascade originated in the period one is completed, while for small values of the damping, it coexists with a stable period two. Can all the orbits interchange their order (in our case, the R value in which they are born)? This would contrast sharply with the behaviour observed in 1-d maps.

Let us have a look at figure 12.1(c). The period-one solution undergoes a period-doubling bifurcation, and, for a higher value of R, the period-two daughter is annihilated with a period-two saddle born in a previous saddle-node bifurcation. Can all the orbits interchange freely their bifurcation companions?

If the answer to these questions were positive, the rigid structure discussed in chapter 11 would fall apart. In what remains of this section we will see that not everything is lost, presenting naturally the kind of invariant that we have to define in order to understand the structure of 3-d flows. The rigorous definitions will be presented in the following section.

In figure 12.2 we show four orbits constituting two saddle-node pairs of solutions of a parameterized system at a certain value μ^* of the parameter. The superindex of the orbit indicates the pair it belongs to, and the subindex indicates whether the orbit is a *regular saddle* (positive Floquet multipliers) or a *flip saddle* (negative Floquet multipliers). Notice the way in which they are organized. Each one of the two orbits O^1 wrap around the O^2 orbits in a different way. The O^2 orbits are *linked* to O_r^1, while they do not link O_f^1 at all. Now let us try to imagine the saddle-node bifurcations in which the orbits were born. There are going to be two parameter values $\mu_a < \mu_b < \mu^*$ that we

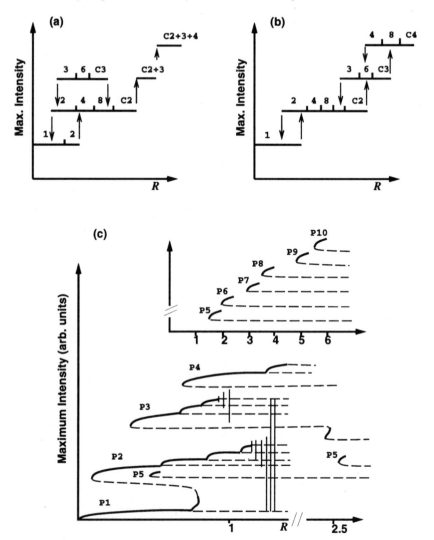

Figure 12.1.   (a) Schematic bifurcation diagram for the parameter values $\gamma/k_0 = 2.7 \times 10^{-4}$ and $\omega/k_0 = 1.35 \times 10^{-2}$. (b) A change of the parameter values shifts the location at which the different branches come to existence. (c) The period two arising in a period-doubling bifurcation is annihilated in an inverse saddle-node bifurcation with another period-two (unstable) orbit.

have to look at closely, namely the values at which each pair is born. It could be the case that at μ_a the pair O^1 is born. At $\mu_b > \mu_a$, the orbits O^1_f and O^1_r are already well separated in phase space and the pair O^2 is born, linked around O^1_r. Finally, at μ^* we arrive at the situation in figure 12.2. This sequence of events was perhaps not hard to imagine.

Let us try to follow a similar reasoning describing the other possibility, i.e., that the O^2 pair is born at $\mu_a < \mu_b$. At μ_b, the two O^2 orbits are well separated in phase space and the O^1 pair should be born in a saddle-node bifurcation. At the bifurcation point, the two orbits of the pair O^1 are identical. They will therefore link both O^2 orbits in the *same* way (i.e., wrapping around either one, both, or none of the orbits). As a consequence, one of the two orbits (at least) would have to change the way it links the pair O^2 as the parameter μ is changed towards μ^*. The problem is that doing so would imply that for an intermediate μ value the orbits intersect, violating the uniqueness of solutions of a system of ordinary differential equations! This sequence of events is hence impossible. Here we can draw our first conclusion: the pair O^1 *had to exist* when the pair O^2 was born, if they coexist for a μ^* value with the organization displayed in figure 12.2 [sola88].

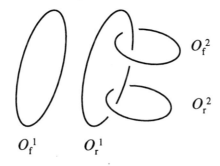

O_f^2

O_r^2

O_f^1 O_r^1

Figure 12.2. The orbits O^2 link one of the orbits O^1 differently from the other.

Let us now have a look at figure 12.3(a). The two closed curves represent periodic orbits of a 3-d flow. Can they be saddle-node companions? In principle, that would require that as the parameters are changed, the orbits deform until they are identical at the bifurcation parameter value μ^*. Is that possible? Compare figure 12.3(a) with figure 12.3(b). The reader is encouraged to try to deform the curves smoothly until they are identical. But we warn her/him: unless the deformations include a self-intersection of the curves, it is impossible! Again, there is a deep reason for ruling out self-intersections of the flow: it would violate the uniqueness of the solutions of ODEs and therefore causality.

In 1-d maps, the relative order in the interval of the points constituting a periodic orbit could not change. That was at the core of the rigidity that exists in 1-d discrete dynamics. In the case of 3-d flows, the two examples illustrated in figure 12.2 and figure 12.3 address the restriction that maintains part of the rigidity exhibited in 1-d maps: *periodic orbits can neither intersect nor self-intersect*. The way in which two orbits wrap around each other or the way orbits are knotted cannot change as parameters are varied. In the following

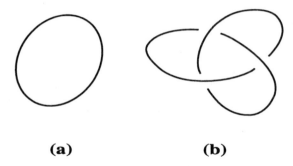

(a)   **(b)**

Figure 12.3. (a) An orbit that can be deformed into a circle. (b) An orbit that cannot be deformed into a circle. These two orbits cannot be saddle-node companions.

section, we will define some topological invariants that quantify these ideas.

12.3 Topological invariants

Using the parameterization in time, one may say that a periodic orbit is a one-to-one smooth function of S^1 in phase space. In fact, as mentioned in the previous section, periodic orbits of a three-dimensional dynamical system are closed one-dimensional curves imbedded[†] in a three-dimensional manifold. This is precisely the definition of a *knot* [kauf91]. Let us consider the concept of knots a little more closely.

If a given closed curve can be continuously deformed into a circle without self-cuttings it is called a *trivial knot* or *unknot*. In general, knots can be classified into *classes* using this operation. All knots that can be continuously deformed into each other without self-intersections belong to the same class. If one knot belongs to two classes, these classes are identical (because the composition of two continuous deformations without self-cuttings is another continuous deformation of the same type). Hence, the knot-classes are *equivalence classes*. The equivalence operation (the deformation) is called *homotopy*, and will be relevant many times through this chapter.

The reason for studying knots in the framework of three-dimensional systems is the following. If an orbit is a solution of a dynamical system that belongs to a parameterized family, the control parameters can be changed, and eventually the orbit might disappear. But as discussed in the previous section, as long as the orbit exists, the knot-class to which it belongs, i.e., the *knot type*, will not change [holm85].

[†] An imbedding of an n-dimensional manifold $\mathcal{M}$ is a smooth one-to-one mapping $\phi : \mathcal{M} \mapsto \mathbb{R}^m$ preserving the differential structure (i.e., with a one-to-one tangent map $D\phi$) where also the inverse map from $\phi(\mathcal{M}) \subset \mathbb{R}^m$ to $\mathcal{M}$ is smooth. m has to be quite large ($\geq 2n + 1$) to assure that there will exist a ϕ for *any* (nice enough) n-d manifold $\mathcal{M}$ (try to imbed a straight line and a circle twisted in the form of a number 8 in $\mathbb{R}^2$ or $\mathbb{R}^3$) [kosi93]. Some authors use the word 'embedding'.

In general, three-dimensional dynamical systems will have many coexisting periodic solutions. For example, if a system has a strange attractor as a solution, there will be infinitely many coexisting periodic solutions (recall chapters 9 and 10). Therefore, it is only natural to define another topological entity: the *link*. A link is an imbedding (on 3-space) of a collection of knots [kauf91].

A crucial question arises: given a knot or link, is it possible to compute some number or object out of it which would indicate to which class it belongs? The question has attracted much research, but the answer is still negative. There are a few *class invariants* (numbers or objects which are the same for all knots within a class) which can do the 'easy part' of the work. If two different knots have different associated class invariants, this indicates that the knots cannot belong to the same class (they are not deformable into each other). But the inverse is not true. For every class invariant proposed so far in the literature, there are examples of knots which are not equivalent and yet are associated with the same invariant. In other words, there are at least two knot-classes where the invariant takes the same value. In the following subsection we will describe one of those invariants: the Conway polynomial.

12.3.1   The Conway polynomial and the linking number

The intuitive idea behind the construction of this invariant is as follows. There is a class of knots which consists of all the curves which are homotopic to a circle. There are also knots which might not be deformed into circles without self-intersections. If we were allowed to cheat, violating the 'no-self-intersection' rule, we could deform any knot into a circle. Some knots would require us to cheat many times, other just a few. The Conway polynomial keeps track of the systematic 'cheating procedure' that would be necessary to reach a trivial (or set of trivial) knot(s) starting from an arbitrary one.

It is then clear that the construction of the Conway polynomial is defined through an algorithmic procedure. The skeleton of that procedure is a set of three axioms:

- Axiom 1: For each oriented knot or link K there exists an associated polynomial $p_k(z) \in Z(z)$, with $Z(z)$ the ring of polynomials in z with integer coefficients. Equivalent knots and links receive identical polynomials: $K_1 \sim K_2 \Rightarrow p_{K_1} = p_{K_2}$.
- Axiom 2: if $K \sim 0$ (the trivial knot) then $p_K = 1$.
- Axiom 3: if three knots or links K_1, K_2 and L differ at the site of one crossing as in figure 12.4, then

$$p_{K_1} - p_{K_2} = z \, p_L. \tag{12.9}$$

This procedure allows us to compute, in an iterative way, a polynomial for a knot diagram.

In figure 12.5(a) we show a *threefoil knot*. Let us compute its Conway polynomial. First, pick up a crossing (anyone will do), and draw the two

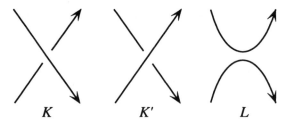

Figure 12.4. Construction of the Conway polynomial.

associated diagrams that are obtained when this crossing is modified in the
ways shown in figure 12.4. Let us call the original knot K, the second diagram
K', see figure 12.5(b), and the third diagram, L (figure 12.5(c)). According to
the third axiom, whatever p_K, $p_{K'}$ and p_L are, they will satisfy

$$p_K - p_{K'} = z\, p_L. \tag{12.10}$$

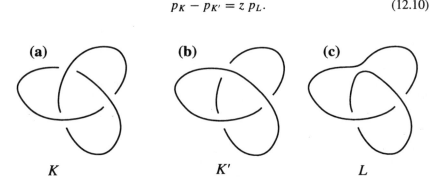

Figure 12.5. (a) Threefoil knot. (b) A knot that differs from the threefoil at one crossing.
(c) A link obtained changing a crossing from the threefoil.

A careful inspection of figure 12.5(b) shows that the knot K' is equivalent
to the unknot (recall that knots are imbedded in 3-space, so you can lift the
rightmost part of the knot, flip it and recover a circle). Therefore, $p'_K = 1$,
and $p_K = 1 + z p_L$. We can proceed as above with p_L, in order to find the
polynomial p_K.

Exercise 12.1: Show that the Conway polynomial for the threefoil is

$$p_K = 1 + z^2. \tag{12.11}$$

Exercise 12.2:

(a) Show that the knot of figure 12.6(a) has the same polynomial that you computed in the last exercise.

(b) Prove that the knot of figure 12.6(b) cannot be deformed into the threefoil.

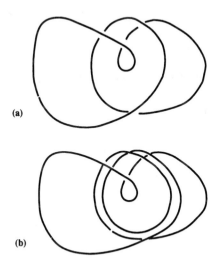

(a)

(b)

Figure 12.6. (a) A periodic orbit (from a real experiment!) knotted as a threefoil. (b) From the same experiment, a period-three orbit.

Given a link of two knots, it is natural to define a number that accounts for how many times they wrap around each other. That number is called the *linking number,* and it is a link invariant [uezu82]. For a given specific pair of knots, it can be computed as

$$L(N_1, N_2) = -\frac{1}{4\pi} \int_{N_1} \int_{N_2} \frac{(\vec{r_1} - \vec{r_2})(d\vec{r_1} \wedge d\vec{r_2})}{\|\vec{r_1} - \vec{r_2}\|^3}. \qquad (12.12)$$

In fact, there is a much easier way to compute this number. Given a *regular projection* of the knots (a diagram like the ones displayed in the previous figures, in which we keep the information of the *over-* and *under-* crossings), we can proceed as follows [mind91]. First, let us define an orientation for the knots. Then, for each cross c between the knots N_1 and N_2, let us assign a number $\sigma_{i,j}(c) = \pm 1$ following the convention of figure 12.7. Then

$$L(N_i, N_j) = \frac{1}{2} \sum_c \sigma_{i,j}(c). \qquad (12.13)$$

Exercise 12.3: Show that the links of figure 12.8(a), (b) have the same linking number. Does this calculation inspire a way to prove that (12.13) is an invariant?

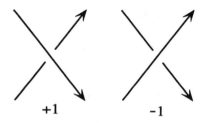

Figure 12.7. Assigning integers to crossings in a regular projection.

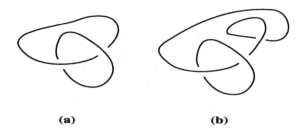

(a)   (b)

Figure 12.8. (a) A link of two orbits. (b) This link looks more complex, though it is just the result of smoothly deforming one of the orbits.

12.3.2 Braids

It is possible to characterize periodic orbits of 3-d flows through their knot polynomials and linking organization even if a global Poincaré section does not exist for the problem. When the section does exist, as in the motivating example of this chapter, we can define new topological invariants closely related to the previous ones which take into account the existence of another invariant: the period.

Whenever a global Poincaré section exists, a periodic orbit will intersect this section in a finite number of points n. One can visualize then a periodic orbit as a set of 'threads' (*strands*) emerging from the Poincaré section at the n points and terminating in the same set of n points after one excursion through the complement of the Poincaré section in phase space. This object is called a *braid* [holm85].

Braids can also be classified into equivalence classes. Two braids are equivalent if we can deform one into the other with the end-points fixed and without making a strand cross through another. We shall see now that it is possible to write an algebraic expression for the braid in terms of *generators*, and that the geometric equivalence restrictions mentioned above can be translated into algebraic relationships between the generators [arti47].

Every braid can be expressed as an ordered set of symbols $\sigma_1, \sigma_2, \ldots \sigma_{n-1}$ and their inverses $\sigma_1^{-1}, \sigma_2^{-1}, \ldots \sigma_{n-1}^{-1}$. Each of the symbols σ_i (σ_i^{-1}) represents a piece of a braid in which strand i crosses under (over) strand $(i+1)$ in a given regular 2-d projection, as displayed in figure 12.9(a),(b). Notice that $\sigma_i \sigma_i^{-1}$ corresponds to two consecutive crossings in which (1) strand i crosses under strand $i+1$ and then (2) strand $i+1$ crosses under strand i. Observing the resulting braid, figure 12.9(c), it is easy to realize that it is possible to deform it by pulling the strand $i \rightarrow i+1 \rightarrow i$ under the strand $i+1 \rightarrow i \rightarrow i+1$. The resulting braid consists of a set of parallel strands with no crossings. Associating this braid with the 'identity' I_n, it is possible to translate our geometric equivalence relationship into the following algebraic one:

$$\sigma_i \sigma_i^{-1} = I_n. \tag{12.14}$$

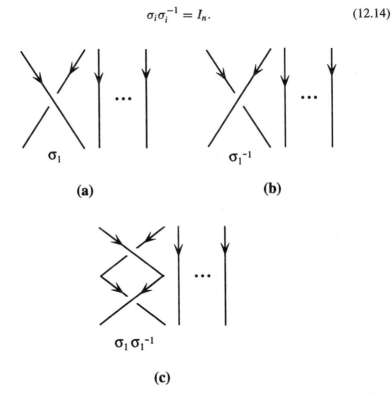

Figure 12.9. (a) The geometrical meaning of σ_1. (b) Same for σ_1^{-1}. (c) The identity (if you play a little bit with the threads it is clear why!).

There are two other algebraic relations that correspond to the geometric moves allowed within a class of braids:

$$\sigma_i \sigma_{i+1} \sigma_i = \sigma_{i+1} \sigma_i \sigma_{i+1} \tag{12.15}$$

$$\sigma_i \sigma_j = \sigma_j \sigma_i, \quad |i - j| > 1. \tag{12.16}$$

Exercise 12.4: Draw the braids corresponding to the left hand sides of the equations written above and compare them with the braids corresponding to the expressions of the right hand sides. What is the geometrical meaning of the operations associated with the algebraic relations?

What happens if one chooses a different Poincaré section? The crossing σ_i that occurs in the first place (following the direction of the flow) after a Poincaré section will occur in the last position for a new section defined immediately after this crossing. Hence, if a given periodic orbit is represented by the braid $B = \sigma_i^p \sigma_k^q \sigma_l^r \ldots$, where $p, q, r = \pm 1$, all braids B' obtained by cyclic permutations of the generators present in B will represent the same periodic orbit 'seen' from another Poincaré section. The reader may recognize behind this property a new equivalence class. Braids can be grouped into equivalence classes with respect to cyclic permutation of the generators. Braid properties which translate into properties of the associated periodic orbits should then be class invariants. The equivalence class of braids associated to a given periodic orbit is called the *braid type* of the orbit [boyl84].

Notice that two braids corresponding to periodic orbits of the same period do not necessarily correspond to the same class of braids. In figure 12.10, two period-three orbits are displayed. The one in figure 12.10(a) (with generators $\sigma_2\sigma_1$) can be thought as a periodic orbit of a rigid rotation of a disc, while the braid in figure 12.10(b) (with generators $\sigma_2^{-1}\sigma_1$) cannot. Also, notice that both orbits are trivial knots. This observation makes it clear that whenever a global Poincaré section exists the braid type of an orbit carries more information that its knot type [sola88].

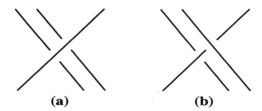

(a)   (b)

Figure 12.10. (a) A 'well ordered' period-three braid. Can you imagine this orbit on a torus? (b) A 'badly ordered' period three. Can you imagine this one on a torus?

12.3.3 Relative rotation rates

Just as braids carry more information than knots, whenever a global Poincaré section can be found it is possible to define a set of numbers, called *relative rotation rates*, which are a decomposition of the linking number through the Poincaré section. These numbers account for 'how much of the linking' took place between consecutive intersections with the control section. More precisely, these numbers are defined as follows. A period-P orbit intersects the Poincaré section in a set of P points $(a_1, a_2, \ldots, a_p)$, while a period-Q orbit does it in a set $(b_1, b_2, \ldots, b_q)$. For a given pair of points (a_i, b_j), we can propagate the difference $(a_i - b_j)$ forward in time along $P \times Q$ periods (if P is not prime with Q the *mcm* will suffice). The vector difference will then have rotated an angle $PQR_{i,j} \times 2\pi$. The relative rotation rate $R_{i,j}$ will therefore be the average rotation per period made by the vector difference. In order to recover the linking number between two periodic orbits we just have to add all the relative rotation rates:

$$\frac{1}{PQ} \sum R_{ij} = L. \tag{12.17}$$

12.4 Orbits that imply chaos

We began this chapter comparing the rigid structure observed in 1-d maps with the behaviour of 2-d maps obtained by a stroboscopic inspection of 3-d flows. We concluded that part of the structure survives, and that this structure could be made evident considering the topological properties of the periodic orbits in the 3-d flow. For example, we saw that some pairs of saddle companions have to be born before others. Maybe one of the most amazing results obtained from the study of 1-d maps is that the presence of orbits of specific periods implies positive topological entropy [bowe76]. There exists also a simple algorithm to compute a lower bound for this entropy[†] [bloc80]. Let us see how these facts translate to the 2-d map/3-d flow case.

For example, if f is a map of the interval, then f has zero topological entropy if and only if the set of periods of its periodic orbits is in one of the following forms:

$$P(f) = \{p = 2^i, i = 0, \ldots, n\} \tag{12.18}$$

or

$$P(f) = \{p = 2^i, i = 0, \ldots, \infty\}. \tag{12.19}$$

The rigid orbit ordering of unimodal maps of the interval guarantees then that as soon as we find an orbit of period 6 or 5 or whatever number which is not an integer power of two, we know that the map has positive topological entropy.

Is it possible to state a similar result for 2-d maps? Are there periodic orbits such that their existence implies that any map (with specific continuity properties) that has them as solutions has also infinitely many other periodic solutions?

[†] Again, you can think of the entropy as a measure of chaos. A precise definition is presented in chapter 13.

Let us go back to the previous section. In figure 12.10 we showed two orbits of period three. One, figure 12.10(a), could be thought as a periodic orbit of a rigid rotation. But the braid of figure 12.10(b) is qualitatively different. Try to imagine a map that has a periodic orbit that is represented by such a braid. For example, draw a circle on a Poincaré section such that it encloses the three points in which the periodic orbit intersects the section. Now if you follow the evolution of this circle, even if you try to do it in the 'less complex way' (just as when we thought of the period three of figure 12.10(a) as the orbit of a rigid rotation), you will find that the image of the disc enclosed by the circle is twisted, stretched and folded. This suggests that a dynamics such as the one described in chapter 10 might be coexisting with this orbit. A braid of the kind of figure 12.10(b) is called *non-rotation-compatible* [gamb89].

The key point is understanding when an orbit will yield 'simple' dynamics or 'complex' dynamics. All the results obtained so far in this field are based on the work of Thurston [thur88] on classification of surface homeomorphisms. Given a compact 2-d manifold M and an invariant set A, let $G(A)$ be the C^1 orientation-preserving diffeomorphisms of M leaving A globally invariant, Thurston showed:

Theorem (Nielsen–Thurston): If $g \in G(A)$, then one and only one of the following three possibilities holds:

(i) g is *periodic*; i.e., g^n is isotopic to the identity map for some n.

(ii) g is *reducible*; i.e., there exists a homeomorphism $h \equiv g$ such that h leaves some finite union of disjoint simple essential closed curves invariant. (A curve is said to be essential with respect to A if it cannot be deformed either to a point in M/A or to the frontier of M.)

(iii) There exists $s_g > 0$ such that for any map $h \equiv g$ one has $s_{\text{top}}(h) \geq s_g$. (In fact, g is isotopic to a *pseudo-Anosov* map.) Here s_{top} stands for the topological entropy.

We recall that h and g are *isotopic*, $h \equiv g$, if there is a continuous deformation (homotopy) of h in g ($f(t)$ with $f(0) = h$ and $f(1) = g$) with $f(t)$ a C^1 map in M/A. So, an isotopy is something like a 'differentiable homotopy'.

From an application-oriented perspective, one would like to determine the conditions assuring that a given braid can be a solution of a zero-entropy 2-d map, or, alternatively, a solution of a positive-entropy map. For example, Gambaudo, van Strien and Tresser [gamb89] showed that if a map *only* has periodic orbits which are rotations or 'rotations around rotations', such as that of figure 12.10(a), then the map has zero entropy. Boyland [boyl84] gave a recipe to determine whether certain orbits will imply positive entropy or not by considering the associated braid. The recipe is as follows. Take an irreducible braid, for example the braid of a prime-period orbit. Now sum the exponents (they are ±1) of all the crossings σ present in the braid[†]. If this sum is not divisible by $n-1$, where n is the period, then the braid implies (or 'has') positive

---

[†] We note in passing that the exponent sum is a class invariant.

entropy. If this is not the case, there is still a chance. If B^n, the nth power of the braid, is *not* a rigid rotation, then the braid implies positive entropy.

Exercise 12.5:

(a) Check that the period-five orbit with braid $B = \sigma_1\sigma_2\sigma_3\sigma_4^{-1}$ has positive entropy.
(b) Show that the braid of figure 12.10(a) has zero entropy while the one in figure 12.10(b) has positive entropy.

12.5 Horseshoe formation

The horseshoe map is a key element to understand in dynamics as the complexity it displays can be observed in any map with transverse homoclinic points (recall chapter 10). In tune with our motivating example, we will stress the comparison between the universal bifurcation sequence of unimodal maps of the line and the horseshoe by studying the Hénon map, which allows us to turn one into the other through smooth parameter changes.

12.5.1 Hénon map and the logistic map

In the first section, some of the solutions of the laser with modulated losses were reported for different values of the damping parameters. We observed that for high damping values, the organization of the solutions looked similar to the organization of a quadratic 1-d map. In this section, we will review the Hénon map, which was presented in chapter 11:

$$F_{\mu,\epsilon}(y, x) = (x, -\epsilon y + \mu - x^2). \tag{12.20}$$

The map stretches and folds phase space (x, y), as the horseshoe map discussed in chapter 10 does, for $\mu > (5 + 2\sqrt{5})(1 + \epsilon^2)/4$ and non-zero ϵ [guck86] (see figure 12.11). For $\mu < -(1 + \epsilon^2)$, all orbits escape to ∞. Therefore, this model is appropriate to study the creation of a horseshoe: as μ is varied, periodic orbits are created until the complete horseshoe is established. Another important feature of this model is that as $\epsilon \to 0$, the map collapses to the unimodal 1-d map $f_\mu(x) = \mu - x^2$, for which the results of chapter 11 apply.

Holmes and Whitley, and Holmes and Williams, studied the bifurcation structure of this map as μ is changed, for different values of ϵ [holm84, holm85]. In figure 12.12, a partial bifurcation diagram is displayed. At $\epsilon = 0$, the usual order predicted by kneading theory holds. As ϵ is increased, the solid lines indicate the (ϵ, μ) values at which the orbits are created. Notice that some lines cross each other, indicating that for different values of ϵ the relative order of appearance changes. Needless to say, some features of this bifurcation diagram

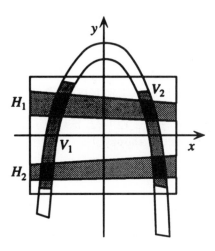

Figure 12.11. A horseshoe map.

depend upon the particularities of the Hénon map, but others will pertain to every flow generated by a horseshoe-like stretching and folding mechanism.

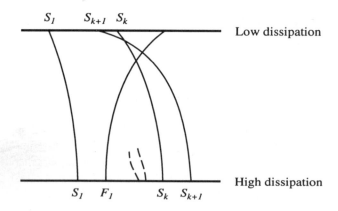

Figure 12.12. A partial bifurcation diagram for the Hénon map. Notice that for different values of ϵ, the order in which orbits are born changes. You might want to have a look at figure 12.1 again. The letter S indicates that the orbit is born in a saddle-node bifurcation, while the F indicates that the orbit is born at a flip bifurcation. The subindices indicate the order of appearance of the orbits in the high-dissipation limit.

12.5.2   Orbit organization

In section 2, we mentioned that the possibility of an interchange of the order in which the bifurcations take place is restricted by the linking organization of the periodic orbits. Let us make this idea more precise. Consider two pairs of orbits A and B, and let us assume that the orbits are saddle-node pairs. Then, a forcing criterion can be stated: if the linking numbers between one of the A orbits with the B pair are equal, but different from the linking number between the other A orbit and the B orbits, then the pair A has to be created before the pair B [sola88]. Of course, if the linking numbers between the four orbits are identical, neither pair forces the other to exist, and whether they might occur in one relative order or the other is not determined by this topological criterion. This 'forcing' criterion is weaker than the ones described in the previous section, where we discussed that some orbits imply infinitely many others. In this section we are restricted to the sets of orbits of a class of diffeomorphisms (the horseshoe), and all we are doing is to order a given set of solutions.

There is a subtle point to be discussed before applying the forcing argument of the previous paragraph. Let us have a look at figure 12.13. Six orbits are illustrated, corresponding to three pairs A, B and C. A linking number analysis may indicate that A forces B to exist, and that A forces C. However, if B and C have the same braid type, they might interchange partners so that the flip of the pair B, with the regular orbit of the pair C (and also the flip of the C pair with the regular orbit B) form a saddle-node pair. This process is called *exchange elimination*. This 'cross association' may occur in some map having these six orbits and it just may not occur in other maps, so nothing general can be said about the relative forcing of these orbits.

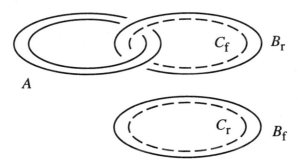

Figure 12.13. Recombination of saddle pairs.

In the case displayed in figure 12.13, the exchange elimination means that A cannot be said to force either orbit of the B, C quartet. Therefore, we learn that linking numbers must be computed between orbit multiplets which might participate in an exchange elimination (that means that they have the same period and the same braid type) rather than on 'individual' pairs. In figure 12.14 we show the result of these forcing restrictions, for every orbit in a horseshoe map up to period eight [mind93].

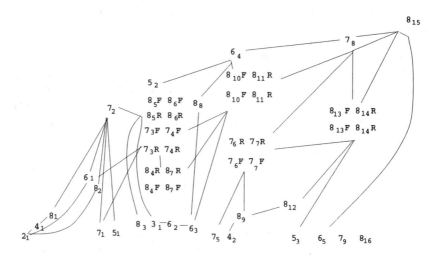

Figure 12.14. Forcing relationships between horseshoe orbits. This table must be read from right to left. If two orbits are connected with a line, the orbit at the right forces the existence of the orbit at the left. The subindices indicate the order of appearance in the highly-area-contracting limit, in which the dynamics can be described in terms of a one-dimensional map.

Notice that the period-three orbit does not force any of the 2^n orbits of the first period-doubling sequence. This explains the observations of the first section: the statement 'period three implies chaos' holds only in the 1-d limit. On the other hand, notice that there is a period-seven orbit that forces the three orbits in the period-doubling sequence included in the diagram. What can be said about these braids? We can compute the entropy associated with the period-three and period-seven orbits using the recipe described above [boyl84]. The former has zero entropy (i.e., it does not imply chaos) while the latter has positive entropy. In order to perform this computation one has to know how to write the braid of a horseshoe orbit.

Exercise 12.6: Write the braid word for the period-three horseshoe orbit
... 001.001 .... Hints [hall94]: (1) Try to approximately place the three points of the orbit in the unit square. The number string to the right of the dot represents the x coordinate of the points. Replace $0, 1$ by L, R and use the unimodal ordering of chapter 11. The same holds for the y coordinate and the string starting leftwards from the dot. (2) Check how the points map to each other by iterating with the horseshoe map. (3) Imagine a torus having the unit square (conveniently deformed) as transverse section and draw flow lines on the torus

representing the periodic orbit.

12.5.3   Subtleties of orbit organization

Perhaps the reader is puzzled since chapter 11 about a subtle point of orbit organization. Šarkovskii's theorem states that there is a rigid orbit order in the sense that if an orbit of a given period p is present in a unimodal map, then there will exist orbits of all periods q below p in Šarkovskii's order. To fix ideas, having an orbit of period three implies that there will exist an orbit of period $2, 4, 5, \ldots$ Now the question comes: what if we have a unimodal map where *many* orbits of, e.g., period five are present? The theorem assures that at least one of these orbits will be forced by the period three but not that *all* of them will!

Exercise 12.7:

- Using the notation of chapter 11, show that if $I(1) = (RLLLC)^\infty$ then there exists a period-five orbit forcing the period-three orbit $(RLR)^\infty$.
- Show that the period-three orbit forces a period-five orbit.

In fact, already in 1-d unimodal maps there is a more subtle ordering (compatible, however, with Šarkovskii's ordering) involving all *types* of periodic orbit of any period [bald87]. The infinitely many periodic orbits ordered by Šarkovskii's result are not many enough! Moreover, for general (not necessarily unimodal) continuous 1-d maps, not every pair of periodic orbits is connected by the order relation [bald87]. This order relation can be called *nonlinear* in opposition to a *linear* order relation, i.e., such that any two elements can be put in relative order.

Considering only the specific orbits of the horseshoe diffeomorphism, there exists also a nonlinear order relation based on the forcing mechanism, in accordance with figure 12.14. Hence, not every pair of orbits is connected by this order, although there are simple rules to decide whether two orbits are related and in such a case which orbit forces the other. One can even obtain lower bounds for the topological entropy of a given orbit [hall94].

12.5.4   Suspensions

The reader might wonder how the actual calculation of the linking numbers of the orbits was performed when constructing the 'tree' of figure 12.14. There are two points to clear up. First of all, given a map, what kind of flows are compatible with it? Second, is there an algorithmic procedure to compute topological invariants once a compatible flow is chosen?

Let us address the first question. In figure 12.15(a) we display the 'simplest' flow compatible with a horseshoe map [holm85]. Its Poincaré map contains a saddle point whose stable and unstable manifolds intersect infinitely many times. The 'time' axis is vertical and the upper and lower xy surfaces represent two consecutive cuts on the Poincaré section. Identifying these two surfaces, one obtains a 3-d flow that can be contained in a torus.

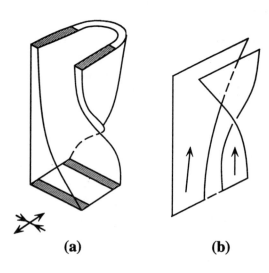

(a)   **(b)**

Figure 12.15. (a) A horseshoe suspension. (b) The associated template.

Now consider the flow obtained by turning the upper xy surface a whole turn (2π radians) around the vertical axis preserving the continuity of the flow lines. This process is called a *full twist*. The new flow has the same Poincaré map as the previous one. Notice that all the flows which differ from this one in an integer number of full twists will be associated with the same Poincaré map. A flow compatible with a map is called a *suspension*. The full twists are a normal proper subgroup of the braid group and hence the braids associated to a given periodic orbit in different 'full twist suspensions' can be grouped into equivalence classes preserving the braid type (an orbit of a given braid type in one suspension has the same braid type in all suspensions).

In fact, we can drive this observation even further. Having noted that many suspensions can have the same Poincaré map and also the same braid types (for their corresponding periodic orbits) could not it be that the braid type of a periodic orbit can be read *directly* from the Poincaré map without using the actual shape of the flow ? The affirmative answer to this question can be obtained by noting that the equivalence classes of braids with respect to full twists are in one-to-one correspondence with the maps of $\mathbb{S}^1$ on the Poincaré surface induced by the periodic orbits of the Poincaré map [nati94].

12.5.5   Templates

Birman and Williams realized that there was a one-to-one correspondence between the periodic orbits in a flow like the one displayed in figure 12.15(a) and the orbits in the branched manifold shown in figure 12.15(b) [birm83]. This simple observation allows us to compute algorithmically the linking and knot properties of the periodic solutions of the 2-d map. Such a branched manifold is called a *template*. Let us review the ideas that led them to define the template or *knot holder* of a flow.

We have mentioned many times that three-dimensional flows can be quite complex. Strange attractors are possible solutions, and we saw in chapter 10 that these objects coexist with infinitely many (unstable) periodic solutions. One could have the idea than the flow confined to a strange attractor should then look like a 'Gordian knot', in other words, a tangle of threads. Yet, take any famous picture of a strange attractor and show it to someone having fewer prejudices. She/he will probably describe the Lorenz attractor as an exotic 'mask'. This was the driving idea behind the work of Birman and Williams. They noticed that although three-dimensional dynamical systems could exhibit non-trivial knots, under certain circumstances it was possible to define an object called a knot holder, a branched manifold that can hold all the coexisting periodic solutions of the system.

Let $\phi_t : M^3 \to M^3$ be a flow on a 3-manifold having a hyperbolic invariant set (see chapter 10) with a neighbourhood $N \in M$. Let $\sim$ denote the equivalence relation $z_1 \sim z_2$ if $|\phi_t(z_1) - \phi_t(z_2)| \to 0$ as $t \to \infty$, and $\phi_t(z_i) \in N$ for all $t \geq 0$. The flow becomes a semiflow on a two-dimensional manifold, see figure 12.15(b). Effectively, this equivalence relationship induces a 'collapse' of the flow along the stable manifold, and identifies the orbits with identical 'future'. What is remarkable about this tremendous collapse is that *the periodic solutions within the invariant set will not change their topological properties under the projection.*

The reason is the following. Let x be a point on a periodic orbit, and $W^s = \{y : d(\phi_t(y), \phi_t(x)) \to 0 \text{ as } t \to \infty\}$. Clearly, $W^s(x)$ will not intersect any other periodic orbit, as two hyperbolic periodic orbits are separated in phase space. Therefore, the collapse does not change the linking or knot type. Figure 12.15 illustrates the procedure. In figure 12.15(a) we show the simplest flow compatible with a horseshoe map. The branched manifold obtained through the collapse along the stable manifold is displayed in figure 12.15(b). In both figures we included a periodic orbit. In fact, using the symbolic dynamics description for the periodic orbits of the horseshoe, it is immediate to see that different periodic orbits have different futures, and hence the map ϕ_t is one-to-one within the set of periodic orbits.

The existence of templates for hyperbolic invariant sets allows us to compute algorithmically the topological features of any orbit in the flow. In figure 12.16 we show the templates corresponding to the Lorenz system [birm83], and the template corresponding to the Rossler [holm85]. In both of them we show the two period-one orbits and the period-two orbit. There is a clear difference in the way the orbits are organized among themselves for both systems. In the Lorenz, the period-two is not linked to the period-one orbits, while in the Rossler system, the period-two orbit links one of the two period-ones. This is the kind of observation that will allow us to classify 3-d flows. A sensible classification scheme cannot accept as equivalent two flows for which the periodic orbits link around each other in a different way! In the following section, this programme is described in more detail.

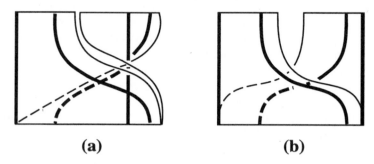

(a)   **(b)**

Figure 12.16. Templates for the Rossler (a) and Lorenz (b) flows.

12.6 Topological classification of strange attractors

We began this chapter pointing out that the existence of chaos poses a problem to the validation of a model proposed to explain the result of an experiment. At the core of the problem lies the sensitive dependence on initial conditions, which rules out the direct inspection of the time-series data. All this chapter is devoted to the idea that the skeleton of a chaotic solution is a set of periodic orbits, which is organized in some way. A handle to solve the problem of classification is suggested by the result of Birman and Williams: for a hyperbolic invariant set, there exists a simple branched manifold that holds all the periodic orbits within the invariant set. There is a problem though: strange attractors are *not* hyperbolic invariant sets!

In figure 12.17(a) we display a saddle cycle. In a Poincaré section, the stable and unstable manifolds are drawn as they approach a tangency. As we saw in chapter 8, a chaotic solution might coexist with such a saddle. If the system parameters are changed, the manifolds might have reached a last tangency, as displayed in figure 12.17(b). In this case, a hyperbolic invariant set exists, and a template can be constructed through the Birman–Williams procedure. The question is: how do the orbits of the flow in figure 12.17(a),(b) compare? The answer is: some orbits which exist in one of the flows might not exist in the other. But the orbits which have survived the parameter excursion must have kept their topological features. This observation suggests the following classification scheme:

Definition: A flow is compatible with a branched manifold when all the periodic orbits of the flow can be associated with periodic orbits in the branched manifold. Two flows will be equivalent if they are compatible with the same branched manifold [mind91].

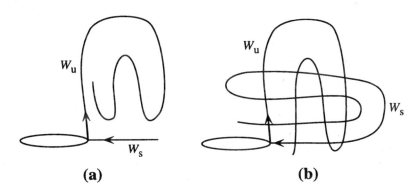

Figure 12.17. The creation of a horseshoe flow in phase space.

This equivalence relationship is useful from the practical point of view. A time-series taken from a dynamical system will 'contain' at most a finite number of periodic orbits. The definition may place within the same class time-series differing in the spectra of orbits, being thus a suitable tool for comparison among different 'runs' of an experiment. In other words, two time-series corresponding to the Rossler system with different parameter values will be in the same class even if they have not the same spectra, but a chaotic attractor solution of the Rossler system and a chaotic attractor solution of the Lorenz equations will be placed in different classes.

Exercise 12.8: Show that a flow that contains all the periodic orbits displayed in figure 12.18 cannot belong to the same class as a horseshoe flow.

A serious drawback of this classification scheme is that the 'boxes' in which we will place our time-series are too big and are not 'well defined' since a time-series, being finite, might be compatible with different (incompatible) templates. This contrasts sharply with the classification proposed by Nielsen–Thurston which assigns to the same class those flows that can be deformed one into each other carrying along the periodic orbits with the deformation. Thurston's classification scheme produces 'smaller boxes'. However, in practical terms this classification presents two drawbacks: first, a given system (say the

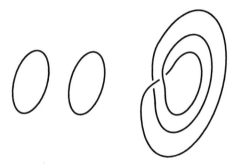

Figure 12.18. Three periodic orbits imbedded in a three-dimensional space which cannot be simultaneously placed in a horseshoe template.

Rossler oscillator for example) might be classified in different classes even for very similar parameter values; and second, if we 'miss' an orbit not implied by other orbits in the description of our system we will place exactly the same flow in two different classes. In any case, this chapter only outlines the state of a field which is still open and in need of much work.

12.7 Summary

In this chapter we have analysed some properties of 3-d flows. The rigid structure of the bifurcation sequences for 1-d unimodal maps (based on the period of their periodic solutions) could be in part preserved, when other topological invariants were taken into account. Looking at the periodic orbits as curves imbedded in a 3-d manifold, the uniqueness theorem for solutions of ODEs allowed us to identify these new invariants (like the knot type, the braid type, the linking number and the relative rotation rates).

One hopes that a similar programme should allow us to understand the structure of higher-dimensional flows, but just as the 'order' fell from 1-d to 2-d maps, knot like invariants for periodic orbits will fall in the passage from 3-d flows to n-d ones (with $n > 3$), except for what can be called the 'simplest' higher-dimensional flows [nati94]. The reason: one can 'unknot' any curve in a manifold of dimension larger than three. Which topological property of the flows should we inspect in order to pursue this programme further? This is still a completely open problem.

Chapter 13

The dynamics behind data

13.1 Introduction and motivation

Despite the fact that the pioneer work of Poincaré is almost one hundred years old, his ideas have filtered quite slowly into the paradigm of the natural sciences. Until recent years, the existence of irregularities in the output of an experiment was likely to be attributed to external sources of noise. On the other hand, a scientist trying to simulate noise is likely to use (even now) random number generators based upon a simple deterministic rule. In chapters 8 to 10 we analysed strange attractors. The high sensitivity to initial conditions allowed us to conciliate, in certain cases, the concepts of irregularity and determinism. In other words, thanks to the elements of nonlinear dynamics already discussed, we have enlarged our list of suspects when facing complex time-series data.

Typically, the output of an experiment is a set of (time-fluctuating) values for a scalar variable sampled at equally spaced intervals of time. We shall call the set a *time-series*. It is important to keep in mind that usually the sampling rate, the precision and the number of members (size) of the time-series are determined by technological limitations. A scientist dealing with the analysis of time-series must frequently find a compromise between the *great ideas* and whatever is *possible*. For example, whether determinism can be detected in the data does not only depend on the *existence* of a simple rule governing the dynamic evolution of the measured variables, but on the size of the time-series data as well.

We will assume hereafter that the data considered carry enough information to allow a meaningful exploration of the deterministic dynamics. In particular, we will assume that all the necessary information for making predictions is contained in the collected data. This is a minimal hypothesis in the sense that if a time-series does not satisfy the hypothesis for any number of data points, then it does not carry all the necessary information to predict the future evolution. Consequently, any test of determinism should fail on such time-series no matter how much data-massaging is performed.

The reader is warned that this hypothesis by no means is the only one assumed throughout this chapter. We are not going to present a list of recipes which would provide the reader with a 'black box' to be used with a time-series in order to get a set of 'significative values'. On the contrary, we will present some questions which we feel are sensible to ask when dealing with a time-series, and define some quantities which allow us to answer them.

13.2   Characterization of chaotic time-series

The sort of questions which we would like to pose are:

- Is there a deterministic rule behind the data?
- If that is the case, what is the dimensionality of the space supporting the rule (the phase space)?
- Are the data chaotic?
- 'How chaotic' are the data?

Before addressing these questions in the context of realistic data, let us discuss what we would find when analysing *ideal* time-series data. By ideal time-series we mean one that is infinitely long, free of noise and consisting of infinitely precise data points.

The condition of carrying enough information to make predictions, imposed in the introduction, translates as the existence of some $n > 0$ such that if $\{x_i\}$ and $\{y_i\}$ are two ideal time-series obtained by the same procedure and $x_j = y_j$ for $j = k, \ldots, k + n - 1$ then $x_j = y_j$ for all values of j. This is to say that the time-series is uniquely determined by n consecutive values of the registered data.

By a deterministic rule we mean, precisely, a function $F : M \to M$. We will have to find a way to associate data points with points in a space M, and a function F from and into the space M that provides the deterministic rule (the time evolution). Actually, very often we will have to settle for less. The situation which is most frequently encountered is that we are able to determine that the space M and the function F exist but without being able to characterize them completely. At best, we can establish the action of F at a finite number of points (those points that will be associated with our finite time-series).

For example, assume the time-series represents a sine function, sampled at intervals of time $\delta t = 2\pi/k$, i.e., $z_i = \sin(t_0 + 2\pi i/k)$, with k a sufficiently large integer to obtain a good sampling, typically $k = 10$ or more. We know that it is not enough to know the value of z for the jth sample, z_j, in order to determine the value of z_{j+1} (the next sample). The problem arises because $\sin(t)$ actually takes each value in $(-1, 1)$ *twice*, once when the wave is going up and another when it goes down. With only one datum value z_j we lack information on whether z is rising or falling. On the other hand, the pair (z_{j-1}, z_j) allows us to predict the pair (z_j, z_{j+1}). Our phase space is therefore two dimensional, actually $\mathbb{R}^2$. The construction of the phase space is addressed in the following subsection.

In most situations we would like to have an *a priori* estimate of the dimension of M, for example, we would like to gauge whether the task in front of us is possible or materially impossible. We address the problem in the subsection 'Dimensions'.

Let us concentrate for a moment on a data set for which we have established that it is deterministic (or that we know is deterministic because of any other external reason). We would like to find out whether the data set is chaotic or not. In such a case we would also like to estimate the rate at which (incomplete) information is lost. To meet this end we will have to give a computable meaning to the defining characteristic of chaotic trajectories, i.e., the divergence of nearby trajectories (*Liapunov exponents*) and to measure the progressive loss of information with some sort of *entropy*. We will address these topics in respective subsections.

13.2.1   Imbeddings

We are going to assume that we are dealing with a time-series provided by an experiment in which the relevant variables are reasonably isolated from the environment. No matter how carefully planned our experiment is, a complete isolation cannot be achieved: therefore it is a natural step to try to quantify the degree of *reasonability* of our assumption.

We would like to know whether the series is deterministic and, in addition, whether it contains enough relevant information to make predictions (these are certainly different issues). If both conditions are fulfilled, a sufficiently long segment of data, say $\{x_i, i = k, \ldots, k + n - 1\}$ should be enough to predict the next segment of length n, i.e., $\{x_i, i = k + 1, \ldots, k + n\}$. Determinism will then require that there is a unique value x_{k+n} for any existing segment $\{x_i, i = k, \ldots, k + n - 1\}$.

In other words, we can represent the time-series, $\{x_i\}$, by points in $\mathbb{R}^n$, each one of the form $X_k = \{x_i, i = k, \ldots, k+n-1\}$. These points uniquely determine the state of the system. In principle, not every point in $\mathbb{R}^n$ will represent a point in the time-series. Hence, the time-series will in general be restricted to a submanifold of $\mathbb{R}^n$. This situation describes an *imbedded* manifold. Note that this imbedding does not need to be unique or optimal.

It is in principle possible to reduce the dimensionality of the imbedding. We could take just some elements within the array $X_k = \{x_i, i = k, \ldots, k+n-1\}$, with the natural restriction that they still determine uniquely the state of the system. The so-called 'time-delay' imbedding [take81], for example, proposes to use the points $Y_k = \{x_{k+d\tau}\}, d = 0, \ldots, m - 1$ (where τ is a small positive integer and $(m - 1)\tau \leq n - 1$) in order to uniquely determine the state of the system. Other alternatives could be to use an approximation to the time derivatives by finite differences [pack80], i.e., to use as imbedding coordinates

x_k, $x_k - x_{k-1}$, etc (intending to approximate with them $x(t)$, $x'(t)$, etc), or to use linearly optimized bases [broo86].

Each of the above-mentioned methods has advantages and disadvantages depending on the features of the time-series. For example, finite differences are likely to help us to interpret the measurements in terms of a dynamical system, but they are also likely to enhance the noise. The linearly optimized basis helps us to deal with noise reduction. Tradition, on the other hand, has 'established' time-delay imbeddings frequently as a first choice. Neither are these imbeddings the only possible ones, nor is there a need to restrict imbeddings to linear combinations of the measured variables.

We want to discuss here what are the conditions imposed to the imbeddings. It is important to realize that what is really accessible to us is the map $f(X_k) = X_{k+1}$ defined on the imbedded manifold. Let us call f the *experimental map*. We can decompose f in the form $f = G^{-1}FG$, where F will be the imbedded map on the manifold M and $G : \mathbb{R}^n \to M$ the realization of the imbedding. If instead of being an imbedding G were one-to-many, $(G^{-1}$ is a projection), we will find that f defines many possible futures (or pasts) for some points in M, since one point in $\mathbb{R}^n$ will correspond to several different dynamical situations or points in M. On the other hand, if G is a projection, i.e., it sends several points in $\mathbb{R}^n$ to the same point in M, the inverse G^{-1} will be multivalued. However conceivable, this case is not often found in practice. Consider for example an experimental set-up in which the detection is performed by the parametric excitation of a resonant circuit. As discussed in chapter 2, the widest Arnold tongue corresponds to the $1/2$ resonance in which the resonant system has a fundamental frequency $1/2$ of the input frequency. In such a circumstance we expect that a periodic orbit of period T in the experiment will be registered by the detector as a periodic orbit of period $2T$, thereafter generating two different points in the reconstructed space for each point in the actual phase space of the experiment, thus G^{-1} maps one point into two points and G is many-to-one. This situation is depicted in figure 13.1.

What happens when the X_k fail to uniquely describe the system? If there is a point X_p that actually corresponds to two different dynamical situations, we expect an intersection of the trajectories drawn in the candidate for imbedded manifold, since two different dynamical situations have associated different past and future[†], see figure 13.2.

The criterion that we have adopted above is very difficult to implement. To check the absence of self-intersections might be an impossible task, especially if the required dimension for the imbedding is larger than three. An alternative approach could be to try a statistical test for determinism. If the system is

---

[†] The present discussion is based on the image of a dynamical system represented by differential equations or invertible maps. If we enlarge our class of dynamical systems to include non-invertible maps, as discussed in chapter 4, only those points showing two different futures in the imbedded dynamics will signal violations of determinism and will highlight the failure of the proposed imbedding.

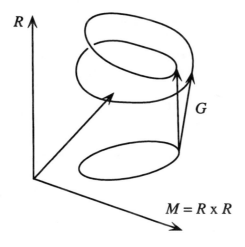

Figure 13.1. A period-T orbit in $M = \mathbb{R}^2$ presented as a period-$2T$ orbit in $\mathbb{R}^3$. G is not an imbedding.

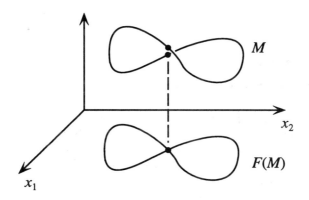

Figure 13.2. A closed curve in the form of a number eight in $\mathbb{R}^3$ has self-intersections when projected to $\mathbb{R}^2$. The projection is not an imbedding.

deterministic, a point in the imbedding should be enough to determine the next one in the time evolution of the system. Roughly speaking, one point cannot have two futures. Considering the finiteness of the time-series we may say that two nearby points cannot have images that are too far away. This can be written as follows:

$$P(|X_i - X_k| < \epsilon \text{ given that } |X_{i-1} - X_{k-1}| < \delta(\epsilon)) \approx 1 \qquad (13.1)$$

where P denotes a probability, and δ, ϵ are adequate small numbers. Experimentally there is a lower bound to how small numbers we can distinguish in a time-series output, namely the error in the measurements. Alternatively, in the case of problems modelled by differential equations, one could propose that two points which are separated more than $\delta(\epsilon)$ will have images that are separated more than ϵ, thus reversing the inequalities above.

Exercise 13.1: Show that the two alternatives are equivalent. Hint: perform the substitution $t \to -t$.

An upper bound for the dimension of the imbedding space can be computed from the dimension of the submanifold where the dynamics actually takes place. For example, in the case of the sine data the imbedding is $\mathbb{R}^2$. However, using as imbedding coordinates the original data set and the time derivative of the data set, we see immediately that the imbedded points lie in a one-dimensional space, the circle $\mathbb{S}^1$. For a manifold of dimension m the imbedding space is not larger than $\mathbb{R}^{2m+1}$ [whit36]. Intuitively, the dimension of the manifold is greater or equal than the 'size' of the invariant set around which the dynamics of our system is organized (the attractor). In the coming subsection we will discuss how to estimate the 'size' of the invariant set and the relation between this size and the imbedding dimension.

13.2.2   Dimensions

We know that a line is a one-dimensional set, a plane is a two-dimensional set, the physical space $\mathbb{R}^3$ has dimension three, but what is the dimension of the Cantor set described in chapter 10? This is certainly not an idle question since we expect some invariant sets to have a structure like the structure of the invariant set in the horseshoe, namely the intersection of two ('horizontal' and 'vertical') Cantor sets as described in chapter 10.

We know that the points belonging to the Cantor set add up to zero length (in fact, we observed that the complement of the Cantor set in the unit interval is a collection of disjoint non-empty open intervals adding up to length one). We also know that the Cantor set is larger than a finite or countable union of isolated points (the typical dimension-zero sets). How can we extend the idea of dimension in such a way that:

- The (local) dimension of any manifold agrees with what we already know (lines have dimension one, planes dimension two, etc).
- The dimension of the Cantor set is neither zero nor one.

There are several ways to extend the idea of dimension conforming to these constraints [abar93]. We are going to develop the most intuitive of these ideas, the *capacity*.

Let us consider a plane, a straight line and a point imbedded in $\mathbb{R}^3$. We will cover each of these manifolds with cubes of side ϵ and count how many of these cubes are needed to cover the manifolds under consideration.

To cover a square piece of the plane of side one with ϵ-sided cubes we need to use $\approx 1/\epsilon^2$ cubes. To cover a segment of length one of a line we need $\approx 1/\epsilon$ cubes. To cover a point we need just one cube. We observe that the exponent of ϵ in each of these cases is -2, -1 and 0 respectively. The exponents are precisely the dimension of the manifolds we considered, with negative sign. We can now revert the relation saying that the *capacity dimension* of a manifold is

$$\dim(M) = -\liminf_{\epsilon \to 0} \frac{\ln(N(\epsilon))}{\ln(\epsilon)} \tag{13.2}$$

where $N(\epsilon)$ is the number of ϵ-sided cubes required to cover a (local) piece of the manifold. We use lim inf to be guarded against the possibility that while $\epsilon \to 0$ the quotient above does not approach a limit value, i.e., it continues to oscillate as, for example, $\cos(\omega/\epsilon)$ does. lim inf picks systematically the lowest estimates of dim.

Exercise 13.2: Show that if $\ln(N(\epsilon))/\ln(1/\epsilon)$ does not approach a limit value when $\epsilon \to 0$ then different choices of ϵ, i.e., different sequences $\{\epsilon_n\}$, with $\epsilon_n \to 0$ when $n \to \infty$, yield different values of the limit for $n \to \infty$ of $\ln(N(\epsilon_n))/\ln(1/\epsilon_n)$. As a result, the *plain* limit in equation (13.2) will not exist as such and lim inf is a necessity.

The above exercise indicates that there is a problem when using lim inf in practice. They have the nice property of being defined even when the usual lim does not exist, but the drawback of being in general very difficult to compute. Fortunately, one seldom needs to exactly compute a lim inf. The origin of our interest in dim was to be able to characterize a time-series. This characterization is by necessity a rough one, since from the very beginning we start with the ambiguity of describing an infinite set (the attractor) with a finite number of points (the time-series). So we can be content with an *estimate* of lim inf. Whatever way we choose of picking boxes $\{\epsilon_n\}$ having radii going to zero with n, will give an upper bound to dim. If your neighbour comes up with a better choice $\{\epsilon_n'\}$, you just drop your old result and adopt the new one. In connection with this (naive) procedure the concept of *boxcounting dimension* has been coined in the physics literature [abar93].

We can now apply naively our rule to the Cantor set. The real line $\mathbb{R}^1$ is a sufficiently large imbedding space for our task, so we will use line segments of size ϵ instead of ϵ-cubes. We can try to cover the Cantor set following the same steps used in the construction of the set. We need one segment of length $\epsilon = 1$ to cover the whole set, to start with. With $\epsilon = 1/3$ we need just two intervals, and in general, at the nth step, we will need 2^n intervals of length $1/3^n$ each to completely cover the set. Now we may let $\epsilon \to 0$ just by letting $n \to \infty$, in order to estimate the dimension.

The estimated dimension turns then to be $0 < d = \ln(2)/\ln(3) < 1$, i.e., a non-integer number between zero and one as desired. Non-integer dimensions are usually referred to as *fractal dimensions*.

Exercise 13.3: Challenge: Show that for the Cantor set our computation actually yields the lim inf.

Cautionary notes:

- The capacity is always defined. The boxcounting dimension coincides with the capacity for those cases where lim inf can be replaced by lim.
- Using the above definition, the capacity dimension depends in principle on the choice of covering. Thus, the limit may not be unique. Taking the sup over all possible coverings [guck86] overcomes this problem, although it renders the new definition difficult to apply in practical situations.
- The association between fractal dimensions and chaotic sets is not one-to-one. It is possible to have chaotic sets that fill a manifold (having therefore integer dimension).
- The information carried by the capacity (and any dimension in general) is of a geometrical nature and as such might not include dynamical information. For example, it might not be possible to distinguish between a time-ordered (deterministic) invariant set and a randomly ordered rearrangement of the same points.

Despite the flags raised by the cautionary notes, the following theorem shows the relevance and usefulness of the capacity dimension.

Theorem (Sauer *et al*) [saue91]: (fractal Whitney imbedding prevalence theorem). Let A be a compact subset of $\mathbb{R}^k$ of capacity d, and let n be an integer greater than $2d$. For almost every smooth map $G : \mathbb{R}^k \to \mathbb{R}^n$,

(i) G is one-to-one on A

(ii) G is an *immersion* on each compact subset C of a smooth manifold contained in A.

Let us analyse the two requirements for G. The first one is closely related to our early requirements about determinism: no two points are mapped to the same image point (i.e., the determinism is not lost in our new state space). But the precise definition of imbedding requires also a differential structure: an imbedding is a map that does not collapse points or tangent directions. The second requirement mentioned in this theorem addresses this issue. A smooth map G on A is an immersion if the Jacobian of G in $x \in A$ is one-to-one for every x. This is equivalent to say that DG has full rank in the tangent space and therefore no differential structure is lost. The subtle difference between imbedding and immersion is that an imbedding $G : M \to N$ is an immersion which is one-to-one and with smooth inverse.

13.2.3   Liapunov exponents

We have discussed that irregular time-series data might be ruled by a
deterministic prescription. That will be the case when dealing with chaotic
data. It is reasonable then to test the eventual chaoticity of the data relying
on the definition of *chaotic*. It was discussed in chapter 8 that sensitivity
to initial conditions is the most relevant feature of chaotic behaviour. The
divergence of nearby trajectories can be measured through a generalization of
Floquet multipliers (see chapter 5).

Given a chaotic trajectory, $x(t)$, we can consider the rate at which
infinitesimally close trajectories depart from $x(t)$ by considering the evolution of
a first-order (linear) perturbation of the trajectory in exact correspondence with
our approach to Floquet multipliers. If the flow of our problem is $F(., t)$, and the
evolution of an orbit $y(0)$ with initial condition close to $x(0)$ $(y(0) = x(0) + \delta)$
is given, up to first order, by $y(t) \approx x(t) + DF(x(t), t)\delta$ we have that
$||DF(x(t), t)\delta||$ measures the distance between both trajectories at a later time
t. Since this distance might grow without bounds we will rather measure the
rate of departure per unit time in the form

$$\lambda = \lim_{t \to \infty} \frac{\ln ||DF(x(t), t)\delta||}{t||\delta||}. \qquad (13.3)$$

The *maximum Liapunov exponent* defined by equation (13.3) can be
obtained for almost all vectors δ. A positive maximum Liapunov exponent
means exponential divergence of nearby trajectories and can be taken as a formal
definition of chaos.

We note that for some special choices of vectors $\delta \in \mathbb{R}^n$ a different limit
can be obtained. For example, if $y(0)$ belongs to the orbit of $x(0)$, i.e., δ is
along the direction of the velocity, we will obtain a neutral departure, i.e., a
zero Liapunov exponent associated with this direction just as we found earlier
a Floquet multiplier one associated with this direction.

If the trajectory $x(t)$ is of saddle type there will be a subspace of vectors
δ with a negative limit for equation (13.3). In such a case we say that they are
associated with a negative Liapunov exponent.

In general we will find a set of subspaces $S_n \subset S_{n-1} \subset \ldots \subset S_1$ such
that almost every vector $\delta_j \in S_j$ has associated a Liapunov exponent λ_j with
$\lambda_n < \ldots < \lambda_1$. The set of values $\{\lambda_j\}$ are the Liapunov exponents of the
trajectory $x(t)$.

If λ_1 is negative the trajectory is stable (linearly attractive). If λ_{n-1}
is positive the trajectory is unstable in all directions, and if some Liapunov
exponents are positive while others are negative we are in the presence of a
saddle. Only saddle-type trajectories can be part of a chaotic attractor.

13.2.4 Entropy

13.2.4.1 Probability entropy

Let us suppose that we are challenged to play a game which we shall call (with 'subliminal' purposes) the *entropy game*. The game consists in the following: given an evolution rule and an initial condition, we have to predict where the initial condition is mapped. As we have to deal with finite precision the following information is made available to us:

- To which box of a grid the initial condition pertains.
- For each box of the grid, what the probability is of being visited, i.e., if a box has probability zero it is never visited, if it has probability one the box is the only one ever visited by our evolution rule and if the probability is $p < 1$ the box will be visited, on average, np times every n steps.

The prediction will consist in picking the box in which the nth iteration falls. Naturally, the key question for us is *which are the chances of winning?*

It is reasonable to expect that our chances depend strongly on the evolution rule: if it is random, our chances will be smaller than if it is deterministic. Yet, even if the rule is deterministic the game can be tricky. We know that if the rule consists of a nonlinear map, it is possible to have exponential divergence of nearby initial conditions. This fact, plus the fact that our initial conditions are given with a finite precision (the diameter of the boxes of the grid) lead us to have to face a risk of losing. We shall try to quantify this risk.

In what follows we will use the notation $P(i_0, i_1, \ldots, i_n)$ to denote the probability that trajectories visit the boxes $i_0, i_1, \ldots, i_n$ and the notation $P((i_{j+1}, \ldots, i_n)/(i_0, i_1, \ldots, i_j))$ to indicate the probability that a trajectory among those that have already visited the boxes $i_0, i_1, \ldots, i_j$, also visits the boxes $i_{j+1}, \ldots, i_n$ i.e., the conditional probability. We will drop the parentheses for one-box sequences. The conditional probability $P((i_{j+1}, \ldots, i_n)/(i_0, i_1, \ldots, i_j))$ is the probability of the (partial) orbit $i_{j+1}, \ldots, i_n$ relative to those with trajectories $i_0, i_1, \ldots, i_j$, thus

$$P((i_{j+1}, \ldots, i_n)/(i_0, i_1, \ldots, i_j)) = \frac{P(i_0, i_1, \ldots, i_j, i_{j+1}, \ldots, i_n)}{P(i_0, i_1, \ldots, i_j)}.$$

First, let us analyse the case in which the evolution rule is regular, i.e., non-chaotic. Pick an initial condition within the box i_0 (with probability $P(i_0) = p$). Ignoring for the moment all complications with measurement errors, we can arrange our grid so that the whole box i_0 is mapped into the box i_1 by the time evolution. This is possible since the evolution law is regular, and hence nearby points evolve into nearby points. We have, thus, $P(i_1/i_0) = 1$ or, equivalently, $P(i_1, i_0) = P(i_0)$. We can say that the information that we were given (that needed to place our initial condition within a box) has not been lost by the time evolution.

In contrast, if the evolution is chaotic, nearby points will be exponentially separated. Therefore, if we pick an initial condition with probability $P(i_0)$, its iteration will fall into *one* of the $N \approx e^\lambda$ boxes in which the initial conditions within the box i_0 can evolve. In other words, $P(i_0, i_1) = P(i_0)/e^\lambda = P(i_0)\, e^{-\lambda}$. For completeness, if the evolution rule is random, $P(i_0, i_1) = P(i_0)P(i_1)$ and $P(i_1/i_0) = P(i_1)$, i.e., knowing that the initial condition was in i_0 does not help to predict where the point will be in the next step.

To summarize our results, *we lost no information in the first case, we lost some information in the second case and we lost all information in the last case.* In order to quantify these observations let us notice that in both the first and third cases the result was independent of the initial point. The second case, though, is more delicate: we might lose more or less information depending on the local expansion rate of the deterministic rule. Let us then define the quantity

$$K_n = - \sum_{(i_0, i_1, \ldots, i_n)} P(i_0, i_1, \ldots, i_n) \ln(P(i_0, i_1, \ldots, i_n)) \qquad (13.4)$$

where the sum extends to all possible trajectories. Then $K_{n+1} - K_n = 0$ for the first case, $K_{n+1} - K_n = \lambda$ for the second one and $K_{n+1} - K_n = K_0$ for the last one.

We can now analyse how the situation changes with the diameter ϵ of the boxes in the grid or, which is equivalent, with the number of boxes with non-zero probability, $N(\epsilon)$. We note that our (rough) estimations do not change with the diameter of the boxes for the deterministic rules (regular or chaotic) but we expect that it will change for the random rule since the probability of the box i will go to zero with the diameter of the box. Moreover, we can compute K_0 for the case of equally probable boxes $P(i) = 1/N(\epsilon)$ obtaining $K_0 = \ln(N(\epsilon))$.

This leads us to associate the quantity

$$K = - \lim_{\epsilon \to 0} \lim_{n \to \infty} (K_{n+1} - K_n) \qquad (13.5)$$

with the average loss of information, and we shall call it the *probability entropy*. The quantity K provides a measure of how our probability of winning deteriorates: our chances are worse with larger K.

The probability entropy allows us to distinguish among the three kinds of rule proposed. If the rule is deterministic and non-chaotic we will find $K = 0$, if it is deterministic but chaotic $0 < K < \infty$, while in the case of a random rule we will find $K = \infty$. Moreover, in the deterministic cases we expect K to be related to the average of the local expansion rates[†].

It is interesting to note that since $K_{n+1} - K_n \to K$ we also have that $\lim_{n \to \infty} \frac{1}{n} K_n = K$.

[†] The situation described for chaotic systems using one expansion rate for all trajectories of length n is certainly a simplification. If we allow each trajectory to have its own expansion rate we have $P(i_0, \ldots, i_{n+1}) = P(i_0, \ldots, i_n)\, e^{-\lambda_{(i_0, \ldots, i_{n+1})}}$ and $K_n = \sum_{(i_0, \ldots, i_n)} P(i_0, \ldots, i_n)\lambda_{(i_0, \ldots, i_{n+1})}$ is the averaged expansion rate.

The interested reader will find a detailed and formal discussion of the probability entropy in [arno68].

13.2.4.2 Topological entropy

We can find a rougher estimate of our chances of winning the game in the following form: compute K assigning to all possible trajectories of length $n+1$ the probability $P(i_0, \ldots, i_n) = \frac{1}{N(n+1, \epsilon)}$ with $N(n+1, \epsilon)$ the number of different sequences of length $n+1$ that can be distinguished with boxes of diameter ϵ.

In other words, the grid of boxes allows us to map trajectories into symbolic sequences $(i_0, \ldots, i_n)$. The resulting entropy, called the *topological entropy h*, is a measure of the exponential growth of the symbolic sequences $N(n, \epsilon) \approx e^{nh}$. The reader may recall that this concept of entropy was the one used in chapter 11 and 12. It is not difficult to see that $K \leq h$. The topological entropy is an upper bound for the probability entropy.

Exercise 13.4: Prove the previous statement.

Remarks:

(i)   The uniqueness of the limits defining the probability and topological entropy is not clear. In principle, different ways of producing smaller and smaller grids might yield different limits. Formally, this problem can be solved by considering all possible ways of covering the attractor with a finite number of 'boxes' all of them of diameter smaller than ϵ and taking K (or h) the largest value obtained among those produced by the different sequences of covers. In other words, we should take $K = \sup_\alpha \lim_{n\to\infty} (K_{n+1} - K_n)$ in the definition of the probability entropy, where α runs over all possible covers. In practice, to consider this infinite number of covers is impossible and one is satisfied with an estimation of the limits for $\epsilon \to 0$ of a single sequence of covers.

(ii)   Both K and h defined as described in the previous remark result in invariants of the dynamics and are independent of the set of coordinates chosen.

13.3   Is this data set chaotic?

Suppose that it is relevant for our studies to be able to answer whether a data set comes from a deterministic system and in such a case whether it is chaotic or not. Furthermore, we may want to rank a number of experimental data sets according to their dynamical complexity. How should we proceed?

We just learned about three *discriminant statistics* (Liapunov exponents, fractal dimensions and entropies). They are statistics in the sense that they are averaged values, and they are discriminant because they evaluate to clearly different values for regular and chaotic behaviour and they also differ from the values obtained for uncorrelated noise.

Our immediate impulse would be to implement algorithms to evaluate these statistics on our (imbedded) time-series, plus—perhaps—a direct test for determinism (see for example [kenn92]) and all the battery of linear tests developed in a couple of centuries of linear statistical analysis (Fourier analysis, correlation functions, statistical regression, filters, ...).

Before abandoning ourselves to this primary impulse it is good to know some of the caveats of these statistics so that we are able to make a proper interpretation of the results.

13.3.1 The well known enemies

Liapunov exponents, fractal dimensions and entropies all rely on the evaluation of limits for long times and infinite precision.

Experimental time-series, however, are finite in both length and precision. The length of a time-series is usually severely limited by storage capabilities (especially in fast experiments) and the impossibility of maintaining a controlled environment for long times (a problem that mostly affects slow experiments). The precision of the measurements is affected by a number of limitations like the finite precision of the measurement apparatus and the unavoidable presence of some level of noise (uncontrolled coupling of the experiment with the environment) both in the data acquisition process and in the experiment itself.

Another inconvenience comes from the uneven distribution of data points in our reconstructed (or constructed) phase space. Boxes of equal diameter covering the imbedded attractor will present rather different probabilities of being visited. As a result of this non-uniformity, when we reduce the diameter of the boxes there will be a large number of empty boxes. Recall that the computation of all our three discriminant statistics is based on the fact that the size ϵ of the boxes goes to zero. At the end (long before the end, in fact), since our number of data points is finite, each box will be either empty or occupied by only one point. We know that in such a circumstance we have gone too far with the smallness of the boxes: so far that we have no possibility of evaluating expansion rates or probabilities for the itineraries. The point is then: where to stop?

To complicate things further, whenever we can expect to have positive Liapunov exponents it is likely that regions of sparse data in our system will be the images of other more populated regions and vice versa. This is a direct consequence of the expanding behaviour we just want to determine. A small region of phase space with high density of points will be mapped in such a way that the points will depart exponentially from each other at a rate controlled by the (yet unknown) Liapunov exponents, landing in a larger region with lower point density. The reverse effect can be associated with the negative Liapunov exponents. We have that the uneven distribution of data points is an image of the unevenness of the expansion rates, thus the undersampled regions are likely to be those with 'negative' expansion rates (this is, where the map is locally contracting) while those well represented regions include the regions with large expansion rates. We would like to be able to describe *both* types of region accurately.

Let us consider as an example of these problems the estimation of an upper bound to the capacity dimension. Assume that you have N points of a data set that can be imbedded in two dimensions. Let us consider for a moment 10^4 iterations of the Hénon map, namely

$$x_{n+1} = a - by_n - x_n^2 \tag{13.6}$$

$$y_{n+1} = x_n \tag{13.7}$$

with $a = 1.4$, $b = -0.3$ and initial condition $x_0 = 0.631$, $y_0 = 0.189$. Since we know the deterministic law that produced the 2-d 'time'-series, we do not actually need to imbed the data. We now produce a sequence of boxes of decreasing size, starting with a square box of size $\epsilon_0 = 1$. We need about nine such boxes to cover the entire data set. Let us now reduce the size of the boxes to 0.9, 0.8, 0.7, 0.6, etc and compute $\ln(N(\epsilon))$ for each case. In figure 13.3 we plot the computed values against $\ln(1/\epsilon)$.

If the behaviour of the curve is good enough, one would expect that the slope of the plot for small ϵ (to the right of the plot) will have a limit value, which we will take as the upper bound to the capacity. Notice however that too small boxes yield statistically non-significant results since 10^4 points may be too few when ϵ gets very small (it is however in that region that we are closest to the limit $\epsilon \to 0$). No more than 10^4 boxes are required to cover all the points. A problem is also encountered with too large boxes. No matter how large we make the boxes, we will always need at least one box to cover the points. Therefore, our curve will sooner or later bend and become a flat line running along the leftmost and rightmost parts of the x axis.

The final result is that in practice only some intermediate ϵ values can be considered to be reliable. We can see that this region of the plot yields a roughly linear dependence of $\ln N(\epsilon)$ on $\ln(1/\epsilon)$. We can therefore attempt to estimate the fractal dimension of the Hénon attractor with the slope of this approximately linear function, computed by a least-squares fitting. We obtain $d \approx 1.23$. The most optimistic error bar we can associate with this number is the standard deviation of the fitting, but we see immediately (see figure 13.3(b) and (c)) that there is a certain unavoidable degree of ambiguity in the whole procedure, since both the result and its error will vary depending on how we choose the 'intermediate' region of linear dependence. For the interested reader we mention that the currently accepted estimation of the capacity for this problem is 1.28 ± 0.01 [gras84].

Although these caveats, once acknowledged, can be partially overcome, they impose limitations on the estimation of the different statistics and make also rather complex the problem of estimating the error bars for the quantities

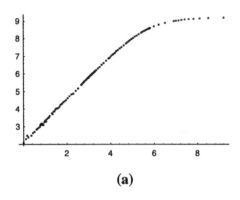

(a)

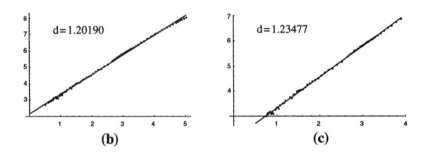

(b)   (c)

Figure 13.3. Estimate of the fractal dimension for the Hénon map. (a) Box sizes ranging from 1 to 0.0001. (b) Central part of the plot in (a) with 85 different box sizes. (c) Same as (b) with 65 different box sizes.

computed. The error bars have to account for statistical as well as experimental errors.

We suggest the interested reader consults [abar93] for further information on methods for chaotic data analysis.

13.3.2   Testing the tests

Can we fool the discriminant statistics? Certainly! So far we have discussed the expected results of measuring dimensions, entropies and Liapunov exponents on deterministic data. The fact that they can discriminate between chaotic and non-chaotic deterministic time-series does not mean that other time-series will pass the tests as 'chaotic' when they are in fact random.

For example, let us assume that we have an (imbedded) deterministic time-series with a fractal dimension. The data points are $\{x_i, i = 1, \ldots, N\}$, $x_i \in \mathbb{R}^n$ and i is the time-order. The same sequence of points can be disordered applying an arbitrary permutation to the N points, thus destroying any trace of determinism. However, the fractal dimension of the set will not change! The dimension of the set is a *geometrical* characterization, completely insensitive to the time-ordering of the points.

The same trick that fools the dimension test will not work with the entropy test, since entropies are precisely geared to discriminate the loss of information. The time-order of the data series is essential to them, but, we can devise other problems to fool the entropy test. For example, if we add to our data noise with a certain amount of 'memory', i.e., such that it adapts according to its previous recent values (this is called *time-correlated noise* and also *coloured noise*, in opposition to the more familiar, memoryless *white* noise), we will be able to alter the way information is lost when we evolve stepwise in our time-series, and thus produce a different estimate for the entropy.

Liapunov exponents are not free of problems (see [krue93] for a method to estimate Liapunov exponents from time-series). They measure the exponential departure of nearby trajectories; hence, we have to focus our interest not in the trajectory itself but in the difference between close-lying trajectories. Each trajectory can be regarded as a 'mean' trajectory (say, a little improperly, the mother trajectory from which our two deviated trajectories depart), plus a small deviation, plus some amount of noise. Computing the difference between the two trajectories we eliminate the 'mean' trajectory which is the largest part of the signal and we remain with the difference of two small deviations plus the noise. Noise may therefore become a significant part of the result (thus rendering the result useless).

13.3.2.1   Testing with surrogate data

In order to be sure that we are giving the correct interpretation to the discriminant statistics we are using, the following method, known as *the method of surrogate data* [thei92], is relevant.

Given a discriminant statistics, a property (like determinism or nonlinearity, for example) and a data set, we would like to determine whether the following statement is true: the estimated value of *discriminant statistics* reflects the *property* of this *data set*.

The method of surrogate data begins by constructing new (surrogate) data sets where the property we are testing for is destroyed, preserving all other properties. Even this goal is hard to achieve: how can we destroy one property without affecting any other?

Producing a very large collection of surrogate data sets (an ensemble), we can use statistical tests to establish the validity of our statement. The final elaboration consists in testing a *null* hypothesis rather than the original statement. The null hypothesis is a proposition of the form: the *data set* cannot be distinguished using the *discriminant statistics* from other data sets generated with the *alternative method*. The word *alternative* here highlights the fact that our surrogate data sets are produced by some or other technique (in short: some recipe; examples will be described below), rather than by sampling the outcome of an experiment.

If the original data set cannot be distinguished with our test from the surrogate data sets we say that the null hypothesis is (statistically) valid and hence that the *alternative method* is a possible explanation of the origin of all the data (including the original data set).

On the contrary, if the discriminant statistics clearly distinguish between the original data sets and the surrogate ones, we say that the null hypothesis is false. In this case, we only have a negative result: the null hypothesis fails to explain the data. This is *not* the same as saying that the estimated value of the *discriminant statistics* reflects a certain *property* which is present in the data and not present in the surrogate data. We have not proved our original assumption, but we can keep our hopes alive.

Let us consider an example. For a given data set, we may want to test the following statement: the estimated value of the *capacity dimension* reflects the *determinism* of the data. We proceed in three steps.

In the first step we produce surrogate data sets which do not have the property in question (determinism). We use the following two recipes (taken from [thei92]) to produce the new data sets:

(a) Take the Fourier transform of the data. This generates a set of complex numbers. The absolute value of these numbers constitutes the Fourier power spectrum of the data set. Produce new data sets with the same Fourier power spectrum (this implies the preservation of mean and variance from the original data set) but changing randomly the phase of the complex numbers in the Fourier transform. Take subsequently the back-Fourier transform.

(b) Compute the mean μ, variance v^2 and first auto-correlation $A(1)$ of the original data,

$$A(\tau) = \frac{\langle x_t x_{t-\tau} \rangle - \langle x_t \rangle^2}{\langle x_t^2 \rangle - \langle x_t \rangle^2} \tag{13.8}$$

where the symbol $\langle \ldots \rangle$ indicates time-average. Produce new data sets by way of an Ornstein–Uhlenbeck process defined by the non-deterministic map

$$y_t = a_0 + a_1 y_{t-1} + \sigma e_t \tag{13.9}$$

where y_0 is a randomly chosen start-point, e_t is a Gaussian random variable of unit variance and the constant coefficients in equation (13.9) are taken so that the mean, variance and auto-correlation of $\{y_t\}$ match the values of the original data: $a_1 = A(1)$, $a_0 = \mu(1 - a_1)$ and $\sigma = v\sqrt{(1 - a_1^2)}$.

The second step consists in estimating the capacity of all the surrogate time-series. We will obtain many different values $x(\{y_t\})$, and we can assign a 'probability' $p(x)$ to these values by counting the fraction of surrogate data sets that yield the same estimate.

The last step consists in estimating the capacity for the original data set and comparing statistically this value against the values obtained with the two sets of surrogate data. Roughly speaking, if the original value is far away from the mean plus/minus standard deviation of the surrogates (a), it is unlikely that the original data set comes from a random process with a given Fourier power spectrum and we proceed to reject the null hypothesis (a). In identical terms, if the estimated capacity dimension of the data is far away from the mean plus/minus standard deviation of the surrogates (b) we can reject the hypothesis that the data come from a Ornstein–Uhlenbeck random process.

More precisely, we can estimate whether the original time-series can be confused with one of the surrogate time-series (in what regards the discriminant statistics) in the following way. If the computed value of the discriminant statistics (capacity) for the original time-series is x_0 we can estimate the fraction P of surrogate data sets where $p(x) > p(x_0)$, using the probability distribution of x obtained from the surrogate data sets. We have

$$P = 1 - \int_{p(x) \leq p(x_0)} p(x) \, \mathrm{d}x. \tag{13.10}$$

This equation says that our data set does not belong to the $100 * P\%$ most frequent random events. The larger P, the more confident we are that our data set cannot be confused with a random process. It will be even better if we are so lucky that we can compute confident error bars for x_0.

13.3.3 Topological tests: how chaotic is our data set?

As stated before, one of the sources of problems with the tests described so far arises from the need to estimate limits for long times or infinite precision. For data sets that can reasonably be thought as originated from one- or two-dimensional maps (or three-dimensional flows) there is however an alternative. We can implement an analysis of the data based on the topological properties of one- and two-dimensional maps discussed in chapters 11 and 12 [mind91, papo91].

The method consists basically in finding pieces of the data set that closely resemble periodic orbits, usually called *close returns* [lath89]. If the attractor is chaotic, we expect that a typical single trajectory will be an open set within the attractor (the attractor itself being the closure of the trajectory), and that there will be infinitely many periodic trajectories belonging to the attractor. The close returns can be considered to mimic the periodic orbits pertaining to the attractor, combined with the measurement noise and the deviations arising from the positive expansion rates.

Once we have a number of candidates for a set of periodic orbits within the chaotic set, we can attempt to compute their braids and to build a (in some sense) minimal template that can hold the orbits together with their linking properties, in accordance with the discussion in chapter 12. If the close returns give a good description of the periodic orbits and if no approximate intersections among the orbits occur, the description has better chances to be accurate.

In this way, topological methods allow, for 3-d flows, to say to some extent *how* chaotic is our data set. The test for chaotic behaviour consists in showing that among the periodic orbits suspected to belong to the attractor there exists one or a set of orbits that are so badly tangled that they imply a positive topological entropy, as the reader may remember from chapter 12.

The topological methods, though believed to be less corruptible by noise and limitations of the data, cannot escape the basic problem of chaotic data analysis: from a finite time segment of an orbit it is not possible to distinguish whether the orbit is chaotic or only has a very long period, nor is it possible to say from a finite number of points pertaining to an invariant set whether the 'holes' we see in phase space (see figure 13.4) are intrinsic or due to insufficient sampling.

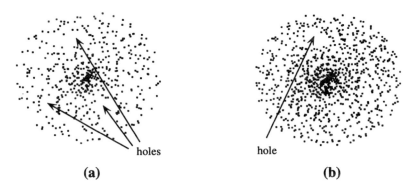

holes	hole
(a)	**(b)**

Figure 13.4. One of the problems of a finite-length sampling. (a) 500 points of a time series imbedded in $\mathbb{R}^2$ showing 'holes' in imbedding space. (b) Adding more points, the holes coming from an insufficient sampling disappear.

13.4 Summary

Motivated by the interest in understanding experimental time-series we have discussed in this chapter three properties of dynamical systems which are related to statistical averages of functions of the data points.

First, we had to place our data points in phase space or rather in some model space which can hold our data in a way that is compatible with the physical properties of the system that we may know in advance. This topic was discussed in the section 'Imbeddings'.

Once we had positioned our points in some kind of *phase like* space, we considered three different statistical averages. Fractal dimensions deal with the position of the points in phase space, Liapunov exponents estimate the departure rate of close-lying trajectories and entropy gauges the way information is lost.

It will never be too repetitive to insist that the finiteness and other limitations of the data introduce a certain degree of ambiguity in these techniques which can never be completely removed. In some cases this may be a question of minor importance, while in others it will destroy all chances of making sensible predictions. The task of the researcher is to render this limitation evident in order to be sure of the extent to which her/his assertions are valid.

Chapter 14

Perturbative methods: averaging

Perturbative methods are at the core of our understanding of the world. They also play an important role in nonlinear dynamics. A problem to be solved using perturbative methods consists of two parts: (1) a problem we can solve exactly but which is not *our* problem; and (2) our problem, that we do not know how to solve exactly but which is 'close enough' to the exactly solvable problem. The immediate temptation is to attempt to carry the solutions of (1) into solutions of (2).

A subclass of perturbation problems consists in those cases in which events develop according to two different time-scales. Let us call these time-scales *fast* and *slow*. If we are able to solve the fast dynamics considering the slow dynamics as 'frozen', we can then look for perturbative solutions. We have already travelled this road in chapter 7 discussing adiabatic elimination of fast-decaying variables, 'reduction' to the centre manifold and normal forms. The techniques presented in this chapter belong to the same family of fast–slow decompositions and, correspondingly, the reader will find some points of contact with chapter 7.

Fast–slow decompositions, though common, are not the only kind of perturbative problem found in dynamical systems. For example, problems involving singular perturbations are of importance and will play an increasingly relevant role as the field moves towards considering questions involving spatially extended systems described by partial differential equations [mise71, chan84].

The discussion of the averaging technique that follows has been largely influenced by the works of Sanders and Verhulst [sand85], Arnold [arno88b] and Guckenheimer and Holmes [guck86].

14.1   The perturbative problem

14.1.1   The problem in standard form

Consider the following problem:

$$dx/dt = f(x, t) + \epsilon g(x, t) \tag{14.1}$$

where ϵ is a small parameter. We will call the system

$$dx/dt = f(x, t) \tag{14.2}$$

the *unperturbed* problem while equation (14.1) is the *perturbed* problem.

According to the idea sketched in the introduction we assume that we have been able to solve equation (14.2) obtaining a flow, $F(\cdot, t)$, such that

$$x(t) = F(x_0, t_0, t) \tag{14.3}$$

which gives the solution x at the time t given any initial condition x_0 at time t_0. This assumption is already quite strong; more likely we will only have approximations to the unperturbed solutions.

In the hope that the perturbed problem will have solutions that, to some extent, resemble the unperturbed ones (the smaller the perturbation, the better the resemblance), we will search for solutions of equation (14.1) letting the initial condition x_0 of equation (14.3) depend on time, i.e., $x(t) = F(x_0(t), t_0, t)$. We get

$$dx/dt = \frac{\partial F}{\partial t} + \frac{\partial F}{\partial x_0}\frac{dx_0}{dt} = f(F(x_0(t), t_0, t), t) + \epsilon g(F(x_0(t), t_0, t), t). \tag{14.4}$$

Since $\partial F(x_0, t_0, t)/\partial t = f(x, t)$, $F(\cdot, t)$ being the flow generated by the unperturbed problem, we obtain an equation for $x_0(t)$ in the form

$$dx_0/dt = \epsilon(\partial F(x_0, t_0, t)/\partial x_0)^{-1} g(F(x_0(t), t_0, t), t) = \epsilon g'(x_0, t) \tag{14.5}$$

from which we are going to launch our approximation.

14.1.2   A first view of averaging

According to Sanders and Verhulst [sand85, p 33] the averaging technique has been used at least for two centuries and was applied (on an intuitive basis!) by Lagrange and Laplace among others. Let us introduce then some intuitive ideas.

We left the previous subsection with an interest for problems of the form (dropping the primes)

$$dx/dt = \epsilon g(x, t). \tag{14.6}$$

Let us assume in addition that the vector field $g(x, t)$ oscillates with period T around a mean value $g^0(x)$ (the periodicity assumption is not strictly necessary as we will see below, but it is one of the most important cases), while ϵ is a small parameter as usual.

We can argue that the variable x will change mostly due to the average vector field $g^0(x)$:

$$g^0(x) = \frac{1}{T} \int_0^T g(x, s)\, ds \qquad (14.7)$$

The effect of the fluctuating term $g(x, t) - g^0(x)$ can be expected to be moderate, since during one period T of oscillation the variable x will have changed very little while the contributions from the fluctuating part will have gone through a complete cycle with positive and negative contributions cancelling each other. Hence, we can write an approximate equation of the form

$$dx/dt \approx \epsilon g^0(x). \qquad (14.8)$$

Moreover, since we are disregarding corrections of order ϵ to approximate $x(t)$ we can expect that the disregarded terms will accumulate an error of order ϵ during a time of order $t \sim 1/\epsilon$ (note that a correction of order ϵ to the variable x will manifest in turn in a correction of only order ϵ^2 to the time-derivative).

The situation is similar in the case of a fluctuating (but not periodic) vector field $g(x, t)$ with the only change that in this case the limit $T \to \infty$ has to be considered.

We can then announce the general theorem of averaging.

Theorem (general averaging theorem) [sand85]: Consider the initial-value problems:

$$dx/dt = \epsilon g(x, t) \qquad\qquad x(0) = x_0$$

with $g : \mathbb{R}^{n+1} \to \mathbb{R}^n$ and

$$dy/dt = \epsilon g^0(y) \qquad\qquad y(0) = x_0,$$

where $x, y, x_0 \in D \subset \mathbb{R}^n$, $t \in [0, \infty)$, $\epsilon \in (0, \epsilon_0]$ and

$$g^0(x) = \lim_{T \to \infty} \frac{1}{T} \int_0^T g(x, t)\, dt.$$

Suppose

(a) g is a Lipschitz-continuous vector field in x and t on $D \subset \mathbb{R}^n$, $t \geq 0$ with time-average g^0
(b) $y(t)$ belongs to an interior subset of D on the time-scale $1/\epsilon$.

Then,

$$x(t) - y(t) = O(\delta^{1/2}(\epsilon)) \text{ as } \epsilon \to 0 \text{ on the time scale } 1/\epsilon$$

with

$$\delta(\epsilon) = \sup_{x \in D} \sup_{t \in [0,1/\epsilon)} \epsilon \left| \int_0^t \left(g(x, s) - g^0(x) \right) ds \right|.$$

We refer the reader to Sanders and Verhulst [sand85] for the proof as well as for higher-order approximations (second-order averaging).

14.2   A non-trivial example

Before moving into the exact results we would like to present an example of the use of averaging techniques. The chosen example is certainly neither a direct application of the method nor an academic example. Although this will lead to additional complications we intend to illustrate what is involved in actual applications of averaging. The fruits of this hard work will be collected in the following section.

We consider a model for a host laser (usually a power laser) with the signal of a (low-power) laser injected in the cavity of the host. Such an experiment usually has the goal of forcing a power laser to work at the frequency of the weak (injected) laser and is usually referred as 'locking' of the signals (a case of one–one resonance). We do not consider here the most general case but only the case of a host laser that, in the absence of injected signal, can be accurately described by the rate equations derived in chapter 3 (for example a CO_2 laser). We refer the interested reader to [sieg86] for more details on lasers and to [sola94] (which we are following) for more details on this particular problem.

14.2.1   Rate equations for the laser with injected signal

Our departing point is the Maxwell–Bloch equations for a laser in which the polarization quickly relaxes to the values determined by the field and is, consequently, adiabatically eliminated following the procedures outlined in chapter 7. The model equations for the laser with injected signal read

$$
\begin{aligned}
\dot{E} &= E\,W + \mathrm{i}\,(\theta\,W + \eta)\,E + \epsilon \\
\dot{W} &= A^2 - \mu\,W\,(1 + q|E|^2) - |E|^2
\end{aligned}
\tag{14.9}
$$

where

- $E \propto$ the complex amplitude of the electric field
- $W \propto$ population inversion
- $\theta \propto$ laser detuning, proportional to the difference between the atomic frequency and the closest eigenfrequency of the cavity
- $\eta \propto$ mistuning of the external signal with respect to the operating frequency of the unperturbed laser
- $\epsilon \propto$ electric field injected in the cavity of the host per unit time
- $A^2 = W_0 - (1 + \theta^2) \propto$ intensity of the unperturbed laser field (W_0 is the corresponding population inversion of the unperturbed laser)
- $\mu \propto$ decay time of the population inversion
- g depends on a combination of various time scales for the laser.

Exercise 14.1: Perform the adiabatic elimination, i.e., obtain equation (14.9) for the case $\eta = 0 = \epsilon$ from equations (3.5)–(3.7). Steps: (1) Substitute $s \mapsto e^{\mathrm{i}w_l t}s'$ and $b \mapsto e^{\mathrm{i}w_l t}b'$. (2) Determine w_l so that the resulting equations can have fixed points ($w_l = \chi\theta/(\gamma_2+\chi)$). (3) Change variables to $E = 2gb'/\sqrt{\gamma_1\gamma_2}$, $P = 2ig^2s'/(\chi\sqrt{\gamma_1\gamma_2})$ and $\Delta = g^2W/(\chi\gamma_2)$. (4) Find the fixed points of the resulting equations. (5) Change P and Δ coordinates to their deviations from the fixed point: $R = P - (1 + i\theta)E$, $\delta = \Delta - (1 + \theta^2)$. (6) Eliminate R. (7) Rescale δ, θ and time adequately.

The unperturbed laser operates with intensity A^2 while the phase of the electric field is undetermined. The lack of an external reference for the phase is reflected in the invariance of the unperturbed equations with respect to a change of phase of the electric field ($\mathbb{S}^1$ symmetry); moreover the equations of motion equation (14.9) (with $\epsilon = 0 = \eta$) are also invariant under the reflection $E \to E^*|E|^{i2\theta}$; hence, the complete symmetry of the unperturbed case is $\mathbb{O}(2)^\dagger$. The operating point of the laser is represented by a circle of fixed points. When the external field is injected it provides the external reference for the phase of the electric field breaking the symmetry.

It is important to notice that the term proportional to θ cannot be neglected if a non-singular (i.e., meaningful) unfolding of the equations is to be achieved. On physical grounds, a perfectly tuned laser cannot distinguish between a positive or negative mistuning of the injected signal since there is only one reference frequency. On the mathematical side, this is reflected in the invariance of the unperturbed equation ($\epsilon = 0$) under the change $(E, \eta) \to (E^*|E|^{i2\theta}, -\eta)$. As a result of this invariance the non-trivial features of the dynamics can only depend on η^2 when $\theta = 0$.

Exercise 14.2: Verify the symmetry property stated above for equation (14.9) with $\epsilon = 0$. Hint: Compute $d|E|/dt$.

14.2.2   Averaged equations

The existence of the invariant circle in the unperturbed equations suggests that we can use an averaging method to extract the slow dynamics associated with the perturbation.

The fast dynamics is associated with an essentially Hamiltonian problem obtained by setting $\mu = \eta = \epsilon = 0$ in equation (14.9). The resulting equation can be written in terms of the logarithm of the amplitude of the electric field, $v = \log(|E|/A)$, the phase of the electric field, ϕ, and the population inversion, W:

$$\dot{v} = W$$
$$\dot{W} = A^2(1 - \exp(2v)) \qquad (14.10)$$

---

† The group $\mathbb{O}(2)$ of symmetry operations is generated by the reflections and the infinitesimal rotations. It takes the name from the orthogonal group in two dimensions. In the present case the rotation is given by a change of phase and the reflection by the operation $E \to E^*|E|^{i2\theta}$.

$$\dot{\phi} = \theta W.$$

The first two equations in equation (14.10) represent a Hamiltonian system with a Toda potential while the last equation is dynamically decoupled. The decoupling of the phase is possible thanks to the $\mathbb{O}(2)$ symmetry of the unperturbed problem.

The Toda oscillator and the harmonic oscillator are topologically orbitally equivalent (it is possible to map orbits of one of them to orbits of the other, although the energy-independent periods of the harmonic oscillator will become energy-dependent in the Toda problem).

In order to set up an averaging problem we need to write the equations in the standard form. To this end we transform equation (14.10) into a harmonic oscillator times the equation of a constant phase (locally) with a canonical transformation[†]. Although the transformation can always be achieved by changing into action-angle variables (a procedure that is always possible in Hamiltonian problems having one degree of freedom) such (canonical) transformations are actually known in exact form only for a few examples. We have to resort to an approximate solution of the equations, since we need to know not only the existence of the transformation but also its actual form and the form of the corresponding inverse transformation.

The change of coordinates reads

$$
\begin{aligned}
v &= V - V^2/3 - 2/3U^2 + O((U^2 + V^2)^{3/2}) \\
u &= W/(A\sqrt{2}) = U + 2/3VU + O((U^2 + V^2)^{3/2}) \\
\phi &= \psi + \theta v + \alpha
\end{aligned}
\qquad (14.11)
$$

(α being a constant to be fixed later) resulting in a set of equations for a fast harmonic oscillator of frequency $f = A\sqrt{2}$

$$
\begin{aligned}
dV/dt &= fU + O((U^2 + V^2)^{3/2}) \\
dU/dt &= -fV + O((U^2 + V^2)^{3/2}) \\
d\psi/dt &= 0.
\end{aligned}
\qquad (14.12)
$$

Note the absence of quadratic terms in equation (14.12) and that the action of the $\mathbb{O}(2)$ symmetry is now simply a shift in the phase ψ (for a pure rotation) and a change of sign of ψ (for a reflection).

The generating function $S = vU + v^2U/3 + 2U^3/9$ provides the near-identity change that is necessary to transform the Toda oscillator into a harmonic oscillator with accuracy up to order $O(U^2 + V^2)$.

The final step consists in restoring the perturbation and averaging the fast motion. We make first the change $(U, V) \to (r\cos(ft - \zeta), -r\sin(ft - \zeta))$

[†] The transformation into a harmonic oscillator is not a *necessary* step. Any other procedure to approximately solve the dynamics of the Toda oscillator would have been equivalent in principle. In practice, different integrals would have to be evaluated.

obtaining an equation similar to equation (14.5) for the initial conditions (r, ζ), which we can average as in equation (14.6) and (14.7). We leave the tedious details of the algebra as an exercise for the non-trusting and write down the final result:

$$dr/dt = -r/2\,(v + \kappa \sin(\psi + 2\alpha)) + O(r^3, \kappa^2 + v^2 + \eta^2)$$

$$d\psi/dt = \eta - \kappa \left(\cos\psi - \frac{r^2}{4\sin^2\alpha}\,(\cos(\psi + 2\alpha) - 2\sin\alpha\,\sin(\psi + \alpha)) \right)$$

$$+ \; O(r^3, \kappa^2 + v^2 + \eta^2) \tag{14.13}$$

$$d\zeta/dt = f - fr^2/6 + O(r^3, \kappa^2 + v^2 + \eta^2)$$

where we have introduced the following constants

$$\kappa = \epsilon/(A^2 \sin\alpha) \tag{14.14}$$

$$\alpha = \tan^{-1}(1/\theta) \tag{14.15}$$

$$v = \mu(1 + qA^2). \tag{14.16}$$

Note that the equations (14.13) are determined up to order r^2, which is fully consistent with the change of coordinates (14.12).

Exercise 14.3: Show that the third-order term in $\dot{r}$ is non-zero.

The averaged equations are obtained dropping the higher-order corrections in equation (14.13). We note that the equation for ζ is dynamically decoupled from the equations for r and ψ and that it represents the fast dynamics (frequency f) of the oscillation. Hence, the reduced averaged flow (in the (r, ψ) variables) is suspended in $\mathbb{S}^1$. We can then interpret the reduced equations as an approximation for the Poincaré map associated to the Poincaré surface given by any constant ζ.

Taking into account that according to equation (14.11),

$$\zeta = \tan^{-1}(U/V) \approx \tan^{-1}(W/(\ln(|E|/A)f)),$$

the surface determined by $\zeta = 0$ is roughly equivalent to the Poincaré plane $W = 0$. Strictly speaking, the Poincaré section has to be taken for $r > 0$ since the three-dimensional flow fails to be transversal to the surface $W = 0$ for $r = 0$, i.e., at the invariant circle of the unperturbed motion.

We notice that some care has to be exercised in the interpretation of the fixed points of equation (14.13). The fixed points of the flow in the (r, ψ) plane correspond to fixed points of the three-dimensional problem only if $r = 0$, while for $r \neq 0$ they correspond to periodic orbits of the full problem that intersect the Poincaré section only once (for this reason they will be called *transversal orbits*). By the same argument, periodic orbits of the reduced bi-dimensional flow correspond to periodic orbits of the full flow only if $r = 0$ while for $r \neq 0$ they are lifted into invariant torus of the three-dimensional averaged flow.

We proceed to analyse the averaged equation equation (14.13) in the following subsection. For this purpose we will restrict our attention to the reduced flow in the (r, ψ) plane. Keeping in mind that it is a suspended flow, we adopt the terminology of the three-dimensional averaged model.

14.2.3   Fixed points and bifurcations

To fix ideas, we will consider the case of positive κ only. The main features of the bifurcation diagram are shown in figure 14.1, which is drawn for $\theta > \sqrt{3}$ (this condition is called 'very large detuning') and $\nu = 1$ (note that $\nu \neq 0$ can be set to one in the averaged version of equation (14.13) by rescaling time, η and κ with $1/\nu$).

For $|\eta| > |\kappa|$ and $r = 0$, $\dot\psi$ is always non-zero. This yields a periodic orbit (the invariant circle $r = 0$) with anticlockwise phase drift for positive η and clockwise drift for negative η. We call this periodic orbit co-planar or *longitudinal* since it lies on (the border of) the Poincaré surface.

The fixed points associated with a laser locked to the external signal exist between the bifurcation lines

$$\eta = \pm\kappa, \qquad\qquad (14.17)$$

at the values $r = 0$ and $\cos\psi = \eta/\kappa$. Both lines correspond to saddle-node bifurcations occurring along the invariant circle (note that there are two values of ψ having the same cosine). Regarding the steering of the powerful laser (one of the main motivations for studying this system) it is crucial to observe that the saddle-node lines separate the region of phase rotations from the region of locked signal, as shown in figure 14.1.

The fixed points can undergo a Hopf bifurcation (which will show up as a pitchfork bifurcation in the radial coordinate of the averaged system[†]) generating periodic orbits that we also label as *transversal orbits* because they are transverse to the Poincaré surface. These bifurcations appear when

$$\sin(\psi + 2\alpha) = -\nu/\kappa$$
$$\cos\psi = \eta/\kappa.$$

Eliminating ψ we obtain a parabolic curve of Hopf bifurcations at

$$\nu^2 + \eta^2 + 2\nu\eta\,\sin(2\alpha) = \kappa^2\cos^2(2\alpha) \qquad\qquad (14.18)$$

which is contained within the region of fixed points and is tangent to the saddle-node line at

$$\begin{aligned}\eta &= -\kappa \\ \kappa &= \nu/\sin(2\alpha).\end{aligned} \qquad\qquad (14.19)$$

[†] Recall the non-zero term in r^3 present in $\dot r$.

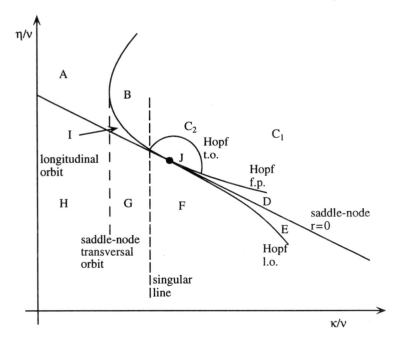

Figure 14.1. Bifurcation set for the averaged equations of the laser with weak injected signal. The singular bifurcation line is drawn for very large detuning.

The point of tangency corresponds to a co-dimension-two bifurcation (the tangency of the bifurcation sets is structurally stable in the presence of a Hopf saddle-node bifurcation) which acts as the organizing centre for the bifurcation diagram. It is important to notice that for α close to $\pi/2$, i.e., a well tuned laser with detuning $\theta \approx 0$, the co-dimension-two bifurcation is reached only with a large injected signal (κ) where the present approximation is no longer valid.

Also the periodic orbit at $r = 0$ undergoes a Hopf bifurcation when the averaged contraction in the radial direction reaches zero, i.e., when $|\eta| \geq |\kappa|$ and

$$\int_0^{2\pi} \dot{r} \, (d\psi/\dot{\psi}) = \int_0^{\tau} \dot{r} \, dt = 0. \tag{14.20}$$

This equation represents a periodic orbit with $r \neq 0$ in the (r, ψ) plane, i.e., a torus in the full system. The bifurcation set is then given by the equations

$$\nu^2 + 2\nu\eta \, \sin(2\alpha) + (\kappa \, \sin(2\alpha))^2 = 0$$
$$\nu + \eta \, \sin(2\alpha) \leq 0. \tag{14.21}$$

This curve is tangent to the line of saddle-node bifurcations at the point of the co-dimension-two bifurcation, figure 14.1. This is not surprising since at that point the period of the orbit approaches infinity and most of the time is spent in the vicinity of the fixed point.

In addition to the bifurcations close to the invariant circle $r = 0$ there are other bifurcations involving transverse periodic orbits (fixed points on the (r, ψ) plane with $r \neq 0$) which cannot be neglected. These orbits are located at

$$\sin(\psi + 2\alpha) = -\nu/\kappa$$

$$r^2 = \frac{4(\cos\psi - \eta/\kappa)\sin^2\alpha}{\cos(\psi + 2\alpha) - 2\sin\alpha\,\sin(\psi + \alpha)}.$$

We note that for κ small enough there are no such transversal orbits. They appear at

$$\begin{aligned} \kappa &= \nu \\ \eta &\leq -\nu\sin(2\alpha) \end{aligned} \tag{14.22}$$

in a saddle-node bifurcation of periodic orbits and furthermore, for $\eta = -\nu\sin(2\alpha)$, a saddle-node bifurcation is produced at the circle ($r = 0$). These orbits are transverse to the Poincaré section.

The transverse periodic orbits can undergo Hopf bifurcations, the bifurcation set being represented by

$$\sin(\psi + 2\alpha) \leq 0$$
$$\cos(\psi + 2\alpha)(\cos(\psi + 2\alpha) - 2\sin\alpha\,\sin(\psi + \alpha)) \geq 0$$
$$\kappa = -\nu/\sin(\psi + 2\alpha) \tag{14.23}$$
$$r^2 = \frac{4\sin\psi\,\sin^2\alpha}{\sin(\psi + 2\alpha) + 2\sin\alpha\cos(\psi + \alpha)} \geq 0$$
$$\eta = \kappa\left(\cos\psi - \frac{r^2}{4\sin^2\alpha}(\cos(\psi + 2\alpha) - 2\sin\alpha\,\sin(\psi + 2\alpha))\right)$$

where we used ψ as a parameter. This bifurcation curve may take different shapes depending on the value of α (in figure 14.1 it separates regions J and C).

Exercise 14.4: Verify that the tori arising from equation (14.21) are different from those of equation (14.23).

The solutions of equation (14.23) do not always lie close to $r \approx 0$. Due to the local character of our approximation (because of the approximation of the Toda potential for small values of r), the behaviour far from the bifurcating circle presents some pathology. There are orbits entering (leaving) the scene from (to) 'infinity'[†], i.e., regions far away from $r = 0$. These orbits however may remain close to the region of interest for certain parameter values and hence it is worth

[†] Also the assumption $\dot{\zeta} \approx f > 0$, on which the Poincaré map interpretation rests, fails far away from $r \approx 0$.

keeping them in mind. The singular bifurcation line given by $r^2 \to \infty$ and $\sin(\psi + 2\alpha) = -\nu/\kappa$ in equation (14.23), i.e., at

$$\kappa = \nu\sqrt{1 + \left(\frac{\sin(2\alpha)}{1 + 2\sin^2\alpha}\right)^2}, \tag{14.24}$$

signals the entrance (or departure) of periodic orbits from or towards infinity. Crossing this line, the number of transverse periodic orbits decreases or increases by one. The maximum number of transverse periodic orbits, however, is always smaller than or equal to two.

The singular line plays also another important role. For $\theta < 1$ ('small' detuning) the intersection of the singular line, equation (14.24), with the upper branch of the Hopf bifurcation for fixed points, equation (14.18), separates the Hopf bifurcations where a periodic orbit appears crossing the line of Hopf bifurcations, equation (14.18), from left to right from those where a periodic orbit disappears crossing the same line in identical form. Above this point, the singular line represents an orbit coming from infinity crossing the line with increasing κ. Below the intersection, the singular line represents an orbit escaping towards infinity.

This situation changes for $\theta > 1$ ('moderately large' detuning). The relevant intersection lies now on the lower branch of the Hopf bifurcation of fixed points. However, the dynamical interpretation remains the same.

The singular line crosses the bifurcation point of co-dimension two (see equation (14.19)) for $\theta = \sqrt{3}$, changing the character of the bifurcation.

These bifurcations have to be completed with the global bifurcations, involving a T^2 torus and heteroclinic orbits, which are expected to be associated with strange sets of the original equations [sola94]. The characteristic flows obtained for the averaged equations with very large detuning are displayed in figure 14.2.

We will stop here our discussion of the laser with injected signal, hoping to share the benefits in the next section.

14.3 The periodic case

In the previous section we have analysed to some extent the problem of a laser with injected signal using our intuitive point of view of averaging. First, we have given an interpretation of the averaged equations (14.13) in terms of intersections with a Poincaré section ($\zeta = 0$). Then, the analysis continued with the existence of fixed points, periodic orbits, tori and bifurcations involving these objects, by associating them with the bifurcations of fixed points and periodic orbits on the coordinates r, ψ of the Poincaré surface. We insisted in presenting a 'real-life' problem rather than an academic one since, except for the very small set of exactly solvable unperturbed problems (which is more or less exhausted in the literature), one can hardly go further than in our example, i.e., one has at best

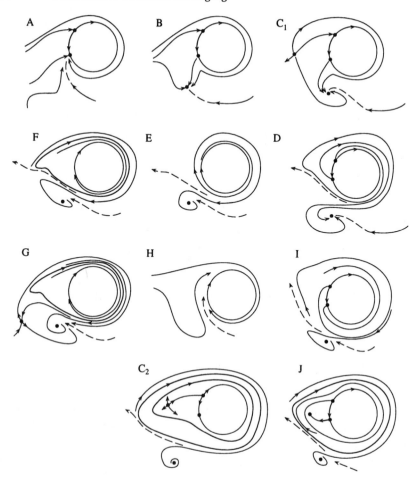

Figure 14.2. Flow types in different regions of the bifurcation diagram for the case of very large detuning. Note that the features of region C_2 and the separatrix of this region have not been discussed in the text.

an *approximate* solution of the unperturbed problem when facing the problem of averaging.

   In this section we would like to know to what extent our solutions and bifurcation sets are 'correct' and reflect the solutions and bifurcation sets of the original problem. We will present without proofs the theorems that establish the scope and validity of the analysis performed. We refer the interested reader to [guck86] for the proofs.

   Through this section we will refer to the periodic averaging problem which we assume to be already stated in the standard form equation (14.6)

$$dx/dt = \epsilon g(x, t; \epsilon)$$
$$g(x, t) = g(x, t + T)$$

with T the period of our problem and x is restricted to a bounded set $x \in U \subset \mathbb{R}^n$.

Averaging g along one period,

$$\frac{1}{T} \int_0^T g(x, s; \epsilon) \, ds = g^0(x; \epsilon), \qquad (14.25)$$

the averaged equation reads (14.8)

$$dx/dt = \epsilon g^0(x; \epsilon). \qquad (14.26)$$

The averaging theorem for the periodic case reads,

Theorem (periodic averaging theorem) [guck86]: There exists a C^r change of coordinates $x = y + \epsilon w(y, t; \epsilon)$ under which equation (14.6) becomes

$$dy/dt = \epsilon g^0(y) + \epsilon^2 g^1(y, t; \epsilon) \qquad (14.27)$$

where g^1 is periodic of period T in t. Moreover

(i)   If $x(t)$ and $y(t)$ are solutions of equation (14.6) and equation (14.26) based at x_0 and y_0 respectively at $t = 0$, and $|x_0 - y_0| = O(\epsilon)$, then $|x(t) - y(t)| = O(\epsilon)$ on a time scale $t \sim 1/\epsilon$.

(ii) If p_0 is a hyperbolic fixed point of equation (14.26), then equation (14.6) possesses a periodic orbit $\gamma_\epsilon(t) = p_0 + O(\epsilon)$ of the same stability type as p_0 for all small enough ϵ.

(iii) If equation (14.26) has a hyperbolic periodic orbit Γ, then equation (14.6) possesses a hyperbolic invariant torus for ϵ small enough.

(iv) If $x^s(t) \in W^s(\gamma_\epsilon)$ $(x^u(t) \in W^u(\gamma_\epsilon))$ is a solution of equation (14.6) lying in the stable (unstable) manifold of the hyperbolic periodic orbit $\gamma_\epsilon = p_0 + O(\epsilon)$, $y^s(t) \in W^s(p_0)$ $(y^u(t) \in W^u(p_0))$ is a solution of equation (14.26) lying in the stable (unstable) manifold of the hyperbolic fixed point p_0 and $|x^s(0) - y^s(0)| = O(\epsilon)$ $(|x^u(0) - y^u(0)| = O(\epsilon))$, then $|x^s(t) - y^s(t)| = O(\epsilon)$ $(|x^u(t) - y^u(t)| = O(\epsilon))$ for $t \in [0, \infty)$ $(t \in (-\infty, 0])$.

The main (first) statement of this theorem is a refinement of the general averaging theorem of section 1.

We can think about the second statement in terms of Poincaré (stroboscopic) sections. Let us regard the averaged and full problems in the form of equation (14.26) and equation (14.27) respectively. Since g^1 is T-periodic, we can consider the equivalent autonomous systems in $\mathbb{R}^n \times \mathbb{S}^1$:

$$\begin{aligned} \dot{y} &= \epsilon g^0(y) & \dot{\theta} &= 1 \\ \dot{y} &= \epsilon g^0(y) + \epsilon^2 g^1(y, \theta, \epsilon) & \dot{\theta} &= 1. \end{aligned} \qquad (14.28)$$

The section $\theta = 0$ will be an adequate Poincaré section. Hence, we see that the averaged flow can be regarded as an $O(\epsilon^2)$ approximation to the Poincaré first return map of the full system.

The averaged flow produces a first-return map in which p_0 is a hyperbolic fixed point. As we have shown in chapter 7, hyperbolicity of a fixed point is a sufficient condition for a fixed point to persist for all members of a parametric family of maps that are sufficiently close to the map considered (in this case the first-return map produced by the averaged flow). We can then regard the averaged flow equation (14.26) and the flow produced by equation (14.6) as members of a parametric family obtained by replacing equation (14.27) by

$$dy/dt = \epsilon(g^0(y) + \lambda g^1(y, t; \lambda)). \qquad (14.29)$$

The corresponding Poincaré maps will differ at most by a function of the order $O(\epsilon\lambda)$ for $\lambda \leq \epsilon$ by the main statement of the averaging theorem. The first-return map associated with equation (14.29) will be of the form

$$x_{n+1} = x_n + \epsilon(F_1(x_n) + \lambda F_2(x_n, \lambda)). \qquad (14.30)$$

The fixed point condition will be

$$\epsilon(F_1(x) + \lambda F_2(x, \lambda)) = 0 \qquad (14.31)$$

which for $\epsilon \neq 0$ is independent of ϵ. By the implicit function theorem if p_0 is a fixed point of equation (14.30) and $\det(DF_1(p_0)) \neq 0$ (i.e., p_0 is a hyperbolic fixed point) there is a parametric family $x(\lambda)$ of fixed points of equation (14.30) for $\lambda_- < \lambda < \lambda_+$, and $\lambda_- < 0 < \lambda_+$ with $x(0) = p_0$ (i.e., for λ in some interval (λ_-, λ_+) around zero). Taking $\epsilon < \min\{-\lambda_-, \lambda_+\}$ we can justify the claim of the theorem.

In addition, we can see from equation (14.30) that the linearized map of equation (14.30) at the fixed point p_λ differs from the linearized map associated with the averaged flow by terms of the order $O(\epsilon\lambda) \approx O(\epsilon^2)$. Hence, if p_0 is hyperbolic we can find an ϵ small enough so that p_ϵ is also hyperbolic (p_ϵ stands for the intersection of the orbit γ_ϵ with the Poincaré section).

Plausibility arguments for the third statements (3) of the theorem require a little more elaboration (see for example [sand85, hale69]), but they evolve in a similar way. A hyperbolic periodic orbit will persist for all members of a family of dynamical systems of the form equation (14.29) for adequately small ϵ and λ. The properties of Poincaré sections and first-return maps guarantee the existence of the torus for the full system. The reader may recognize this situation in the periodic orbits in the plane (r, ψ) for $r \neq 0$ described by equation (14.20) in the laser example.

We can interpret statement (4) in the following way. We know that the fixed points of the Poincaré maps associated with equation (14.26) and equation (14.6) are $O(\epsilon)$ close. Moreover, the linearized maps differ by $O(\epsilon^2)$, hence the eigenvalues of the full problem differ by order $O(\epsilon^2)$ and the eigenvectors by order[†] $O(\epsilon)$. Furthermore, by the centre manifold theorem (chapter 5) the manifolds are tangent to the stable subspaces at the fixed point. Hence, for the λ in our family, we can locally map the stable manifold of p_λ, $W^s(p_\lambda)$, in the form

$$W^s(p_\lambda) = h(y) = p_\lambda + (y - p_0) + \lambda T(y - p_0) + \lambda \text{HOT} \qquad (14.32)$$

with $y \in W^s(p_0)W$ and where T, $|T| = O(1)$, is a linear transformation. Then, we can find a neighbourhood $U \subset W^s(p_0)$ where $|h(y) - y| = O(\epsilon)$ for all λ considered.

We now split the problem into what happens inside U and what happens outside U.

For $t = 0$ we 'synchronize' the trajectories taking an initial condition $y(0)$ at the border of U for the averaged problem and $x(0) = h(y(0)) + O(\epsilon)$ for the full problem. The two initial conditions differ by $O(\epsilon)$. By the main statement of the averaging theorem for periodic problems we can continue the trajectories for negative times of order $1/\epsilon$ (iterations of the map of order $O([T/\epsilon])$) without increasing their distance beyond $O(\epsilon)$.

We need to see that, within U, the ϵ-agreement of the stable manifolds holds for $t \in [0, \infty)$. The situation is depicted in figure 14.3. We have that

$$|x_s(t) - y_s(t)| = |x_s(t) - h(y_s(t)) - (y_s(t) - h(y_s(t)))|$$
$$\leq |x_s(t) - h(y_s(t))| + |y_s(t) - h(y_s(t))|. \qquad (14.33)$$

We also know that $|y_s(t) - h(y_s(t))| = O(\epsilon)$ for all $t \in [0, \infty)$ while $|x_s(t) - h(y_s(t))|$ is $O(\epsilon)$ at $t = 0$ and goes to zero for $t \to \infty$. Moreover, $x_s(t)$ goes to p_λ exponentially and also $h(y_s(t))$ goes to p_λ exponentially, however the exponents are $O(\epsilon)$ and differ by $O(\epsilon^2)$ according to our discussion of the stability of the fixed points. The fact that $|x_s(t) - h(y_s(t))| \sim O(\epsilon)$ for all t is discussed in detail in [sand85, p 70].

Remark: the fact that the stable (unstable) manifold of the fixed point of equation (14.26) approximates the stable (unstable) manifold of the periodic orbit of equation (14.6) does not mean that heteroclinic connections between fixed points in the averaged problem translate immediately to heteroclinic connections for the periodic orbits of the full problem.

This theorem justifies part of our intuitive procedure with the laser (Poincaré section and fixed-point interpretation), for η and ϵ small enough. Let us move to the next point.

[†] The zero-order map is just the identity and does not determine eigendirections. The eigenvectors are determined by the $O(\epsilon)$ map, correspondingly, adding the second-order correction to the map produces only a first-order correction to the eigenvectors.

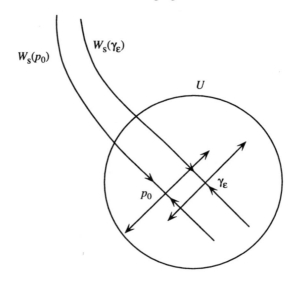

Figure 14.3. Approximation of the stable manifold by the stable manifold of the averaged system.

14.3.1 Bifurcations

It is also reasonable to expect that bifurcation lines of the averaged system will translate into bifurcation lines of the full system. For example, if a fixed point of the averaged problem changes stability in a Hopf bifurcation when crossing the corresponding bifurcation line we will have by the averaging theorem that *away from the bifurcation line* the fixed point translates into hyperbolic periodic orbits of the full system with the same stability. Correspondingly, the stability of the periodic orbit of the full system will also have to change somewhere near the bifurcation line of the averaged problem.

Theorem (averaging and bifurcations) [guck86]: If at $\mu = \mu_0$ equation (14.26) undergoes a saddle-node or a Hopf bifurcation, then, for μ near μ_0 and ϵ sufficiently small, the Poincaré map of equation (14.6) also undergoes a saddle-node or a Hopf bifurcation.

14.4 Perturbation of a Hamiltonian problem

Let us try the averaging method again in another familiar problem.

In chapter 3 we introduced a model for the experiment of a laser with modulated losses in the form of equation (3.11)

$$dI/dt = I(W - R\cos(\Omega t)) \tag{14.34}$$

$$dW/dt = 1 - \epsilon_1 W - (1 + \epsilon_2 W)I. \tag{14.35}$$

We will consider ϵ_1, ϵ_2 and R as the small parameters. The unperturbed equation is obtained by letting the small parameters be zero, giving

$$dI/dt = IW \tag{14.36}$$
$$dW/dt = 1 - I. \tag{14.37}$$

Letting $X = \ln I$ we have the Hamiltonian system

$$dX/dt = W = \partial\mathcal{H}/\partial W \tag{14.38}$$
$$dW/dt = 1 - \exp(X) = -\partial\mathcal{H}/\partial X \tag{14.39}$$

where $\mathcal{H} = W^2/2 + \exp(X) - X$ is the Hamiltonian.

According to our prescription we first put the problem in standard form letting the initial values of the unperturbed problem to depend on time. Let $(X, W) = (Q, P)(t)$ be a solution of equation (14.38) and equation (14.39), and (p, q) the initial condition. Then (p, q) will obey

$$\frac{\partial Q}{\partial q}dq/dt + \frac{\partial Q}{\partial p}dp/dt = f_Q(P, Q, t)$$

$$\frac{\partial P}{\partial q}dq/dt + \frac{\partial P}{\partial p}dp/dt = f_P(P, Q, t) \tag{14.40}$$

where $(f_Q, f_P) = (-R\cos(\Omega t), -(\epsilon_1 + \epsilon_2\exp(Q))P$ is the periodic perturbation. The matrix

$$M(t) = \begin{pmatrix} \frac{\partial Q}{\partial q} & \frac{\partial Q}{\partial p} \\ \frac{\partial P}{\partial q} & \frac{\partial P}{\partial p} \end{pmatrix} \tag{14.41}$$

has an inverse matrix of the form

$$M^{-1}(t) = \begin{pmatrix} \frac{\partial P}{\partial p} & -\frac{\partial Q}{\partial p} \\ -\frac{\partial P}{\partial q} & \frac{\partial Q}{\partial q} \end{pmatrix} = -\begin{pmatrix} 0 & 1 \\ -1 & 0 \end{pmatrix} M^\dagger \begin{pmatrix} 0 & 1 \\ -1 & 0 \end{pmatrix} \tag{14.42}$$

as a result of the Hamiltonian nature of the unperturbed problem[†].

The equations we are seeking are

$$\begin{pmatrix} dq/dt \\ dp/dt \end{pmatrix} = M^{-1}(t)\begin{pmatrix} f_Q \\ f_P \end{pmatrix}. \tag{14.43}$$

We immediately realize that equation (14.43) is not a periodic equation in general. It is periodic only if the period of the matrix $M^{-1}(t)$ is in a rational relation with the period of the perturbation, i.e., if $(P, Q)(t) = (P, Q)(t + T)$ and $T * \Omega/(2\pi) = n/m$ for some integers n and m. Notice that the period T depends (in general) on the energy $\mathcal{H}(P, Q)$ of the orbit (P, Q).

[†] The matrix $M^{-1}(t)$ can also be seen as the time-evolution matrix U associated with the trajectory that begins at (P, Q) and reaches (p, q) at the time t. Recall the discussion in chapter 2.

We also notice that equation (14.43) represents a time-dependent correction to the energy $\mathcal{H}$ of the system since

$$d\mathcal{H}/dt = \frac{\partial \mathcal{H}}{\partial Q} f_Q + \frac{\partial \mathcal{H}}{\partial P} f_P \qquad (14.44)$$

which can be obtained by direct computation or also by multiplying from the left equation (14.43) by $(\partial \mathcal{H}/\partial q, \partial \mathcal{H}/\partial p)$, i.e., a vector perpendicular to the velocity, and using that

$$\left(\frac{\partial \mathcal{H}}{\partial q}, \frac{\partial \mathcal{H}}{\partial p} \right) M^{-1}(t) = \left(\frac{\partial \mathcal{H}}{\partial Q}, \frac{\partial \mathcal{H}}{\partial P} \right). \qquad (14.45)$$

Apart from the correction to the energy, (f_Q, f_P) produces a correction term that is parallel to the unperturbed velocity, i.e., a time-dependent correction to the frequency of the orbit (P, Q). This correction can be obtained by left-multiplying equation (14.43) by $(\partial \mathcal{H}/\partial p, -\partial \mathcal{H}/\partial q)$.

Our goal is to average equation (14.43) and to infer the dynamical properties of the system using the averaged equations. We will divide our discussion into three complementary cases, namely when we are close to a fixed point of the unperturbed system, close to a resonant periodic orbit of the unperturbed system and the general case (when neither of the previous cases occur).

14.4.1   General position

In the general case we can average directly equation (14.43) or equation (14.44). In the case of equation (14.44) we obtain

$$d\mathcal{H}/dt = \lim_{T \to \infty} \frac{1}{T} \int_0^T \left(\frac{\partial \mathcal{H}}{\partial Q} f_Q + \frac{\partial \mathcal{H}}{\partial P} f_P \right) dt \qquad (14.46)$$

which after replacing the values of f_P and f_Q reads

$$\frac{d\mathcal{H}}{dt} = \lim_{T \to \infty} \frac{1}{T} \int_0^T (1 - \exp(Q)) R \cos(\Omega t) - P^2 (\epsilon_1 + \epsilon_2 \exp(Q)) \, dt$$

$$= \lim_{T \to \infty} \frac{1}{T} \int_0^T \left(-P^2 (\epsilon_1 + \epsilon_2 \exp(Q)) \right) dt < 0. \qquad (14.47)$$

The last equality is a consequence of the assumption that the period of the orbit (Q, P) with initial condition (q, p) is not in resonance with the forcing term. Actually, the necessary condition in this case is that the period, T, of (Q, P) is not of the form $T = 2n\pi/\Omega$ for any positive integer n. When this non-resonant condition is fulfilled we can expand $\exp(Q)$ in Fourier series and verify that term by term integration yields zero in all cases. In contrast, when $T = 2n\pi/\Omega$, the n-harmonic term in the Fourier expansion of $\exp(Q)$ produces a non-zero contribution. In this case we say that we are at a *subharmonic* resonance, i.e., when the system oscillates with a period that is n times that of the external driving force $2\pi/\Omega$ in this case, with $n > 1$ an integer (recall the presentation of this concept in chapter 1).

According to the general averaging theorem our approximation is valid for times of the order $O(1/\epsilon)$. We have imposed the additional requirement that the system does not move in the proximity of a subharmonic resonance.

14.4.2 Resonant layer

In a neighbourhood of a resonance the general analysis just performed is not valid. We have to adopt a slightly different strategy to treat this region of phase space.

We will consider a band of size $\epsilon^{1/2}$ around a subharmonic resonance and write

$$\begin{pmatrix} P(t) \\ Q(t) \end{pmatrix} = \begin{pmatrix} P_0(t) \\ Q_0(t) \end{pmatrix} + \epsilon^{1/2} M(t) \begin{pmatrix} x \\ y \end{pmatrix} \tag{14.48}$$

where (P_0, Q_0) is an approximate solution to the unperturbed problem and $|(x, y)| = O(\epsilon^0)$ are the first-order corrections. The equations for (x, y) are obtained from equations (3.10) and (3.11). We perform a Taylor expansion of the unperturbed problem around (P_0, Q_0) and then add the perturbation. Since $M(t)$ is the solution of the linearized problem, after using the method of variation of constants, we have

$$\begin{pmatrix} dx/dt \\ dy/dt \end{pmatrix} = \frac{1}{\epsilon^{1/2}} M^{-1}(t) \begin{pmatrix} f_{Q_0} + QT \\ f_{P_0} + QT \end{pmatrix} + \text{HOT} \tag{14.49}$$

(we will encounter this *ansatz* again later). Note that since the perturbation is of size ϵ, the right hand side is of order $\epsilon^{1/2}$ and can hence be averaged. Also note that QT stands for the quadratic terms in x, y from the Taylor expansion around (P_0, Q_0) of the unperturbed velocity field. These terms are also of size ϵ. The corresponding periodic-averaged equation is

$$\begin{pmatrix} dx/dt \\ dy/dt \end{pmatrix} = \frac{1}{\epsilon^{1/2} T} \int_0^T \left(M^{-1}(t) \begin{pmatrix} f_{Q_0} + QT \\ f_{P_0} + QT \end{pmatrix} \right) dt + \text{HOT} \tag{14.50}$$

where $T = 2n\pi/\Omega$ is the period of the n-subharmonic resonance.

One of our main interests at this point is to see whether (or under which conditions) an orbit of period T exists and approximates an orbit of the unperturbed problem. For such an end it is desirable to go backwards in the change of coordinates equation (14.48) and write the approximate condition for the existence of a periodic orbit

$$\begin{pmatrix} Q(0) \\ P(0) \end{pmatrix} = \begin{pmatrix} Q_0(T) \\ P_0(T) \end{pmatrix} + \epsilon^{1/2} M(T) \begin{pmatrix} x \\ y \end{pmatrix} (T) = \begin{pmatrix} Q(T) \\ P(T) \end{pmatrix}; \tag{14.51}$$

hence we have the condition

$$\begin{pmatrix} x \\ y \end{pmatrix}(0) = M(T)\begin{pmatrix} x \\ y \end{pmatrix}(T) \approx M(T)\left(\begin{pmatrix} x \\ y \end{pmatrix}(0)\right.$$
$$\left. + \epsilon^{-1/2}\int_0^T \left(M^{-1}(t)\begin{pmatrix} f_{Q_0}+QT \\ f_{P_0}+QT \end{pmatrix}\right)dt\right) \qquad (14.52)$$

since (Q_0, P_0) is T-periodic and $M(0)$ is the identity.

We recall that $M(T)$ has one eigenvalue equal to one associated with the velocity direction (recall chapter 2) and a second eigenvalue one if (and only if) the period T does not depend on the energy at the lowest order, i.e., $dT/dE = 0$. Multiplying equation (14.52) from the left with the corresponding eigenvector, see equation (14.45), and recalling the discussion on affine equations in chapter 2, we obtain the necessary and sufficient condition for the existence of solutions up to first order (this condition is also known as the Fredholm alternative [cour53, p 115])

$$0 = \mathcal{H}(T) - \mathcal{H}(0) = \int_0^T \left(\frac{\partial\mathcal{H}}{\partial Q_0}f_{Q_0} + \frac{\partial\mathcal{H}}{\partial P_0}f_{P_0}\right)dt. \qquad (14.53)$$

We finally observe that by the implicit function theorem the first-order solutions can be extended to the nonlinear equation and that equation (14.53) can also be read as a condition for the phase of the periodic orbit since, if $(P, Q)(t)$ is periodic for the unperturbed problem, $(P, Q)(t + \phi)$ is also periodic, hence equation (14.53) is an equation that selects (if any) the correct phase of the orbit that survives as periodic under the perturbation.

The condition equation (14.53) produces none or a multiple of n solutions. For, if $(P_0, Q_0)(t)$ satisfies equation (14.53) for a given perturbation, then $(P_0, Q_0)(t + m2\pi/\Omega)$ satisfies exactly the same equation. Actually, changing integration variables from t to $t + \phi$ the condition for the persistence of the orbit $(P, Q)(t + \phi)$ reads

$$0 = \int_0^T \left(\frac{\partial\mathcal{H}}{\partial Q_0}f_{Q_0}(t - \phi) + \frac{\partial\mathcal{H}}{\partial P_0}f_{P_0}(t - \phi)\right)dt. \qquad (14.54)$$

which is clearly periodic in ϕ with period $2\pi/\Omega$.

For our case study the condition reads

$$\int_0^T \left(-P^2(\epsilon_1 + \epsilon_2\exp(Q))\right.$$
$$\left. + R(1 - \exp(Q))\cos(\Omega t - \phi)\right)dt = 0 \qquad (14.55)$$

$$\int_0^T -P^2\left(\epsilon_1 + \epsilon_2\exp(Q)\right)dt$$
$$= \int_0^T R\exp(Q)\cos(\Omega t - \phi)dt \qquad (14.56)$$

and represents the intersection of a constant function (left side of the last equality) with a periodic function of ϕ. By increasing R the amplitude of the oscillation increases and the curves will become tangent. Immediately after, they will intersect in $2n$ points. At the tangency, a pair of saddle-node orbits of period n is created, see figure 14.4. The analysis of subharmonic resonances can be parallelled, with the appropriate changes, to study the case when $T = (q/p)(2\pi/\Omega)$ with q/p rational.

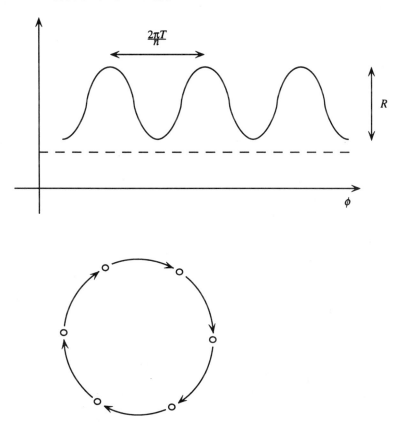

Figure 14.4. Creation of saddle-node pairs in the subharmonic resonance.

Although it is possible to study the stability and the first bifurcation of the orbits created [guck86, schw88] we will leave our discussion at this point.

14.4.3   Perturbation of fixed points

The last element for our analysis of equation (3.11) consists in finding out what happens in the vicinity of the fixed points of the unperturbed system.

The equations (14.38) and (14.39) have a fixed point for $(X, W) = (0, 0)$.

For the Hamiltonian system, the fixed point is the centre of the complete family of periodic solutions and from the point of view of linear stability theory it is a centre with undefined asymptotic stability. The frequency of oscillation around the centre is unity, i.e., the time-scale in equation (3.11) has been taken precisely to make the period of the oscillations around the unperturbed fixed point 2π.

Once again we can consider a neighbourhood of the fixed point of size ϵ and we approximate the full equation in such a neighbourhood to obtain

$$dX/dt = W - R\cos(\Omega t) \qquad (14.57)$$
$$dW/dt = -X - X^2/2 - (\epsilon_1 + \epsilon_2)W, \qquad (14.58)$$

i.e., a perturbed nonlinear oscillator. According to our general recipe, we move to the reference frame of the solution of the unperturbed problem (harmonic oscillator) with the transformation

$$X = q\cos t + p\sin t \qquad (14.59)$$
$$W = -q\sin t + p\cos t \qquad (14.60)$$

i.e., a *Van der Pol transformation* [wigg90] (the name comes from its application in a 'classical' nonlinear oscillation problem). Using again our original recipe of letting the initial conditions of the approximate solution depend on time, we obtain the equations in standard form for the pair of variables (q, p):

$$dq/dt = -R\cos(\Omega t)\cos t + (\epsilon_1 + \epsilon_2)(p\cos t - q\sin t)\sin t$$
$$+ (q\cos t + p\sin t)^2 \sin t/2 \qquad (14.61)$$
$$dp/dt = -R\cos(\Omega t)\sin t - (\epsilon_1 + \epsilon_2)(p\cos t - q\sin t)\cos(t)$$
$$- (q\cos t + p\sin t)^2 \cos t/2 \qquad (14.62)$$

which after averaging results in

$$dq/dt = -q(\epsilon_1 + \epsilon_2)/2 \qquad (14.63)$$
$$dp/dt = -p(\epsilon_1 + \epsilon_2)/2 \qquad (14.64)$$

saying that there exists a stable periodic orbit of period one for ϵ small enough.

Figure 14.5 presents an approximated idea of the flow and main bifurcations of the CO_2 laser constructed with the information obtained in this section. This picture compares well with the numeric results presented in chapter 8 and the experimental results commented on in chapter 1.

Let us close this section saying that what has been shown for the specific case of the CO_2 laser with modulated losses can be achieved by identical methods for any other Hamiltonian system having one degree of freedom.

14.5 Perturbation of homoclinic loops

We have left for the end of our discussion of the perturbation of Hamiltonian systems with one degree of freedom the perturbation of an orbit homoclinic to

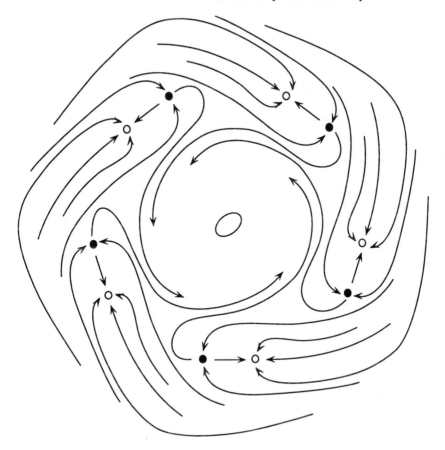

Figure 14.5. Sketch of the invariant sets in the stroboscopic map of the CO_2 laser.

a fixed point. Such an orbit is not present for the system studied in the previous section but it might be present in other important cases. The canonical example is that of the Duffing oscillator equation (3.15)

$$dQ/dt = P \tag{14.65}$$
$$dP/dt = -\beta Q - Q^3 - \mu P + A\cos(\omega t) \tag{14.66}$$

which, for $\beta = -1$ and $\mu = 0 = A$, presents a fixed point of saddle type for $(Q, P) = (0, 0)$. This point is the limit of periodic orbits of two different types (the ones confined to one side of the well and those *not* confined to one side of the well). Both types of periodic orbit are separated by a homoclinic orbit that connects the inset and outset of the fixed point.

For identical reasons to those discussed in section 3 the fixed point of the unperturbed system will become a periodic orbit of the periodically forced system and the intersection of this periodic orbit, $\gamma(t)$, with a stroboscopic section will be ϵ close to the value of the fixed point. Also, the stable and unstable manifolds can be approximated to order $O(\epsilon)$ by the stable and unstable manifolds of the unperturbed system during times $(-1/\epsilon, \infty)$ and $(-\infty, 1/\epsilon)$.

Let us then work once again keeping only the $O(\epsilon)$ corrections in all of our approximations.

Let the stable and unstable manifolds of the orbit $\gamma(t)$ be

$$\begin{aligned}
(Q^s, P^s)(t) &= (Q_0^s, P_0^s)(t) + (q^s, p^s)(t) \\
(Q^u, P^u)(t) &= (Q_0^u, P_0^u)(t) + (q^u, p^u)(t)
\end{aligned} \tag{14.67}$$

where (q, p) are of size ϵ.

The reader may have noticed the formal similarity (apart from some factors of ϵ) between the above equations and equation (14.48). We will hence repeat here the procedure of section 4.2 being now a little more explicit. Up to order ϵ the first-order corrections obey the equation

$$d/dt \begin{pmatrix} q \\ p \end{pmatrix} = \begin{pmatrix} \frac{\partial^2 \mathcal{H}}{\partial Q \partial P} & \frac{\partial^2 \mathcal{H}}{\partial P \partial P} \\ -\frac{\partial^2 \mathcal{H}}{\partial Q \partial Q} & -\frac{\partial^2 \mathcal{H}}{\partial Q \partial P} \end{pmatrix} \begin{pmatrix} q \\ p \end{pmatrix} + \begin{pmatrix} f_q \\ f_p \end{pmatrix} (t, \phi) \tag{14.68}$$

where $\mathcal{H} = P^2/2 + \beta Q^2/2 + Q^4/4$ in the case of the Duffing oscillator. Note that we have approximated $\partial \mathcal{H}/\partial Q$ and $\partial \mathcal{H}/\partial P$ by their first-order expansions in (q, p) around (Q_0, P_0).

Using the method of variation of the constants, we find the general solution of the equation (14.68):

$$\begin{pmatrix} q \\ p \end{pmatrix}(t) = M(t, t_0) \left(\begin{pmatrix} q \\ p \end{pmatrix}(t_0) \right.$$
$$\left. + \int_{t_0}^t M(t_0, s) \begin{pmatrix} f_q \\ f_p \end{pmatrix}(s, \phi) \, ds \right) \tag{14.69}$$

for a time-dependent perturbation $(f_q, f_p)(t + \phi)$. $M(t, t_0)$ is again a time-evolution matrix for the linearized unperturbed problem, where t_0 is the initial time and t the final time. Note the relation with its matrix inverse: $M^{-1}(t, t_0) = M(t_0, t)$. The above equation can be compared with the solutions for $x(t)$, $y(t)$ used in equation (14.52) recalling that since $M(t_0, t_0) = I$, the identity matrix, and (q, p) above correspond to $M(t, t_0)$ times (x, y) in equation (14.52), we have $x(t_0) \equiv q(t_0)$ and $y(t_0) \equiv p(t_0)$.

In order to find the conditions for the existence of a homoclinic orbit to the periodic orbit $\gamma(t)$ we will proceed as follows (see figure 14.6):

(i)   We will consider a local Poincaré section, Σ, to the unperturbed flow intersecting the unperturbed homoclinic orbit (the normal of Σ parallel to the unperturbed velocity at the intersection point). For example, the surface defined by $P = 0$ in our example.

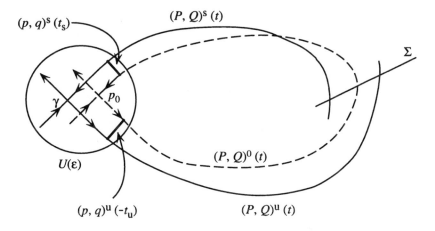

Figure 14.6. Perturbation of a homoclinic orbit.

(ii) Letting $t = 0$ be such that $(Q_0^s, P_0^s)(0)$ and $(Q_0^u, P_0^u)(0)$ lie on Σ, the times t_s and t_u are taken so that Q and P are close enough (see next point to see how close) to the fixed point along the stable (s) and unstable (u) manifolds.

(iii) Within a neighbourhood (that scales with ϵ) of the fixed point, the perturbed and unperturbed stable (unstable) manifolds differ by the amount $(q^s, p^s)(t_s)$ $((q^u, p^u)(-t_u))$ by the arguments given in section 3. Moreover, $(q^s, p^s)(t_s)$ $((q^u, p^u)(-t_u))$ can be taken in the direction of the unperturbed unstable (stable) manifold.

(iv) We seek conditions for the equality $(q^s, p^s)(\tau) = (q^u, p^u)(0)$ to hold for some $\tau = O(\epsilon)$.

Using the solutions (14.69) the condition to be satisfied is

$$\begin{pmatrix} q^s \\ p^s \end{pmatrix}(0) = M(0, t_s) \left(\begin{pmatrix} q^s \\ p^s \end{pmatrix}(t_s) + \int_{t_s}^0 M(t_s, s) \begin{pmatrix} f_q \\ f_p \end{pmatrix}(s, \phi)\, ds \right)$$

$$= M(0, -t_u) \left(\begin{pmatrix} q^u \\ p^u \end{pmatrix}(-t_u) \right.$$

$$\left. + \int_{-t_u}^0 M(-t_u, s) \begin{pmatrix} f_q \\ f_p \end{pmatrix}(s, \phi)\, ds \right)$$

$$= \begin{pmatrix} q^u \\ p^u \end{pmatrix}(0). \tag{14.70}$$

Any difference along the normal of Σ can be accounted for by adjusting τ introducing only a perturbation of order ϵ^2 in the perpendicular direction that we proceed to disregard consistently with our previous approximations.

The difference along Σ reads

$$\left(\frac{\partial \mathcal{H}}{\partial Q}, \frac{\partial \mathcal{H}}{\partial P}\right)(0)\{M(0, t_s)\left(\begin{pmatrix} q^s \\ p^s \end{pmatrix}(t_s)\right.$$

$$+ \int_{t_s}^0 M(t_s, s)\begin{pmatrix} f_q \\ f_p \end{pmatrix}(s, \phi)\, ds\left.\right) - M(0, -t_u)\left(\begin{pmatrix} q^u \\ p^u \end{pmatrix}(-t_u)\right.$$

$$+ \int_{-t_u}^0 M(-t_u, s)\begin{pmatrix} f_q \\ f_p \end{pmatrix}(s, \phi)\, ds\left.\right)\} = 0 \qquad (14.71)$$

and using equation (14.45) written in the form

$$\left(\frac{\partial \mathcal{H}}{\partial Q}, \frac{\partial \mathcal{H}}{\partial P}\right)(0)M(0, t) = \left(\frac{\partial \mathcal{H}}{\partial Q}, \frac{\partial \mathcal{H}}{\partial P}\right)(t) \qquad (14.72)$$

and the composition law

$$M(t, t')M(t', s) = M(t, s)$$

equation (14.71) simplifies into

$$0 = \int_{t_s}^0 \left(\frac{\partial \mathcal{H}}{\partial Q}, \frac{\partial \mathcal{H}}{\partial P}\right)(s)\begin{pmatrix} f_q \\ f_p \end{pmatrix}(s, \phi)\, ds$$

$$- \int_{-t_u}^0 \left(\frac{\partial \mathcal{H}}{\partial Q}, \frac{\partial \mathcal{H}}{\partial P}\right)(s)\begin{pmatrix} f_q \\ f_p \end{pmatrix}(s, \phi)\, ds$$

$$+ \left(\frac{\partial \mathcal{H}}{\partial Q}, \frac{\partial \mathcal{H}}{\partial P}\right)(t_s)\begin{pmatrix} q^s \\ p^s \end{pmatrix}(t_s)$$

$$- \left(\frac{\partial \mathcal{H}}{\partial Q}, \frac{\partial \mathcal{H}}{\partial P}\right)(-t_u)\begin{pmatrix} q^u \\ p^u \end{pmatrix}(-t_u) \qquad (14.73)$$

or

$$\left(\frac{\partial \mathcal{H}}{\partial Q}, \frac{\partial \mathcal{H}}{\partial P}\right)(t_s)\begin{pmatrix} q^s \\ p^s \end{pmatrix}(t_s) - \left(\frac{\partial \mathcal{H}}{\partial Q}, \frac{\partial \mathcal{H}}{\partial P}\right)(-t_u)\begin{pmatrix} q^u \\ p^u \end{pmatrix}(-t_u)$$

$$- \int_{-t_u}^{t_s} \left(\frac{\partial \mathcal{H}}{\partial Q}, \frac{\partial \mathcal{H}}{\partial P}\right)(s)\begin{pmatrix} f_q \\ f_p \end{pmatrix}(s, \phi)\, ds = 0. \qquad (14.74)$$

Finally, since $(p^s, q^s)(t_s)$, $(p^u, q^u)(-t_u)$ and (f_q, f_p) are bounded for $t_u, t_s \to \infty$ and $\left(\frac{\partial \mathcal{H}}{\partial Q}, \frac{\partial \mathcal{H}}{\partial P}\right)$ decreases exponentially towards zero in the same limits, the limit of the expression (14.74) is

$$0 = \int_{-\infty}^{\infty} \left(\frac{\partial \mathcal{H}}{\partial Q}, \frac{\partial \mathcal{H}}{\partial P}\right)(s)\begin{pmatrix} f_q \\ f_p \end{pmatrix}(s, \phi)\, ds. \qquad (14.75)$$

The zeros of the *Melnikov function* equation (14.75) (if they exist) determine the values of ϕ for which the homoclinic orbits in the perturbed system exist. In the case in which the Melnikov function is positive (or negative) for all values of ϕ there are no homoclinic orbits in the perturbed system.

We also notice that the Melnikov function represents the limit case of equation (14.53) which in turn determines the subharmonic periodic orbits that can be extended to the perturbed system. As discussed in chapter 9, these subharmonic orbits accumulate at the homoclinic orbit, a result that we recover in this particular case.

14.6   Hopf bifurcation at a strong resonance

In our last discussion in terms of averaging we will complete the analysis of the strong-resonant Hopf bifurcations that we started in chapter 7. Instead of considering the problem of Hopf bifurcations for maps we will approach the question by way of the averaging theorem.

Consider the bifurcation of a periodic orbit in a periodically forced problem having two eigenvalues of the monodromy matrix $\exp(\pm 2\pi i\lambda)$. Let $X(t)$ be a periodic solution of a flow $dx/dt = f(x,t) = f(x,t+T)$. T needs not necessarily be the minimum period of the vector field. We will perform our study in a neighbourhood of the periodic orbit. Hence, we change coordinates to

$$x = X(t) + M(t,t_0)\exp(-\Lambda t)y \qquad (14.76)$$

where $M(t_0+T, t_0) = \exp(\Lambda T)$ i.e., the components of x are conveniently taken as displacements in the eigendirections of the monodromy matrix $M(t_0 + T, t_0)$ and Λ is the diagonal matrix with the logarithm of the eigenvalues of the monodromy matrix. With this *Floquet transformation* the equations for y read

$$dy/dt = \Lambda y + g(y,t) \qquad (14.77)$$

where $g(y,t)$ is of size $O(y^2)$ or higher and T-periodic.

In order to further simplify equation (14.77) we need to review the theory of normal forms (chapter 7) making the appropriate extension to the case when the vector field depends explicitly on time.

As usual, we seek an almost identity change of coordinates to get rid of as many nonlinear terms as possible; hence, we write

$$y = u + h(u,t)$$

$$\left(1 + \frac{\partial h(u,t)}{\partial u}\right)\frac{du}{dt} + \frac{\partial h(u,t)}{\partial t} = \Lambda(u + h(u,t)) + g(u + h(u,t), t)$$

and look for solutions to the linear equation

$$\frac{\partial h(u,t)}{\partial u}\Lambda u + \frac{\partial h(u,t)}{\partial t} - \Lambda h(u,t) = g(u,t), \qquad (14.78)$$

in the same spirit as the corresponding equation in chapter 7. Instead of taking h simply as a polynomial in u, we introduce now the time dependence via a Fourier series expansion. Formally,

$$h_m^l(u, t) = \sum_{k=1}^{\infty} h_{mk}^l \left(\prod_j u_j^{m_j} \right) e^{iwkt} \tag{14.79}$$

where m is the polynomial degree of the l-component of h, $w = 2\pi/T$ and j runs over all components of the variable $\{u\}$ with non-negative integer exponents m_j such that $\sum m_j = m$. We will be able to eliminate all terms except when the resonant condition

$$\sum_j (m_j \lambda_j) + iwk = \lambda_l \tag{14.80}$$

is fulfilled.

For example, consider a periodic orbit in a bi-dimensional flow at a Hopf bifurcation. The eigenvalues are $e^{i\Omega T}$, $e^{-i\Omega T}$ and consequently the matrix Λ has diagonal elements $i\Omega$, $-i\Omega$. The resonant terms are of the form

$$i(m_1 - 1 - m_2)\Omega + iwk = 0 \tag{14.81}$$
$$i(m_1 + 1 - m_2)\Omega - iwk = 0 \tag{14.82}$$

where we have used the coordinates z, z^* to represent the directions which are transverse to the flow. Note that in equation (14.82) we have used the negative-k part of the Fourier expansion associated with z^*. The normal form is

$$dz/dt = i\Omega z + a|z|^2 z \tag{14.83}$$
$$dz^*/dt = -i\Omega z^* + a^*|Z|^2 z^* \tag{14.84}$$
$$d\phi/dt = 1. \tag{14.85}$$

ϕ is the time introduced as a dependent variable to make the system autonomous. Note that equation (14.81) and equation (14.82) describe the same resonance condition we have already found for the Hopf bifurcation for maps, equation (7.75). We will concentrate in the sequel on the z-equation, since the equation in z^* can be obtained from it by complex conjugation.

In the case in which w/Ω is a rational number p/q we have to add the corresponding resonant terms. Hence, according to our discussion in chapter 7, for the strong resonances at $p/q = 3$ and $p/q = 4$ the normal forms are

$$dz/dt = i\Omega z + a|z|^2 z + b(z^*)^2 e^{iwt} \tag{14.86}$$
$$d\phi/dt = 1 \tag{14.87}$$

for the resonance at $\Omega = w/3$ and

$$dz/dt = i\Omega z + a|z|^2 z + b(z^*)^3 e^{iwt} \tag{14.88}$$
$$d\phi/dt = 1 \tag{14.89}$$

for the resonance at $\Omega = w/4$.

If we change coordinates once again $z \rightarrow e^{i\Omega t}\zeta$ we obtain the averaged equations

$$\begin{aligned} d\zeta/dt &= a|\zeta|^2\zeta + b(\zeta^*)^2 \\ d\phi/dt &= 1 \end{aligned} \tag{14.90}$$

for the resonance at $\Omega = w/3$ and

$$\begin{aligned} d\zeta/dt &= a|\zeta|^2\zeta + b(\zeta^*)^3 \\ d\phi/dt &= 1. \end{aligned} \tag{14.91}$$

With the equations (14.90) and (14.91) we have achieved a decoupling of the evolution in the transversal direction from the evolution in the velocity direction, i.e., equation (14.90) and (14.91) represent a direct product of two flows or a suspended flow. We only need to study the evolution in the transverse direction which will provide the approximation for the Poincaré section.

Note that equation (14.90) and equation (14.91) are invariant under a change of phase in ζ by $2\pi/3$ and $2\pi/4$ respectively.

14.6.1 The 1/3 resonance

After a proper rescaling of ζ in equation (14.90) and the reintroduction of the linear terms (unfolding) the problem to study reads

$$d\zeta/dt = \epsilon\zeta + A|\zeta|^2\zeta + (\zeta^*)^2. \tag{14.92}$$

We already know from chapter 7 that there is an orbit of period three of the original problem which is reflected here in three fixed points related by a change of phase of $2\pi/3$.

We transform $\zeta = |\epsilon|y$ and scale time as $t = t'/|\epsilon|$. Disregarding terms of higher order in $|\epsilon|$ we obtain

$$dy/dt' = e^{i\theta}y + (y^*)^2 \tag{14.93}$$

with $\epsilon/|\epsilon| = e^{i\theta}$.

There are four special cases that can be analysed without difficulty, i.e., when $\theta = \pi/2, 3\pi/2$ and $\theta = 0, \pi$. In the first two cases, the system responds to a Hamiltonian $H = \pm(p^2+q^2) - p^3/3 + q^2 p$ for $y = q + ip$ (the minus sign corresponds to $\theta = \pi/2$).

For $\theta = \pi/2$ there are fixed points at $(q, p) = (0, -1)$ and at the points obtained rotating $(0, -1)$ by $\pm 2\pi/3$, i.e., $(\sqrt{3}/2, 1/2)$, $(-\sqrt{3}/2, 1/2)$. These points are saddles while there is another fixed point at $(0, 0)$ which is a centre. The lines of constant energy $H = 1/6$ connect the three non-zero fixed points in a cyclic heteroclinic connection (see figure 14.7).

For $\theta = 3\pi/2$ the same description holds except that the direction of the rotations has been reversed and the triangle with the fixed points has vertices at

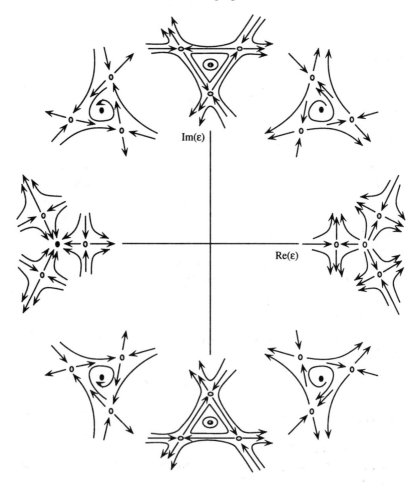

Figure 14.7. Bifurcation diagram for the resonant Hopf bifurcation at a 1/3 resonance.

$(0, 1)$, $(\sqrt{3}/2, -1/2)$ and $(-\sqrt{3}/2, -1/2)$ while the energy of the heteroclinic connection is $H = -1/6$.

For $\theta = 0$ the fixed point at the origin is a source. The system responds to a gradient system of the form

$$\mathrm{d}q/\mathrm{d}t = \partial W/\partial q$$

$$\mathrm{d}p/\mathrm{d}t = \partial W/\partial p$$

with $W = (p^2 + q^2)/2 + q^3/3 - qp^2$. The function W is a sort of Liapunov function for the system and increases continuously (decreases for negative time). Every trajectory is born 'at' a fixed point and ends up either at another fixed point or goes towards infinity. The saddle fixed points are located at $(-1, 0)$,

$(1/2, \sqrt{3}/2)$ and $(1/2, -\sqrt{3}/2)$. The axis $(q, 0)$ is invariant, as well as the lines connecting $(0, 0)$ with the saddle-points (see figure 14.7).

For $\theta = \pi$ the flow is the time-reversed version of the flow for $\theta = 0$ with the saddle fixed points located at $(1, 0)$, $(-1/2, \sqrt{3}/2)$, $(-1/2, -\sqrt{3}/2)$, and the flow is now the gradient of $W = -(p^2 + q^2)/2 + q^3/3 - qp^2$.

For general values of θ the flow represents a smooth transition between the special situations considered as depicted in figure 14.7.

Finally a word about the effect of the neglected terms. The terms neglected are of two kinds: time independent and time dependent. In the first case, the terms will modify the exact values of the transitions between different kinds of flow in figure 14.7. The overall result will be a slightly deformed picture. The time-dependent terms have the potential ability of making transverse homoclinic connections of the homoclinic cycles associated with the Hamiltonian cases by the Melnikov mechanism. There is then, according to our discussion in chapter 9, a possibility of *chaos* in these regions.

14.6.2   The 1/4 resonance

We rewrite the suspended flow equation (14.91), as in the previous subsection, restoring the linear terms and with the scalings $\zeta = |\epsilon|^{1/2} y$ and $t = t'/|\epsilon|$:

$$dy/dt' = e^{i\theta} y + A|y|^2 y + (y^*)^3. \qquad (14.94)$$

We will consider only the case $\mathrm{Re}(A) < 0$, $\mathrm{Im}(A) < 0$ to fix ideas. All the other cases can be analysed in the same form and correspond to two reflections: $A \to -A$ is the inverse-time flow and $A \to A^*$ is the same flow with y replaced by y^*, i.e., reflected.

As discussed in chapter 7 there are two different sub-cases, $|A| > 1$ where we can find none or eight fixed points in addition to $y = (0, 0)$ and $|A| < 1$ where there is always one fixed point present.

The flow far from the origin is dominated by the cubic terms. By far from the origin we really mean $Y > |y| \gg 1$ for some large Y. We can actually make $|\epsilon|^{1/2} Y > |\zeta|$ as small as we want by taking ϵ small enough.

Disregarding the linear terms and letting $w = y^2$, far from the centre we have

$$dw/dt = 2|w|^2 (Aw + w^*) \qquad (14.95)$$

which away from $w = 0$ is the flow of a vector field parallel to $Aw + w^*$. In other words, the flow of equation (14.95) is topologically orbitally equivalent to the linear flow

$$dw/dt = (Aw + w^*) \qquad (14.96)$$

which has an attractor at $(0, 0)$ when $|A| > 1$ and a saddle if $|A| < 1$. Then, globally, the region around the origin is attractive if $|A| > 1$ and of saddle type if $|A| < 1$.

We now continue the analysis for these two cases independently.

14.6.2.1 Case |A| > 1

Apart from $(0, 0)$, there are fixed points for those θ such that $-e^{i\theta}/|y|^2$ intersects a unit circle centred at A. The fixed points will exist for ϵ within a triangular region in the first quadrant $(0 < \theta_0 < \theta < \theta_1 < \pi/2)$. For $\theta = \theta_0$ and $\theta = \theta_1$ the fixed points collapse in saddle-node bifurcations.

Exercise 14.5: Show without performing the whole calculation that there will be eight fixed points for $\theta \in (\theta_0, \theta_1)$. Hint: Start by writing $y = r\, e^{i\phi}$.

Let us study the fixed point at the origin *locally* (recall that the analysis at the beginning of this subsection was *global*, i.e., for y far from zero). The eigenvalues associated with $(0, 0)$ are $e^{\pm i\theta}$, and hence for $-\pi/2 < \theta < \pi/2$ the origin is unstable while for $\pi/2 < \theta < 3\pi/2$ it is stable. For $\theta = 0, \pi$ the direction of rotation around the origin changes.

Note that for $\theta \in (-\pi/2, \pi/2)$ and $\theta \notin [\theta_0, \theta_1]$ the origin is locally unstable, globally stable and there are no other fixed points. This leads to the existence of a periodic orbit circulating around zero in y, i.e., a *torus* in the full system. Tori may also be present when $|A| < 1$ (see below).

The flow in the different zones determined by the values of ϵ or θ is depicted in figure 14.8.

14.6.2.2 Case |A| < 1

The stability of the origin changes at $\theta = \pm\pi/2$ as in the previous case, since this fact is independent of the size of $|A|$. For $|A| < 1$, apart from the origin there are other fixed points present for all values of θ.

Exercise 14.6: Show that there are four non-zero fixed points.

We can also attempt to analyse the existence of tori (periodic orbits in the transverse section) circulating around the fixed point at zero. It is convenient to rewrite equation (14.91) using $y = \rho\, e^{i\phi}$

$$d\rho/dt = \rho(\cos(\theta) + \rho^2(\mathrm{Re}(A) + \cos(4\phi))) \qquad (14.97)$$
$$d\phi/dt = \sin(\theta) + \rho^2(\mathrm{Im}(A) + \sin(4\phi)). \qquad (14.98)$$

We are facing the problem of finding a torus (i.e., a periodic orbit of small radius going around the origin in the system above) *without* being able to solve the problem in closed form. The first thing we need is to assure ourselves that a whole revolution in ϕ around the origin is possible. A condition guaranteeing this is that $d\phi/dt$ is never zero (hence it is monotonically positive or negative).

For the special case of $\theta = 0, \pi$ we observe that it is not possible to have any periodic orbit since the sign of $d\phi/dt$ changes according to whether

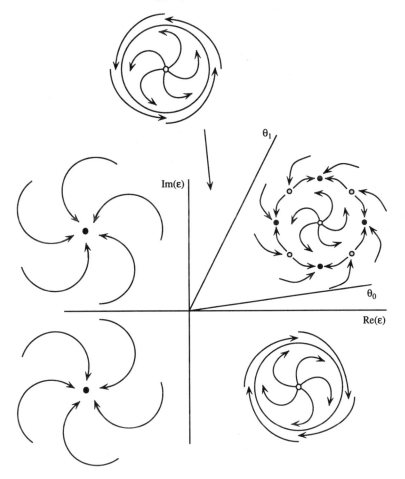

Figure 14.8. Bifurcation diagram for the resonant Hopf bifurcation at a 1/4 resonance, $|A| > 1$.

$\mathrm{Im}(A) + \sin(4\phi)$ is positive or negative, and this expression is independent of ρ. Note on passing that on the turn-around points for ϕ (the zeros of the expression above) we find the phase of the fixed points of the flow.

In contrast, if $\theta = \pm\pi/2$, ϕ increases monotonically for $\rho^2(1 - \mathrm{Im}(A)) < 1$. Here we have a chance. Note also that (apart from the periodic orbit we are looking for) the equation $\mathrm{Re}(A) + \cos(4\phi) = 0$ determines in this case the phase of the fixed points of the system.

To find the periodic orbit, we need to show also that ρ 'closes itself' after a turn of 2π in ϕ. When θ is increased passing $\theta = \pi/2$ we can argue that a periodic orbit of small radius is present by averaging $d\rho/d\phi$. This (incomplete)

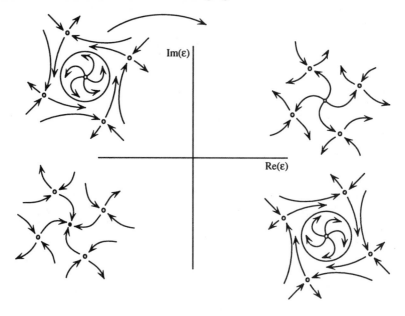

Figure 14.9. Bifurcation diagram for the resonant Hopf bifurcation at a 1/4 resonance, $|A| < 1$.

bifurcation diagram is illustrated in figure 14.9.

Exercise 14.7: Study the saddle-node, the flip and the weakly resonant Hopf bifurcations for maps in terms of suspended flows following the ideas of our discussion of Hopf bifurcations at a strong resonance.

Exercise 14.8: Challenge: Study the bifurcation at a double 1 eigenvalue with Jordan canonical form

$$\begin{pmatrix} 1 & 1 \\ 0 & 1 \end{pmatrix}. \tag{14.99}$$

Can you argue the possibility of chaos in this bifurcation in terms of Melnikov's method? Hints: Use the normal form derived in chapter 7. Look for parameter values where the flow in the transverse section is Hamiltonian.

14.7 Summary

What did we learn in this chapter? We started by noting how to rephrase a problem consisting of an easy part plus a perturbation into a new problem having only the perturbation, and showing further how to get an approximate solution by averaging the perturbation.

In the laser examples, we learned that the 'easy part' is seldom so easy. We often have to be content with an approximate solution to the unperturbed problem.

Later we addressed the case of periodic perturbations and its derivation: how the unperturbed periodic solutions change because of the perturbation, and in what conditions the (unperturbed) homoclinic orbits are preserved. Consistently with the examples, we approached the problem assuming that an approximate solution to the unperturbed problem was available. The reader may have noticed that having an *exact* solution of the unperturbed problem would have simplified much of our work, but life is never so simple....

Our last application of averaging addressed the final part of the Hopf bifurcation for maps that we left open in chapter 7.

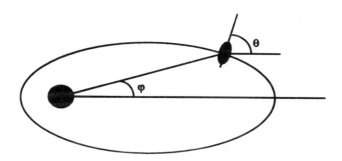

Figure 14.10. Schematic picture of an ellipsoidal satellite orbiting around a planet.

14.8   Additional exercise

Exercise 14.9: The present exercise is based on a model that consists of an orbiting satellite in the shape of a triaxial ellipsoid, with axes of inertia $C > B > A$. Assume that the largest axis of inertia is in the plane of the orbit. The orientation θ of this axis will now be a relevant physical variable (together with the coordinates (r, φ) of the orbit). A dynamical equation for the system (see figure 14.10) is

$$\theta'' + \frac{w^2}{r^3} \sin(2(\theta - \varphi)). \tag{14.100}$$

The parameter w is a measure of the deviation from sphericity, $w^2 = 3(B - A)/C$. In the case of circular orbits, $r = a$ and $\varphi' = \Omega$, with both a and

Ω constants. In this case, one can write $\phi = 2(\theta - \varphi)$ and the equation for this system will read

$$\phi'' + \frac{2w^2}{a^3} \sin(\phi) = 0. \tag{14.101}$$

If, instead of a circular orbit, the satellite follows an elliptical one, both r and φ' become known periodic functions of time (see a textbook on mechanics). Using the expressions for these functions, analyse the effect of this perturbation using the method proposed by Melnikov.

Chapter 15

Bifurcations and symmetries

15.1 Introduction

Symmetries have long been held dear to the hearts of natural scientists. One of the most elegant achievements of theoretical physics was to realize that behind every conservation law there is a symmetry involved. There is a profound reason for symmetries being so frequently found in physics: the isotropy of space and time for the absolute physical laws. This fact makes the symmetries of $\mathbb{R}^3$ (and more generally of $\mathbb{R}^n$) of primary importance in physics. In addition, it is usual to isolate a phenomenon in order to study it in the most simple configuration compatible with it. Typically, this leads to a restriction to subsymmetries of $\mathbb{R}^n$, for example through boundary conditions.

We have mentioned in chapter 1 that a nonlinear system can display a set of coexisting solutions. If this nonlinear system eventually has a symmetry, not all the solutions will necessarily reflect it. For example, it can be the case that for certain values of the parameters the system has a stable symmetric solution, and by changing the parameters one reaches a situation in which a bifurcation takes place yielding stable asymmetric solutions. Recall the discussion around the Lorenz equations in chapter 3.

Can we predict any of the qualitative features of the bifurcating solutions of a symmetric system using only symmetry considerations? This question will be the focus throughout this chapter. As the reader may expect, the answer to this question is positive. Here lie the advantages of taking symmetries into consideration...and the disadvantages as well. These problems will be discussed at the end of the chapter.

The natural language of symmetries is group theory. Therefore, the following section will deal with the presentation of some of the mathematical tools to be used throughout the chapter.

15.2 Symmetries and groups

Let us cut a piece of paper into the shape of an equilateral triangle. Which is the set of operations that leaves the triangle *invariant*? There is a trivial operation that leaves it invariant: 'leave it as it is'. We are going to call this operation the identity, I. What about non-trivial operations? We can rotate the triangle by an angle $\psi = 2\pi/3$ around an axis perpendicular to the plane of the paper. This is a second operation which we shall call r. If we apply r twice, we get another operation such that if applied to the original triangle, it also gives rise to a triangle indistinguishable from it. Apparently, we have obtained a prescription for finding infinite operations that leave the triangle invariant: successive applications of r.

The reader may have noticed that what we more precisely mean by invariant is that the triangle after the operation still remains a plane triangle (i.e., it does not break apart or twist) and that the vertices remain—as a set—in the same place (although the individual vertices may switch position with each other). Clearly, if the vertices were numbered, we would distinguish between the original triangle and the result of applying e.g., r^2 to it. Now, even if you numbered the vertices you would not be able to distinguish the original triangle from the result of applying r^3 to it: if the r operation is applied successively three times, the triangle comes out in the same position as it was originally. In other words, $r^3 = I$.

Yet, there are more operations. We could interchange two vertices leaving the third one untouched (this requires us to lift our piece of paper from the table). We shall call this operation R. Can we generate new operations with successive applications of R? It does not work, as two times R is already the identity I. What we could do is to begin to play with combinations of R and r. This gives rise to two new operations Rr and rR. In summary, there are six operations that leave an equilateral triangle invariant. A natural composition law between these operations exists, namely the successive application of the operations. What happens if we compose any two operations within this set of six? We get another operation of the set. The set of operations that leave an equilateral triangle invariant with the composition rule mentioned above constitutes an example of the algebraic structure called a *group*.

Definition: A group G is a set of elements with a law of composition $\times$ such that

- For any pair of elements a, b in G, $a \times b$ is in G
- For any set of three elements a, b, c it holds that $(a \times b) \times c = a \times (b \times c)$
- There is an element I in G such that $a \times I = I \times a = a$
- For any a in G there is a unique element a_{inv} such that $a \times a_{inv} = a_{inv} \times a = I$.

Notice that all the operations of the group we have described so far in this section have been obtained composing r and R. Therefore we shall call these two operations the *generators* of the group.

Exercise 15.1:

- Show that the set of operations and composition rule described above constitutes a group. This group is called $\mathbb{D}_3$.
- Show that the set of operations which leaves a square invariant ($\mathbb{D}_4$) has eight elements.

Exercise 15.2:

- Show that in $\mathbb{D}_3$, the set (I, R) constitutes a group.
- If the set B with the operation '$\times$' constitutes a group, and the subset $A \subset B$ with the same operation also constitutes a group, then we say that A with '$\times$' is a *subgroup* of B with '$\times$'. Are there other subgroups in $\mathbb{D}_3$?

Suppose that we cut another piece of paper with a circular shape. Now the number of operations that leave the piece of paper invariant is infinite, as any rotation is an element of the set, and the rotations can be parameterized in terms of an angle $\theta \in [0, 2\pi)$. Just as in the case of $\mathbb{D}_3$, we can pick an axis passing along a diameter of the circle and flip the piece of paper. This operation leaves the circle invariant as well. The group generated by the continuous rotations and the flip is called $\mathbb{O}(2)$.

In the following section, we shall present a physical system with a set up which has a $\mathbb{O}(2)$ symmetry. This system will be our source of questions throughout the rest of the chapter. The reader interested in getting a sound background in group theory can find a variety of texts in which this topic is treated. Among them, we suggest [gilm74] and [hame62]. Short, yet comprehensive, chapters reviewing group theoretical concepts can be found in [blum89] and [golu88].

15.3 Intensity patterns in lasers

The appearance of intensity pulsations at the output of a laser was noted immediately after the first successful operation of the ruby laser. A qualitative change in the understanding of laser physics occurred when it was realized that those oscillations were due to the nonlinear interaction between the electromagnetic field and the atoms in the cavity. Since then, lasers have become a source of questions for dynamicists as well as a test-bench for nonlinear theories [molo90].

The intensity output of a laser can not only display a complex temporal behaviour: it might also have a complex spatial structure. The pattern shown in figure 15.1 corresponds to the intensity output of a CO_2 laser, as observed when part of the output beam is sent to an infrared imaging plate [gree90].

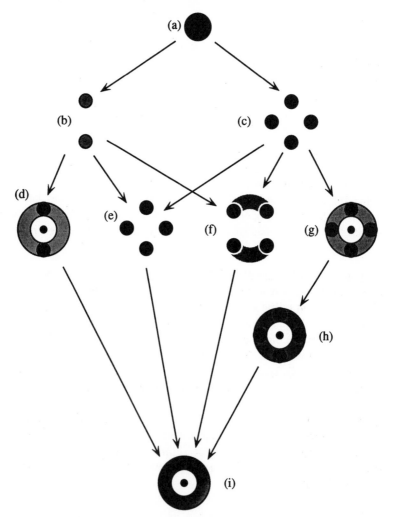

Figure 15.1. Intensity patterns (schematic) observed as part of the laser beam is sent to an infrared plate. (a) Pattern with $\mathbb{O}(2)$ symmetry. (b) Standing wave based on the mode $l = 1$. (c) Standing wave based on the mode $l = 2$. (d) Travelling plus standing waves based on the mode $l = 1$. (e) Superposition of the standing waves based on the modes $l = 1$ and $l = 2$. (f) Superposition of a 1 standing wave and a 2 standing wave. (g) Travelling plus standing waves based on the mode $l = 2$. (h) A pattern with three different frequencies and reflection symmetry. (i) Pattern without symmetry. Arrows indicate sequences of patterns reported in the experiment.

Notice that this pattern has *less* symmetry than the experimental set-up, which consists of a cylindrical tube (containing the active medium) placed between two mirrors. In order to make this statement more precise we will make use of more elements from group theory.

The setup has an obvious $\mathbb{O}(2)$ symmetry generated by the following geometric operations:

(a) $\theta \to \theta + s$ with $s \in \mathbb{R}$ and θ angular variable
(b) $\theta \to -\theta$.

The intensity pattern is invariant under the dihedral group ($\mathbb{D}_n$), which is generated by

(a) $\theta \to \theta + j\frac{2\pi}{n}$ with $j = 0, 1, \ldots, n - 1$.
(b) $\theta \to -\theta$.

$\mathbb{D}_n$ being a subgroup of $\mathbb{O}(2)$, we say that the pattern has 'less symmetry' than the set-up. This observation triggers a first question: is it in general the case that we shall observe solutions of a symmetric system which posses a symmetry that is a subgroup of the symmetry of the system? As stated, the question cannot be answered, but it can be restated in the following way. The physical system under study has an $\mathbb{O}(2)$ symmetry. The trivial solution (laser off) is invariant under this group. As parameters are changed, the solution with dihedral symmetry is born. For which subgroups of $\mathbb{O}(2)$ is it generic to find bifurcating branches of solutions that are invariant under the action of the subgroups? This is almost a well posed question. Before trying to answer it, we should point out that the symmetry to be considered is slightly larger than $\mathbb{O}(2)$. We obtained $\mathbb{O}(2)$ by 'natural' geometric arguments. Measuring the intensity in one of the bright spots, it is observed that the amplitude grows from zero with the parameters, and it oscillates with a well defined optical frequency. There are hence strong reasons to expect that the solution with dihedral symmetry is in fact a standing wave which is born in a Hopf bifurcation. Therefore, we should enlarge our geometrical symmetries to take into account the phase of the standing wave: the periodicity of the solutions we seek implies that two phase shifts will have the same effect if they differ in a period. Therefore the symmetry operations for periodic solutions bifurcating from a trivial solution of a system with geometric $\mathbb{O}(2)$ symmetry will be pairs of operations $(\gamma, \phi) \in \mathbb{O}(2) \times \mathbb{S}^1$, with $\mathbb{S}^1$ the circle group (also called $\mathbb{U}(1)$) [golu85].

Now we can state the following (answerable) question: for which subgroups of $\mathbb{O}(2) \times \mathbb{S}^1$ can we find bifurcating branches of periodic solutions (from a trivial state) having the symmetry of these subgroups? In the following section we will introduce some elements which are necessary to answer this question.

15.4   Symmetries and bifurcation theory: the tricks

In chapter 7 we outlined a programme for studying bifurcations. It included the identification of a number of marginally stable modes, the construction of nonlinear terms for those modes through a centre manifold reduction and the reduction of the equations to a normal form. What we are trying to do now is to identify, given a bifurcation problem in a system with symmetries, how much of the resulting behaviour is due to the symmetries and not to the details of the physics.

Let us assume that our bifurcation problem has been already reduced to its appropriate unfolded normal form

$$\dot{x} = f(\lambda, x) \tag{15.1}$$

with $x \in \mathbb{R}^n$, $\lambda \in \mathbb{R}$ and the Jacobian $Df(0,0)$ with eigenvalues in the imaginary axis (recall that we expect a Hopf bifurcation).

Definition: We say that the system of equations (15.1) is *equivariant* under the group Γ iff

$$f(\lambda, \gamma x) = \gamma f(\lambda, x) \tag{15.2}$$

for each $\gamma \in \Gamma$.

The equivariance of a vector field imposes strong restrictions on a bifurcation problem. First, its effect on the linear part of equation (15.1) may be that more than one eigenvalue becomes critical (i.e., its real part becomes zero) at the same parameter value(s). Second, it can force some nonlinear terms to be absent. From the point of view of the analysis, the normal form of a symmetric system is more difficult than the normal form of a non-symmetric one. Yet, symmetries can make a problem simpler. The reason is that questions as the one stated in the previous section can be solved using *algebra* [golu88], as we will see throughout the chapter.

In the next two subsections, the effect of symmetries in the linear and nonlinear parts of equation (15.1) will be discussed.

15.4.1 Symmetries and linear theory

Let us call $Df(\lambda, x)$ the Jacobian of equation (15.1). The equivariance implies that

$$\gamma Df(\lambda, x) = Df(\lambda, x)\gamma \tag{15.3}$$

for every[†] $\gamma \in \Gamma$. This commuting relationship strongly affects the structure of $Df(\lambda, x)$. Let us see in which way, making use of the example that motivates the chapter.

We are interested in analysing a Hopf bifurcation in a physical problem with $\mathbb{O}(2)$ symmetry. The spatial modes which lose stability in this bifurcation will be parameterized by amplitudes in terms of which the normal form equation (15.1) is written. We can first conjecture that, our bifurcation being a Hopf, the dimension of the centre manifold will be two. The normal form of this bifurcation was presented in chapter 7. At the bifurcation,

---

[†] If $x \in \mathbb{R}^n$ and $f(\cdot, \cdot) \in \mathbb{R}^n$ we may identify γ with its $n \times n$ matrix representation. The crucial point is that γ acts *linearly*.

$$Df = \begin{bmatrix} 0 & -w \\ w & 0 \end{bmatrix}.$$ (15.4)

This fact will help us to isolate a problem with our conjecture. The action of $\mathbb{O}(2)$ on $\mathbb{R}^2 \equiv \mathbb{C}$ is generated by

(a)   The planar rotations, i.e.

$$R_\theta = \begin{bmatrix} \cos\theta & -\sin\theta \\ \sin\theta & \cos\theta \end{bmatrix}$$ (15.5)

(b)   The reflection along one axis, say the x axis:

$$k = \begin{bmatrix} 1 & 0 \\ 0 & -1 \end{bmatrix}.$$ (15.6)

This last matrix will not commute with equation (15.4) unless $w = 0$. In other words, a system

$$\dot{x} = f(\lambda, x)$$ (15.7)

with $x \in \mathbb{R}^2$ cannot undergo a Hopf bifurcation if f commutes with $\mathbb{O}(2)$ [golu88]. This is a property of the action of $\mathbb{O}(2)$ on $\mathbb{C}$. The only 2×2 matrices that commute with every $\gamma \in \mathbb{O}(2)$ will be written as

$$\sigma = \begin{bmatrix} \lambda & 0 \\ 0 & \lambda \end{bmatrix}$$ (15.8)

with $\lambda \in \mathbb{R}$. Whenever this happens, we say that the action of the group on the vector space is *absolutely irreducible*. Hence, for our $\mathbb{R}^2$ or $\mathbb{C}$ problem, no matrix with complex eigenvalues will belong to the set of the matrices that commute with every matrix of the group.

Our first conclusion from the study of how symmetries affect a normal form is that the dimensionality of the critical space can be forced to be *bigger* than in a problem without symmetries, if the symmetry is to be broken at the bifurcation. It is also possible for the group to act trivially on the parameters of the bifurcating mode. In terms of our example, this would happen if the mode to bifurcate is the *Gaussian mode*, a mode such that its intensity pattern looks like a spot.

Exercise 15.3: Let us consider $\mathbb{O}(2)$ acting on $\mathbb{C}^2$ in the following way:

$$\gamma(z_1, z_2) = (e^{i\gamma} z_1, e^{-i\gamma} z_2)$$ (15.9)
$$k(z_1, z_2) = (z_2, z_1).$$ (15.10)

Construct the most general matrix commuting with all elements of $\mathbb{O}(2)$. Show that Hopf bifurcations are now possible.

15.4.2   Invariant theory

The equivariance condition does not only impose restrictions on the linear term of equation (15.1). It also constrains the nonlinear terms that can be included in the Taylor expansion of the vector field.

Going back to our motivating example, we are interested in the possible structure of the normal form of a Hopf bifurcation for an $\mathbb{O}(2)$-symmetric problem. We showed that unless the symmetry is acting trivially on the bifurcating modes, we need at least four real dimensions for the centre manifold. A possible action of $\mathbb{O}(2)$ on $\mathbb{R}^4 \equiv \mathbb{C}^2$ was proposed in the last exercise (in section 6 it will be clear why we are so enthusiastic about this action). The first few terms of the most general Taylor expansion of our normal form candidate up to third order (third order is enough for a Hopf bifurcation) will look like

$$\dot{z}_1 = (\sigma + i\omega)z_1 + az_1z_1 + bz_1z_1^* + \ldots \tag{15.11}$$

$$\dot{z}_2 = (\sigma + i\omega)z_2 + cz_1z_1 + dz_1z_1^* + \ldots. \tag{15.12}$$

It is clear that many of the terms written in (or implied by) the last equations will have to be absent, since they do not allow the vector field to be equivariant. For example, under the action of γ (15.9), the left hand side of equation (15.11) becomes $e^{i\gamma}\dot{z}_1$, while the first nonlinear term becomes $e^{2i\gamma}z_1z_1$. As they do not transform in the same way, a must be zero if the equation is to be equivariant under the action of the symmetry group. Instead, a term like $z_1^2 z_1^*$ in equation (15.11) can survive to the action of γ. To continue this procedure is easy though tedious. The reader can convince herself/himself by solving the following exercise.

Exercise 15.4: With the action of $\mathbb{O}(2)$ on $\mathbb{C}^2$ proposed in the last exercise, show that the equivariant normal form, to third order, for a Hopf bifurcation must be written as

$$\dot{z}_1 = (\sigma + i\omega)z_1 + (az_1z_1^* + bz_2z_2^*)z_1 \tag{15.13}$$

$$\dot{z}_2 = (\sigma + i\omega)z_2 + (bz_1z_1^* + az_2z_2^*)z_2. \tag{15.14}$$

Needless to say, if a higher order is needed, this procedure can get to be dangerously tedious. Fortunately, there is a systematic way to address the problem. The following lemma gives the key for constructing equivariant vector fields [golu88].

Lemma: Let $h : \mathbb{R}^n \to \mathbb{R}$ be a Γ-invariant function (i.e. $h(\gamma x) = h(x)$ for all $\gamma \in \Gamma$ and $x \in \mathbb{R}^n$), and let $f : \mathbb{R} \times \mathbb{R}^n \to \mathbb{R}^n$ be an equivariant vector field. Then hf intended as the product of a vector $v \in \mathbb{R}^n$ with a real number, i.e., $(hf)(\lambda, x) = h(x)f(\lambda, x)$ is equivariant.

In general, it can be proven [golu88] that for a given representation of a compact Lie group[†] Γ, there always exists a *finite* set of *invariant polynomials* $(u_1, u_2, ..., u_m)$ such that any other smooth invariant function can be written in terms of them. It can also be proven in a similar way that there always exists a finite set of equivariant polynomials $(v_1, v_2, ..., v_n)$ such that *any* equivariant vector field $f(\lambda, x)$ can be written as

$$f(\lambda, x) = \sum_1^n h_j(\lambda, x) v_j(x) \qquad (15.15)$$

where the *h*s are appropriately chosen invariant functions of $(u_1(x), u_2(x), ..., u_m(x))$.

When both the order of the normal form sought and the dimension of the amplitude space are small, it might be easy to just write down all the terms of the Taylor expansion and eliminate the terms that are not equivariant under the action of the group. If that is not the case, it is better to invest the effort in constructing the amplitude equations within the lines of the results described above.

Exercise 15.5: Consider the action of $\mathbb{O}(2)$ on $\mathbb{C}^2$ described in equations (15.9) and (15.10). Write down the most general equivariant vector field.

15.5   The treats

So far we have only faced complications. The critical eigenvalues are typically degenerate, increasing the number of modes in terms of which we have to write down the normal forms. We also saw that the nonlinear terms of the normal form will be degenerate, in the sense that not all terms will generically be present. It seems that the price to be paid for the introduction of symmetries in a problem is just too high. Fortunately, there are ways of using the symmetries of a problem to simplify its analysis. We will induce them by trying to answer the very first question that we stated at the beginning of this chapter. The question was: for a subgroup Σ of the symmetry of a problem, should we expect Σ-symmetric bifurcating solutions?

Let us begin analysing steady-state bifurcating solutions (i.e., bifurcation of fixed points, but *not* of Hopf type). As usual we assume that $x = 0$ is our fixed point and $\lambda = 0$ the bifurcation point. The subspace of $\mathbb{R}^n$ that actually 'suffers' the bifurcation is $\text{Ker}(\mathcal{L}) = \text{Ker}(\dot{y} - (Df(0,0))\, y)$, i.e., the zero subspace of the linearized system at the bifurcation values. For steady-state bifurcations this subspace is just $\text{Ker} Df = \{x : Df(0, 0) \cdot x = 0\}$. If we are interested in solutions with a prescribed symmetry Σ, we will be dealing with a subset of $\text{Ker}(\mathcal{L})$ such

[†] A Lie group is a continuous group with an analytic composition law.

that $\sigma x = x$, for all $\sigma \in \Sigma$. Notice that this vector subspace has a remarkable property: it is invariant under the dynamics. The reason is simple. Let us call $\text{Fix}(\Sigma) = \{x \in \text{Ker}(\mathcal{L}) : \sigma x = x \text{ for all } \sigma \in \Sigma\}$. Then we have:

Lemma: For all $\lambda \in \mathbb{R}$, f maps $\text{Fix}(\Sigma)$ into itself.

Proof: Let $x \in \text{Fix}(\Sigma)$, and $\sigma \in \Sigma$. Using the equivariance of f, and the definition of $\text{Fix}(\Sigma)$

$$\sigma f(\lambda, x) = f(\lambda, \sigma x) = f(\lambda, x). \tag{15.16}$$

This result shows that we can identify a vector space that is invariant under the dynamics ruled by equation (15.1). We can therefore work on this subspace of dimensionality in principle smaller than the dimensionality of $\text{Ker}(\mathcal{L})$ (or $\text{Ker}Df$ in this case).

The implications of this simple observation are deep. It is not only saying that the calculations can be made easier. It is suggesting something more powerful: we could translate our problem from the arena of the analysis to the algebraic arena as follows. We could identify the relevant marginal modes (in the language of chapter 7: choose the coordinates that reduce Df to Jordan canonical form), and compute the action of the symmetry group of our problem into the amplitudes of those modes. Then we could compute $\text{Fix}(\Sigma)$ for a given symmetry subgroup Σ. If $\dim(\text{Fix}(\Sigma)) = 1$, let us say, we should then expect Σ-symmetric steady-state bifurcating branches. In other words, the bifurcation process (or part of it at least) will develop along a one-dimensional subspace of $\mathbb{R}^n$ invariant under Σ. The beautiful result is the *equivariant branching lemma*, which we will present now as in [golu88].

Theorem (equivariant branching lemma): Let Γ be a group acting on a vector space V. If

- $\text{Fix}(\Gamma) = (0)$
- $\Sigma \subset \Gamma$ is a subgroup leaving a non-zero $v \in V$ invariant, such that $\dim(\text{Fix}(\Sigma)) = 1$
- $f : \mathbb{R} \times V \to V$ is a Γ-equivariant bifurcation problem satisfying

$$Df_\lambda(0, 0) \cdot (v_0) \neq 0$$

where $v_0 \in \text{Fix}(\Sigma)$ is non-zero, then there is a branch of solutions $(tv_0, \lambda(t))$ to the equation $f(x, \lambda) = 0$.

The first condition of the theorem is slightly less restrictive than requesting the action of Γ on V to be irreducible. Df_λ is the derivative of Df with respect to λ. The third condition can be regarded as requesting that the eigenvalues of Df cross zero with non-zero speed.

Exercise 15.6: Show that $\text{Fix}(\Sigma)$ is a linear vector space.

Exercise 15.7: Prove the equivariant branching lemma. Hint: Use that a 1-d subspace is of the form $tv_0 = v$, with $v_0 \neq 0$ and then use the implicit function theorem.

The main idea behind this theorem is that, prescribing *a priori* the symmetry of the solution sought, we could restrict the dynamics to a subspace of low dimensionality. In that subspace, the treatment of a bifurcation problem does not differ from the one of a bifurcation problem without symmetries (as discussed in chapter 7). The same programme can be carried out for the Hopf bifurcation, adapted slightly.

What we have learned so far about Hopf bifurcations in symmetric systems is that the imaginary eigenvalues could be degenerate. Let us see how symmetries can help us to tackle this problem. In the case of our motivating example (the appearance of a pattern consisting of a set of spots in a Hopf bifurcation), for every time t there is an operation (rotation of $2\pi/N$) which leaves the pattern invariant. This operation generates a subgroup of the group of symmetries of the problem ($\mathbb{O}(2)$ in the example). We shall call this subgroup a *spatial symmetry* of the solution.

On the other hand, we do not know (until we revisit the laser model in the next section) what is the dimensionality n of the vector space in terms of which the problem can be described. We just know that a two-dimensional space is not big enough. We need at least four dimensions, with the linear part of the vector field at the bifurcation point such that (1) it commutes with every $\gamma \in \mathbb{O}(2)$, and (2) the eigenvalues are complex. Regarding the action of $\mathbb{O}(2)$ on $\mathbb{R}^4$ as being the analogue of equation (15.5) and (15.6), a possibility could be

$$Df = \begin{bmatrix} 0 & -I_{2\times2} \\ I_{2\times2} & 0 \end{bmatrix}. \tag{15.17}$$

We leave to the reader to check that with the action of $\mathbb{O}(2)$ on $\mathbb{R}^4 \equiv \mathbb{C}^2$ described in (15.9) and (15.10) the most general commuting matrix looks quite different.

Another remark related to the application of the equivariant branching lemma to Hopf bifurcations deals with $\text{Ker}(Df)$ and $\text{Fix}(\Sigma)$. In the case of steady-state bifurcations (i.e., saddle-node, etc) the linearized dynamical system $\mathcal{L} \equiv \dot{y} - (Df)y$ had as zero eigenspace exactly $\text{Ker}(Df)$. The zero eigenspace of $\mathcal{L}$ is now $\text{Ker}(Df - iwI)$ instead, w being the imaginary part of the eigenvalues going through the Hopf bifurcation. However, it is still $\text{Fix}(\Sigma) \subset \text{Ker}(\mathcal{L})$ that is relevant for us.

If the subgroup Σ of the symmetry group that leaves the pattern invariant has a *two-dimensional* $\text{Fix}(\Sigma)$, it is enough to request that the eigenvalues cross at non-zero speed the imaginary axis to show the existence of a branch of solutions with spatial symmetry Σ. The reason is that the vector field f leaves $\text{Fix}(\Sigma)$ invariant, and, in that two-dimensional space, the Hopf bifurcation theorem (see chapter 7) can be applied.

Notice that the symmetric periodic solutions that arise in the Hopf bifurcation are spatially symmetric, i.e., symmetric for every t. But symmetries in the context of periodic solutions can be richer. Imagine the following set of operations:

$$(\gamma,\theta)\in\Gamma\times\mathbb{S}^1 : \gamma v(t)=v(t-\theta).$$

In order to understand their meaning, an example can help. Imagine the circular piece of paper of section 2, and an insect walking around its border. The operations described above are saying that if we take the disc at time t and we perform instantaneously a rotation γ, we shall obtain the same configuration as if we had waited certain amount of time θ. This induces us to think that a more general equivariant branching theorem for Hopf bifurcations can be stated, now not with conditions on a subgroup $\Sigma\in\Gamma$, but with conditions on $\Sigma\in\Gamma\times\mathbb{S}^1$.

Theorem (Golubitsky, Stewart and Schaeffer) [golu88]: Let $f:\mathbb{R}^n\times\mathbb{R}\to\mathbb{R}^n$ be a Γ-equivariant vector field, with linear part as in equation (15.17) at bifurcation, and eigenvalues that cross the imaginary axis at non-zero speed. Then, if the subgroup $\Sigma\in\Gamma\times\mathbb{S}^1$ has a two-dimensional $\mathrm{Fix}(\Sigma)$, there exists a unique branch of periodic solutions.

When we dealt with symmetries $\Sigma\in\Gamma$, the proof of the equivariant branching theorem for Hopf bifurcations was a simple extension of the steady-state version of the theorem: the dynamics is restricted to an invariant subspace. The theorem which we have just stated is more subtle to prove: it deals with subgroups $\Sigma\in\Gamma\times\mathbb{S}^1$, and f is equivariant under Γ, and not $\Gamma\times\mathbb{S}^1$. The problem is solved realizing that the normal form of this bifurcation (another vector field which reproduces part of the dynamics of the original one, in particular the existence of periodic solutions) does commute with $\Gamma\times\mathbb{S}^1$. The details are carefully indicated in [golu88].

In the coming section, the ideas presented so far will be used in order to revisit our motivating example.

15.6   Revisiting the laser

Our system consisted of a cylindrical laser tube placed between two mirrors. Part of the output beam was projected into infrared imaging plates, and patterns with dihedral symmetry emerged. A measurement of the intensity of the pattern in one point revealed that the pattern arose in a Hopf bifurcation.

Such a field can be expressed in terms of simple modes as follows [gree90, dang92]:

$$E = P_l(r)\,e^{iwt}(z_1\,e^{il\theta}+z_2\,e^{-il\theta}) \tag{15.18}$$

where r and θ are polar coordinates on a section of the laser tube, $P_l(r)$ is a Laguerre polynomial, z_1, z_2 are the amplitudes of the clockwise and anticlockwise circulating modes and w is the frequency of the modes.

Notice that the geometric operations that generate the symmetry group $\mathbb{O}(2)$ are

$$\theta\to\theta+\gamma$$
$$\theta\to-\theta$$

which translate into the amplitudes (z_1,z_2) as

$$(z_1, z_2) \rightarrow (e^{il\gamma} z_1, e^{-il\gamma} z_2)$$
$$(z_1, z_2) \rightarrow (z_2, z_1).$$

The action of $\mathbb{S}^1$, the phase symmetry, on the amplitudes (z_1, z_2) is given by

$$(z_1, z_2) \rightarrow (e^{i\phi} z_1, e^{i\phi} z_2).$$

Let us call (γ_1, γ_2) the elements of $\mathbb{O}(2) \times \mathbb{S}^1$. It is not difficult to check that the two-dimensional subspace $(0, z)$ is invariant under the action of the subgroup of $\mathbb{O}(2) \times \mathbb{S}^1$ which consists of (γ, γ). Another two-dimensional subspace, (z, z), is left invariant by the dihedral group $\mathbb{D}_l$.

Exercise 15.8: Show the previous assertions. Hint: $(\theta, \phi)(z_1, z_2) \rightarrow (e^{i(l\theta+\phi)} z_1, e^{i(-l\theta+\phi)} z_2)$ and $(R, \phi)(z_1, z_2) \rightarrow (e^{i\phi} z_2, e^{i\phi} z_1)$.

We leave to the reader to show that these subgroups are the only ones which leave two-dimensional subspaces fixed and are not contained in any other subgroup[†] of $\mathbb{O}(2) \times \mathbb{S}^1$.

In terms of our laser problem, the subspace $z_1 = z_2$ implies for the electric field

$$E = P_l(r) \, e^{iwt} z \, \cos(\theta) \tag{15.19}$$

i.e., a standing wave; while the subspace $z_1 = 0$ represents a travelling wave. Notice that we take one travelling wave (the one that travels to the left, in this case) as a *representative* of all the travelling waves. This is a time-saving strategy based on a simple observation: if two solutions are related by a symmetry operation, their symmetries will be also be related in a specific way by this symmetry operation. We say that the symmetries are *conjugate*. We let the reader discover the content of this concept by herself/himself in the following exercise.

Exercise 15.9: Let x_1 be a solution of equation (15.1), a Γ-symmetric system. Show that γx_1, with $\gamma \in \Gamma$ is a solution. Assume that γx_1 is different from x_1, and call Σ_{x_1}, $\Sigma_{\gamma x_1}$ the symmetries of x_1 and γx_1 respectively. Find the relation between Σ_{x_1} and $\Sigma_{\gamma x_1}$.

Here we can see the scope and limitations of the theory developed in this chapter. From the laser-off solution, the existence of a standing wave with dihedral symmetry is within the predictions of a general theory based only on the observation of the symmetries of the problem. Why is the travelling wave not stable? There is no answer to this question in the framework of the theory.

[†] The subgroups of $\mathbb{O}(2) \times \mathbb{S}^1$ are: $\mathbb{Z}_2 \times \mathbb{S}^1$, $\mathbb{S}^1 \times \mathbb{S}^1$, $\mathbb{O}(2)$, $\mathbb{D}_n \times \mathbb{S}^1$, $\mathbb{Z}_n \times \mathbb{S}^1$, $\mathbb{S}^1 \times \mathbb{Z}_n$, $\mathbb{Z}_n \times \mathbb{Z}_n$, $\mathbb{D}_n$, $\mathbb{S}^1$, $\mathbb{Z}_n$, $\mathbb{I}$. To find all the subgroups of a given group is by no means trivial.

Only the computation of the normal form coefficients from the Maxwell–Bloch equations can give a satisfactory answer to that question.

Exercise 15.10: Let $(z_1, z_2) \in \mathbb{C}^2$. Study the solutions bifurcating from $(0, 0)$ for the following system:

$$\dot{z}_1 = (\sigma + i\omega)z_1 + A(z_1 z_1^* + 2z_2 z_2^*)z_1 \qquad (15.20)$$

$$\dot{z}_2 = (\sigma + i\omega)z_2 + A(2z_1 z_1^* + z_2 z_2^*)z_2 \qquad (15.21)$$

with $\mathrm{Re}(A) < 0$. Prove that these are the normal form equations for the Hopf-bifurcating modes $\phi_1 = g(r)\, e^{i\theta}$ and $\phi_2 = g(r)\, e^{-i\theta}$ for our laser example [sola90]. Are the standing waves stable? The careful reader should get very nervous, and read the next section.

15.7   Imperfect symmetries

The existence of symmetries in a problem will imply degeneracies in any system of equations proposed to model it. This has been the bottom line of section 3. However, the reader should keep in mind that a *real* physical system will be at most *reasonably* symmetric. A systematic study, based on symmetry properties for a real system, should therefore include an investigation of the way in which *imperfections* unfold those degeneracies [dang91].

In the case of the laser, misalignments of the mirrors are a common source of imperfections. The continuous component of the symmetry $\mathbb{O}(2)$ generated by

$$R_\theta = \begin{bmatrix} \cos\theta & -\sin\theta \\ \sin\theta & \cos\theta \end{bmatrix} \qquad (15.22)$$

will not survive under the misalignments. Yet it is possible to keep the reflection symmetry $\mathbb{Z}_2$, which consists of the identity and k, where k is the operation such that $(z_1, z_2) \rightarrow (z_2, z_1)$ [dang92, lope94]. Figure 15.2 shows a set-up where this occurs.

It is possible to study qualitatively the effect of this *external* breaking of the symmetry by introducing terms in the normal form which leave the equations equivariant under the symmetries of the perturbed problem. For example, the system of equations

$$\dot{z}_1 = (\sigma + i\omega)z_1 + A(z_1 z_1^* + 2z_2 z_2^*)z_1 + \epsilon z_2$$

$$\dot{z}_2 = (\sigma + i\omega)z_2 + A(2z_1 z_1^* + z_2 z_2^*)z_2 + \epsilon z_1 \qquad (15.23)$$

is equivariant under $\mathbb{Z}_2$, since the system of equations (15.23) is in fact unchanged if $z_1 \leftrightarrow z_2$, while the equivariance under the continuous component of $\mathbb{O}(2)$ has been lost.

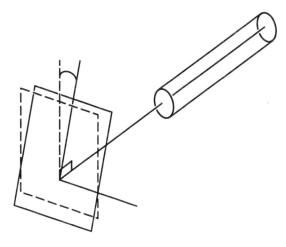

Figure 15.2. Set-up leading towards an imperfect 𝕆(2) symmetry in the laser problem.

Exercise 15.11: Study the solutions of equation (15.23). Notice that if $|\epsilon| \neq 0$, the pure travelling wave solutions do not exist. Check that for $|\epsilon|$ large enough, the standing wave solution is stable (by the way, the reader who got nervous solving the previous exercise can relax now!). Hints: (1) Substitute $z_i = \rho_i \, e^{i\phi_i}$. (2) Perform the additional change of variables: $\rho_1 = B \cos(\phi/2)$, $\rho_2 = B \sin(\phi/2)$. See exercise 7.16.

Equations (15.23) are the highlight of the chapter: they were obtained by the combined action of symmetry considerations and nonlinear techniques. These equations represent a dynamical model that captures the features of the experimental set-up that motivated it.

15.8 Summary

In this chapter we have addressed the question of how symmetries affect a bifurcation problem. The main ideas are that

- Symmetry affects the multiplicity of the critical eigenvalues.
- Symmetry forbids some nonlinear terms in the normal form.
- Symmetry allows us to predict from algebraic considerations a restricted scenario of possible symmetric solutions bifurcating from a solution with *more* symmetry.

It is worth remarking that the search for features which are generic to problems sharing a symmetry can be enlightening. Having prepared an experiment (or a model) with a symmetric set-up, how much of the observed behaviour is imposed by the scientist and how much is *intrinsic* to the phenomenon? This question must be addressed before extracting conclusions from the experiment (or model).

A last remark: symmetries imply degeneracies, and therefore it is very important to take into account the effect of the eventual imperfections which unfold the degeneracies, as they might change qualitatively the features of the solutions of the system under study.

15.9 Additional exercises

Exercise 15.12: Discuss the Hopf bifurcation in the presence of spherical symmetry. Assume that the imaginary eigenspace is a direct sum of two copies of three-dimensional irreducible representations of $SO(3)$ on the space of spherical harmonics of order one. Hints:

(a) Suppose that a complex field Φ is expanded in the *associated Legendre polynomials*:

$$\Phi = z_0 \, \cos(\theta) + z_1 \, e^{i\phi} \sin(\theta) + z_2 \, e^{-i\phi} \sin(\theta). \tag{15.24}$$

Find the actions of the following geometric operations on (z_0, z_1, z_2):

$$\phi \to \phi + \alpha$$
$$\theta \to \pi - \theta.$$

(b) Assuming that the action of the temporal symmetry is

$$\psi(z_0, z_1, z_2) \to (e^{i\psi} z_0, e^{i\psi} z_1, e^{i\psi} z_2). \tag{15.25}$$

Find the number of branches and the symmetries of the bifurcating solutions that are expected according to the equivariant Hopf theorem.

Exercise 15.13: In chapter 1 we discussed the Rayleigh–Bénard convection. A simple bi-dimensional model can be used as a test bench in order to understand important aspects of the mechanisms responsible for the onset of convective instabilities. In this simplified model, one assumes that

$$v_y(x, t) = v_{y,0}(t) \, \cos(kx) \tag{15.26}$$

and the equation ruling the behaviour of $v_{y,0}(t)$ is

$$v'_{y,0} = \epsilon v_{y,0} - v^3_{y,0} \tag{15.27}$$

with $\epsilon = R_a - R_c$, the difference between the Rayleigh number at the given temperature difference between the plates and the critical value of the Rayleigh number at which convection begins. Interpret the physical meaning of the symmetry $v_{y,0} \to -v_{y,0}$.

Exercise 15.14: When convection experiences are performed in open containers, there is an additional mechanism driving the fluid: as colder regions of the free surface experience larger surface tensions, fluid particles tend to escape from hot regions to move towards colder ones.

Based on this observation, show that a general dynamical equation for a convection mode in system with an upper free surface reads as follows:

$$A' = \epsilon A - A^3 + \alpha_1 + \alpha_2 A^2 \qquad (15.28)$$

where A denotes the amplitude of the convection mode and ϵ is a function of the temperature of the lower plate.

Exercise 15.15: We continue our discussion of the convection in open containers (upper free surface). In actual experiments, the lower plate is in contact with a source that provides heat at a constant rate S. The dynamics of the temperature ϵ can be modelled by

$$\epsilon' = -\mu_1 \epsilon + C_1(S - \beta A^2) \qquad (15.29)$$

where A denotes the amplitude of a convection mode, and μ_1, C_1 and β are constant positive parameters. This equation can be interpreted as follows. The temperature increases if there is no convection, and as soon as the convection is established it tends to decrease (this effect is due to the fact that convection takes hot fluid away, and replaces it with cold fluid).

(a) Show the existence of α_1 and α_2 such that the coupled system

$$A' = \epsilon A - A^3 + \alpha_1 + \alpha_2 A^2 \qquad (15.30)$$
$$\epsilon' = -\mu_1 \epsilon + C_1(S - \beta A^2) \qquad (15.31)$$

has periodic solutions.
(b) Draw the periodic solutions for $\mu_1 \to 0$.

Exercise 15.16: Let $x(t)$ be a solution of a Γ-equivariant autonomous problem. For $\gamma \in \Gamma$ show that

- $\gamma(x(t))$ and $(\gamma x)(t)$ are solutions of the problem.
- Let $\Sigma \subset \Gamma$ be the subgroup of symmetry actions, σ, that leave the orbit of x invariant, i.e., $x(t) = \sigma(x(t_\sigma))$ for some t and t_σ. Show that Σ is a subgroup of Γ and that $t - t_\sigma$ depends only on σ.
- Show that if σ is of finite order ($\sigma^n = Id$) and $(t - t_\sigma) \neq 0$ then $x(t)$ is periodic, with period T, and $(t - t_\sigma)$ divides the period.

Chapter 16

Global bifurcations: II

16.1 Introduction

In chapters 9 and 10 we studied the effects of having a transverse crossing of the stable and unstable manifolds of a fixed point on a 2-d map. We observed that under certain general conditions the map behaved like a horseshoe, and that the dynamics of a horseshoe is complex enough to keep us busy for a long time. The natural question is: can we translate this phenomenon to flows?

What was needed in chapters 9 and 10 to obtain a horseshoe was a map having a point such that forward and backward iterates of the point converged to the (saddle, hyperbolic) fixed point, thus organizing the system. Does an analogous orbit exist in flows? In such a case, will we always find horseshoes or some similarly complex structure associated with such an orbit?

These questions are the motivation for the present chapter. We will review the simplest case, two-dimensional flows, where we already know that the possible limit sets are fixed points, periodic orbits and eventually lines connecting fixed points. We know therefore in advance that there will be no horseshoes present, but we will gain more experience in dealing with manifolds that (unlike those of maps) cannot cross. Then we will go over to three dimensions, finishing with a new, deeper, visit to our laser system.

16.2 Homoclinic orbits of maps and flows

As mentioned above, the horseshoe construction on a two-dimensional map required the existence of a point such that both forward and backward iterations of the map based on that point would tend to the hyperbolic saddle fixed point organizing the system. What is the *analogue* to such a point in flows?

One answer to this question could be just to take the sentence above and replace 'forward iterations of the map' by 'positive time-evolution of the flow', and the same for 'backward' and 'negative time-evolution'. We will (naively) adopt this answer throughout this chapter, although this choice is not free from subtleties.

What we obtain is an *homoclinic orbit* to a fixed point, i.e., an orbit such that all points on it have as positive-time limit (this is also called ω-limit, recall chapter 6) and also as negative-time limit (α-limit) the fixed point in question.

In the spirit of chapter 9 we will study homoclinic orbits that are present in parametric families of flows. The orbit may perhaps be present only for a given value μ_0 of the parameter, so we will study the dynamics in a small region $\mu_a < \mu_0 < \mu_b$ in parameter space around the critical value.

16.2.1   Manifold organization around a homoclinic orbit

It may be argued that a fixed point of a map corresponds in fact to a *periodic orbit* of a flow (this is the subtlety mentioned above). This is true at least for Poincaré maps: the suspended flow will have a periodic orbit associated to the fixed point of the map. In the case of flows, the stable and unstable manifolds of a fixed point cannot *cross*. A transverse crossing such as those depicted in chapter 9 and 10, if it were possible, would violate the unicity of solutions of differential equations. We came across this fact also in chapter 12. This property constitutes the main difference between homoclinic orbits to fixed points of flows and homoclinic orbits to fixed points of maps.

What on the other hand may happen is that the stable and unstable manifolds of a fixed point coincide, i.e., that they are just the *same* curve that comes 'out of' the fixed point along the local unstable direction and re-enters along the local stable direction. This is exactly the case for homoclinic orbits in two-dimensional flows, see figure 16.1. Fixed points have in this case two associated eigenvalues. A hyperbolic saddle will hence have one eigendirection associated with the stable eigenvalue and the other associated with the unstable one. Homoclinic connections are thus possible.

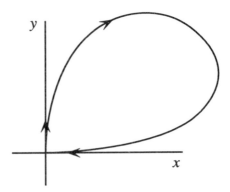

Figure 16.1. A homoclinic orbit on a two-dimensional flow.

A further complication arises in three-dimensional flows. Fixed points will have now *three* associated eigenvalues. A hyperbolic saddle has then either two eigenvalues with negative real parts and one positive real eigenvalue or the other way around (one real negative and two with positive real parts). Still, the one-dimensional manifold cannot cross transversely the 2-manifold, but it may become a *part* of it. This situation is depicted in figure 16.2. Because of the unicity requirement when two manifolds (or in general two orbits) coincide at one point (which is not a fixed point), they coincide along the whole (positive and negative time) orbit based at that point.

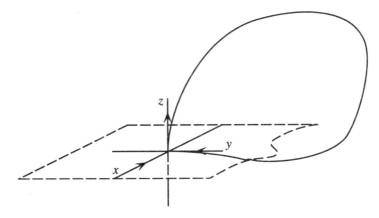

Figure 16.2. A homoclinic orbit on a three-dimensional flow.

16.2.2 Manifold organization around periodic orbits of flows

The manifold organization for homoclinic orbits to a fixed point of flows is clearly quite different from the 'analogue' for maps. The behaviour of map manifolds can be recovered when we consider the stable and unstable manifolds associated with *saddle periodic orbits* of 3-d flows.

Consider a small disc transverse to the orbit at one point. If the disc is small enough and the flow is good enough, initial conditions on the disc will return to the disc after one excursion with the flow along the periodic orbit. The local return map thus defined on the disc has a fixed point associated with our periodic orbit. The eigenvalues of the monodromy matrix associated with the orbit control the stability of the fixed point of the map (in fact, the linearization of the return map is given by the monodromy matrix, see chapter 5). Different points along the orbit yield different (but conjugated) return maps, with the same linearization (in suitable coordinates).

What do the stable and unstable manifolds of the periodic orbit look like? At each point of the orbit, we will have three directions: the direction of the flow, tangent to the orbit, and the stable and unstable directions on the disc. So the stable and unstable manifolds of the periodic orbit are in fact *both*

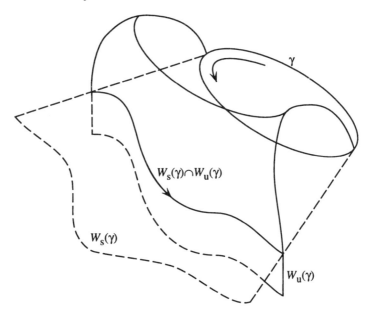

Figure 16.3. Possible organization of the manifolds of a periodic orbit on a three-dimensional flow.

two-dimensional! They are, locally, the direct product of the orbit times the eigenvector along the stable (or unstable) direction at each point.

16.2.2.1 Intersection of manifolds

In 3-d spaces, a 2-d manifold is a hypersurface. Having *two* such manifolds may have interesting consequences. It may exceptionally happen that the manifolds never cross, as it would be the case for a periodic orbit on the xy plane having the z axis as the stable direction and the plane of the orbit as the unstable direction. The other, more general, possibility is that the manifolds cross, as in the example of figure 16.3.

Let us verify that this situation does not violate the unicity requirement for differential equations. If the manifolds cross at one point, unicity demands that the whole orbit of that point will have to belong to both manifolds. The intersection of the manifolds is, hence, a one-dimensional curve, and this curve tends to the periodic orbit both for positive and negative times. Whenever this curve crosses the control disc, we will have a point of intersection between the stable and unstable manifolds of the fixed point of the return map.

Forcing our imagination beyond the simplicity of figure 16.3 we may realize that this first 'innocent' crossing implies that the manifolds have to fold wildly among themselves and have in fact infinitely many crossings. The intersection

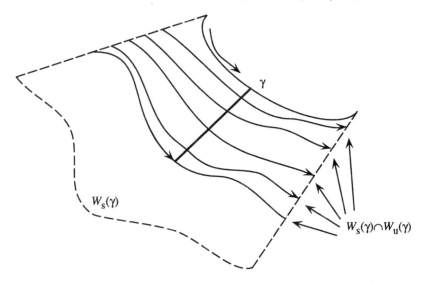

Figure 16.4. Detailed view of the intersection of manifolds of a periodic orbit in a 3-d flow.

curve tends to the periodic orbit for $t \to \infty$. Consider a straight line along the stable manifold drawn from the orbit to our intersection curve, such as the one shown in figure 16.4. The line will in fact cross the intersection curve infinitely many times, since the curve approaches the periodic orbit along the stable manifold. Since *each* point of the curve belongs to the unstable manifold as well, this manifold has to fold in order to cross the stable manifold infinitely many times along the line. This discussion can be repeated switching the roles of the manifolds and considering $t \to -\infty$ instead. We recover, hence, the infinitely many crossings of the manifolds in the return map.

16.3 Homoclinic orbit in two-dimensional flows (to a fixed point)

What happens to a family of flows having a homoclinic orbit such as the one depicted in figure 16.1 ? Let us consider the problem in detail. After a proper change of coordinates, we can assume that the fixed point is located at the origin, that the expanding direction associated with the positive eigenvalue λ is the y axis, and that the contracting direction with eigenvalue $\gamma < 0$ is the x axis. We can safely assume that our parametric family of flows $f(x, y, \mu)$ is such that for $\mu = 0 (\equiv \mu_h)$ there is a homoclinic orbit.

By a continuity argument, we can expect that (at least part of) a small line segment $\Pi_1 = (-\epsilon_x < x < \epsilon_x, y = \epsilon_y)$ will be mapped by the flow in the surroundings of an equivalent segment perpendicular to the x axis, i.e., $\Pi_0 = (x = \eta_x, -\eta_y < y < \eta_y)$. All ϵ and η are assumed to be small positive numbers. It is clear that at least the homoclinic orbit is mapped approximately in this way. Moreover, taking ϵ and η very small, we make no serious approximation by describing the flow in the region between Π_0 and Π_1 with the (hyperbolic) linear part and hence some points in Π_0 will pass through Π_1 as well.

We notice that the flow is 'organized' so that points in Π_0, in a small (x, y, μ) neighbourhood of the homoclinic orbit, flow through Π_1 and back to Π_0 again. We may hence represent the flow by an approximate return map on Π_0. We can describe this map by the composition of a *local map* between Π_0 and Π_1 and a *global reinjection* mapping Π_1 back to Π_0.

The local map will be approximately given by the linearized flow, ($\dot{x} = \gamma x, \dot{y} = \lambda y$). Let us integrate the linearized system to obtain

$$x(t) = x_0\, e^{\gamma t} \tag{16.1}$$

$$y(t) = y_0\, e^{\lambda t}. \tag{16.2}$$

Hence, any point with initial coordinate (η_x, y) on Π_0 will cross Π_1 after a time of flight $F = (1/\lambda)\ln(\epsilon_y/y)$, appearing at the point $x = \eta_x\, e^{\gamma F}$ on Π_1.

Since we have the homoclinic connection, we may expect that the flow outside the linear part will approximately map the positive-x part of Π_1 onto the positive-y part of Π_0[†] stretching it by a positive factor B and displacing it by the quantity μ, i.e., the bifurcation parameter. One usually says that the bifurcation parameter μ *controls* the global reinjection of the unstable manifold.

Exercise 16.1: Show that the factor B cannot be negative for a two-dimensional flow.

What does our approximate map look like? Points y on Π_0 map to $x = \eta_x(y/\eta_y)^\delta$ where $\delta = |\gamma/\lambda|$ and then back to Π_0, B-stretched and μ-shifted:

$$y_{n+1} = \mu + B\eta_x(y_n/\eta_y)^\delta. \tag{16.3}$$

After a suitable rescaling of y to $y' = cy$ and setting $\mu' = \mu/c$ with $c = (\eta_x/\eta_y^\delta)^{1/(\delta-1)}$, we have, dropping the primes,

$$y_{n+1} = \mu + y_n^\delta. \tag{16.4}$$

The fixed points of this map will correspond to periodic orbits of the whole flow. We see that fixed points occur whenever $f(y) = y - y^\delta$ equals μ, i.e., at the intersection of the curve $f(y)$ with a straight line. The two possible cases, for $\delta > 1$ and $\delta < 1$, are shown in figure 16.5.

[†] What happens to the points that land on Π_0 with negative y is not interesting, since the flow takes them far away from our zone of interest.

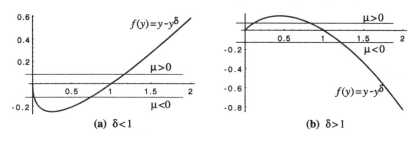

Figure 16.5. Return map near a homoclinic orbit of a 2-d flow. (a) $\delta < 1$. (b) $\delta > 1$.

The general conclusion that can be extracted is that for parameter values that are sufficiently close to the homoclinic condition $\mu = \mu_h$ the system will have a periodic orbit. The orbit occurs for $\mu > \mu_h$ when $\delta > 1$ and for $\mu < \mu_h$ when $\delta < 1$. Note that in this derivation we took $\mu_h = 0$. For other choices of the bifurcation parameter, the shift in the return map equation (16.4) should be changed to $\mu - \mu_h$.

The derivative of the approximate return map at the fixed point value yields the stability of the created orbit. For $\delta < 1$ and sufficiently small μ the created orbit will be unstable and, conversely, it will be stable for $\delta > 1$.

Exercise 16.2: Prove the previous statements about stability.

Exercise 16.3: How does this procedure (and the result) change if the system has reflection symmetry along the y axis or inversion symmetry around the origin?

A couple of remarks are proper before finishing this section. First, the possibility of using a linear approximation of the flow near the fixed point is established by the Hartman–Grobman theorem (chapter 5) and is valid for y sufficiently small. Secondly (but related to the previous remark), following an excursion in $\mu \geq 0$ for $\delta > 1$, one encounters the following scenario: no periodic orbits for large values of μ. A pair of periodic orbits is born in a saddle-node bifurcation for smaller μ. One orbit drifts towards large y ('infinity') while the other approaches zero, becomes homoclinic and disappears at $\mu = 0$. The sequence is reversed with the condition on δ. Note that much of this picture is a result of the approximations we have made. The only thing we can say with certainty is that our approach is valid sufficiently close to $y = 0$ and therefore only the disappearance of a periodic orbit after becoming homoclinic can be assured without doubt. For large y, the model may be more or less valid according to the particular global shape of our dynamical system. It is not determined in this context what is the 'past' of the periodic orbit.

16.4   Homoclinic orbit in three-dimensional flows (to a fixed point)

Encouraged by the (hopefully) simple approach of the 2-d case, let us jump one step in the dimension universe and consider a 3-d problem. The first thing that comes to the (author's) mind is that here the fixed point where the homoclinic orbit is based will have three eigenvalues determining its stability. Instead of having a *unique* way of describing a hyperbolic saddle (as in 2-d) we have now four possibilities:

(i)   One real positive eigenvalue and two real negative eigenvalues.

(ii) One real positive eigenvalue and two complex conjugated eigenvalues with negative real part.

(iii) One real negative eigenvalue and two real positive eigenvalues.

(iv) One real negative eigenvalue and two complex conjugated eigenvalues with positive real part.

The last two cases follow from the first two by making the substitution $t \mapsto -t$, so we will only consider two different examples. The curious reader may want to give a second thought to the word 'different'.

16.4.1   Homoclinic orbit with real eigenvalues

The appearance of a homoclinic orbit to a saddle fixed point with real eigenvalues was discussed in chapter 8 in connection with the Lorenz system. Lorenz' problem has an additional feature, namely that *two* homoclinic orbits appear simultaneously, each one connected to one of the (two) branches of the local unstable manifold of the fixed point. Let us try first the more general case of the existence of only one homoclinic orbit in order to build later on it towards a better understanding of the Lorenz system.

To begin with, the generic case will be that one of the negative eigenvalues is larger than the other (smaller in modulus). We will call it λ_2 while λ_1 will be the positive eigenvalue. The homoclinic orbit will hence approach the origin tangentially to $\hat{e}_2$, the eigenvector associated with λ_2, since the larger negative exponent in the third direction forces trajectories to a faster collapse towards zero.

Let us try what we learned in the previous section. We consider a parametric family of flows having a fixed point at the origin and a homoclinic orbit for $\mu = 0$. Again, we propose a partition of the problem in a detailed local (linearized) flow and a typical global reinjection. Figure 16.6 displays the scenario.

We consider two small planes Π_0 and Π_1 perpendicular to the directions $\hat{e}_2$ and $\hat{e}_1$ respectively, and at distances ϵ, η of the fixed point that are small enough to describe the local part of the flow with its linearization around the fixed point.

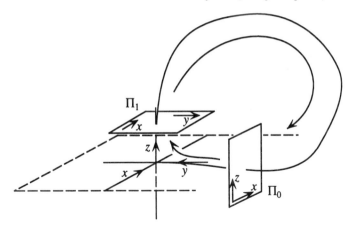

Figure 16.6. Set-up for the local/global approximation near a homoclinic orbit.

Points in Π_0 will have coordinates (x, ϵ, z). The time of flight F is such that $z(t = F) = z\, e^{\lambda_1 F} = \eta$, just as in the 2-d case. We obtain $F = (1/\lambda_1)\ln(\eta/z)$ and consequently a local map P_0:

$$P_0 : \Pi_0 \mapsto \Pi_1 \tag{16.5}$$

$$P_0(x, z) = (x(z/\eta)^{\delta_3}, \epsilon(z/\eta)^{\delta_2}) \tag{16.6}$$

where $\delta_i = |\lambda_i/\lambda_1|$, $i = 2, 3$.

Sufficiently close to the homoclinic orbit (both in phase space and in parameter space) we may expect that the flow will transport a small disc on Π_1 onto Π_0 again. This will define a global map $P_1 : \Pi_1 \mapsto \Pi_0$. By continuity, we can assume that the smaller the disc the less it will be deformed by this excursion, so we can just approximate P_1 with an affine transformation. The image of the disc will hence be stretched, rotated and shifted. The homoclinic orbit occurs for $\mu = 0$, so for that value the eigendirection $\hat{e}_1$ maps to $\hat{e}_2$, so we can safely assume the shift to be proportional to the bifurcation parameter. The general shape of P_1 reads

$$\begin{pmatrix} x' \\ z' \end{pmatrix}_0 = \begin{pmatrix} a & b \\ c & d \end{pmatrix}\begin{pmatrix} x \\ y \end{pmatrix}_1 + \begin{pmatrix} e\mu \\ f\mu \end{pmatrix} \tag{16.7}$$

where a, b, c, d, e, f are real constants describing the affine map. This map has to be one-to-one since it represents a flow, so the only *a priori* requirement on these constants is that $ad - bc \neq 0$. Our goal is to establish what happens with the homoclinic orbit for *any* return map of the form (16.7).

Exercise 16.4: Can the determinant of the affine map P_1 be negative?

Combining P_0 and P_1 we obtain an approximate map $P = P_1 \circ P_0$ on Π_0 describing the surroundings of the homoclinic orbit. From equation (16.6) and equation (16.7) we obtain the general form of P:

$$x' = Axz^{\delta_3} + Bz^{\delta_2} + E\mu \tag{16.8}$$

$$z' = Cxz^{\delta_3} + Dz^{\delta_2} + \mu \tag{16.9}$$

where the constants have been redefined and rescaled to unity). The fixed points of this map describe the periodic orbits that are born or die in the surroundings of the homoclinic orbit. We solve equation (16.8) for x and replace in equation (16.9):

$$x = \frac{Bz^{\delta_2} + E\mu}{1 - Az^{\delta_3}} \tag{16.10}$$

$$z = \mu + Dz^{\delta_2} + Cz^{\delta_3}\frac{Bz^{\delta_2} + E\mu}{1 - Az^{\delta_3}}. \tag{16.11}$$

We will use a graphical method again to estimate the solutions to equation (16.11). The periodic orbits will occur when

$$\mu = \frac{(1 - Az^{\delta_3})(z - Dz^{\delta_2}) - CBz^{\delta_2+\delta_3}}{1 + (CE - A)z^{\delta_3}} \equiv g(z). \tag{16.12}$$

We note that the derivative dg/dz near the origin is either equal to one or diverges when δ_2 is respectively greater or smaller than one (we assume throughout $D \neq 0$; the special case $D = 0$ will be lightly commented below). Hence, for fixed values of the parameters A, B, C, D, E, sufficiently close to $z = 0$ we will have a periodic orbit for either $\mu > 0$ or $\mu < 0$ depending on the value of δ_2 (and on the sign of D for $\delta_2 < 1$). An example of the possible situations is depicted in figure 16.7.

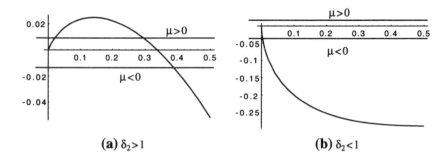

(a) $\delta_2 > 1$ **(b)** $\delta_2 < 1$

Figure 16.7. Location of the fixed points of the real 3-d problem. The parameters are chosen as: $A = 1.1, B = 0.3, C = 0.4, D = 1.2, E = 0.7$. (a) $\delta_2 > 1$. (b) $\delta_2 < 1$.

Whenever we have a solution of equation (16.11) we can compute the stability of the resulting periodic orbit for sufficiently small values of z, via the eigenvalues of the 2×2 Jacobian matrix (Jac) of equation (16.8) and equation (16.9), i.e.,

$$\gamma_{1,2} = \text{Tr(Jac)}/2 \pm \sqrt{(\text{Tr(Jac)}^2/4 - \det(\text{Jac}))} \qquad (16.13)$$

where

$$\text{Tr(Jac)} = Az^{\delta_3} + D\delta_2 z^{\delta_2 - 1} + C\delta_3 \frac{Bz^{\delta_2 + \delta_3 - 1} + E\mu z^{\delta_3 - 1}}{1 - Az^{\delta_3}}.$$

$$\det(\text{Jac}) = \delta_2(AD - CB)z^{\delta_2 + \delta_3 - 1}.$$

Sufficiently close to $z = 0$ the eigenvalues may be very large or very small when the values of δ_2, δ_3 and $\delta_2 + \delta_3$ are smaller or larger than one. We have

(i)   Stable periodic orbit for $\delta_2 > 1$ (consequently δ_3 and $\delta_2 + \delta_3$ are both larger than one). Note that if $D = 0$ it is enough with $\delta_3 > 1$ to obtain a stable periodic orbit.

(ii) Saddle periodic orbit for $\delta_2 < 1$. Note that if $D = E = 0$ we may have a situation where $|\det(\text{Jac})| \sim |\text{Tr(Jac)}| \gg 1$ and hence have a source periodic orbit.

We also notice that the sign of the eigenvalues is determined, for z very small, by the coefficient in front of $z^{\delta_2 - 1}$ in equation (16.13), i.e., by the sign of D. If $D > 0$ the homoclinic orbit lies on a regular band while if $D < 0$ the homoclinic orbit lies on a Möbius band with points with $z > 0$ mapped into points with $z < 0$.

Exercise 16.5: Verify the previous statement.

The general conclusion is that a periodic orbit is born from the homoclinic orbit and persists for small adequate values of μ [siln66, wigg90]. This result is in direct correspondence with the 2-d problem, with the additional feature that saddle periodic orbits may now occur.

16.4.1.1   Parameter versus period plots

In order to render more evident the fact that a periodic orbit is approaching the homoclinic condition when parameters are varied (or the other way around if the variation in μ is done in the reverse direction), we can replace z in equation (16.11) or in $g(z)$ by $z = \eta \, e^{-\lambda_1 F}$, where F is the time of flight (uniquely defined for any possible z) and produce a μ versus F plot where it is seen that $F \to \infty$ when $\mu \to 0$, see figure 16.8. We may call these plots *parameter versus period plots* although we must bear in mind that F is *not* the period of the periodic orbits satisfying $\mu = g(z)$ but rather the time that the orbits spend in the local region. The true period is larger than F and it must include the time spent travelling along the reinjection. The homoclinic orbit at $\mu = 0$ appears in figure 16.8 as a periodic orbit of 'infinite' period. The logarithmic transformation between z and F transforms the fixed point analysis near $z = 0$ on $g(z)$ into an analysis for large F on $h(F) = g(z(F))$.

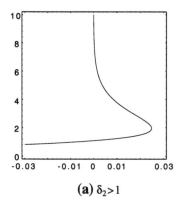

(a) $\delta_2 > 1$

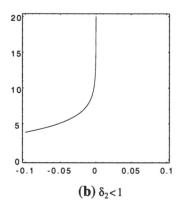

(b) $\delta_2 < 1$

Figure 16.8. Parameter versus period plots for the fixed points of the return map P. The fixed parameters are chosen as in the previous figure.

Exercise 16.6: Derive the different cases depicted in figure 16.7 analytically, by studying the function $h(F) \equiv g(z(F))$.

16.4.2 Homoclinic explosions: homoclinic orbit with symmetry

Let us now consider the Lorenz problem again. We know that for $\sigma = 10$ and $b = 8/3$ the unstable manifold of the fixed point at the origin changes character for $\rho = \rho_0 \simeq 13.926\ldots$ [spar82]. Each branch of the manifold circles around one of the symmetry-related fixed points for $\rho < \rho_0$, while for $\rho > \rho_0$ they 'exchange' partners and circle around the *other* fixed point. It may be convenient to go back some chapters and review figure 8.11 at this point. At ρ_0 the unstable manifold becomes homoclinic [spar82]. Actually *both* branches of the unstable manifold become homoclinic.

When studying parametric families of flows where the stable and unstable manifolds of a fixed point may cross, we can say that in general one parameter 'controls' one manifold. What we mean is the following. Imagine that the 1-d unstable manifold of a fixed point is, for a given value of the parameter, very close to the 2-d stable manifold (this may happen only for one branch of the manifold, while the other branch may be far apart in phase space and far apart from the 2-d manifold). The relative distance between the manifolds will change when the parameter is varied and it may eventually become zero. It would be a rare coincidence (which in addition could be destroyed by a small perturbation of the system) if both manifolds would move with the parameter remaining always equally far apart! It will in general happen that one of the manifolds changes comparatively much less than the other or that it does not change at all with the parameter. We say that the parameter *controls* the distance between the manifolds or, more loosely, that it controls the rapidly varying manifold.

Two different manifolds or the two branches of a 1-d manifold may however become homoclinic 'simultaneously' (along a parameter excursion) in the presence of symmetries. If the position of e.g., one branch of the unstable manifold is related by the symmetry to the position of the other branch, then whatever deformation one branch suffers will be 'symmetrically' experienced by the other branch as well. In the same sense as above, in symmetric systems one parameter may control two manifolds or that two branches of a manifold become homoclinic at the same parameter value.

It is clear that symmetric systems are rare in nature, since a small perturbation on a system may leave no symmetries left (depending on the nature of the perturbation). Nevertheless, whenever there *is* a symmetry, it leaves deep traces in the dynamics, even for perturbed non-symmetric systems.

Let us try to determine how symmetry plays its crucial role exemplifying on the Lorenz system. The symmetry of the system is $(x, y) \rightarrow (-x, -y)$. The homoclinic orbit comes out of the fixed point at the origin through one branch of the unstable manifold and returns to the fixed point along the eigenvector associated with the largest (stable) eigenvalue (the one closest to zero, since they are negative). In this case, it is the eigenvector tangent to the z axis[†].

If we act with the symmetry on the homoclinic orbit, we obtain another orbit that comes out of the same fixed point (since it is symmetric) through the *other* branch of the unstable manifold (since the two branches are related by the symmetry) and re-enters tangent to the z axis (since it is symmetric). Hence, two homoclinic connections occur at the same parameter value because of the symmetry! In the previous subsection, we were concerned only with one branch of the stable and unstable manifolds of the fixed point. The choice of the branch was given by the presence of the homoclinic orbit. Now both branches of the unstable manifold will be simultaneously 'active' rendering a pair of homoclinic orbits as in figure 16.9. The case (*a*) resembles what is known from the Lorenz system (note however the different orientation of the flow: see below and figure 16.10).

16.4.2.1 *Local and global maps: a geometric approach.*

Recalling the approach of the previous subsection, we can now define *two* planes Π_1 and Π_1' perpendicular to the branches of the local unstable manifold of the fixed point, which after an excursion through the global region of the system will be reinjected onto Π_0, a plane perpendicular to the z axis at a height $z = z_0$

[†] Recall that the z axis has the associated eigenvalue $-b$ ($\simeq -2.666\ldots$ for our parameter choice), while the other negative eigenvalue is smaller than $-\sigma$ ($= -10$) for positive ρ. The reader may want to go back to chapter 3 and review the stability analysis of the fixed point at the origin.

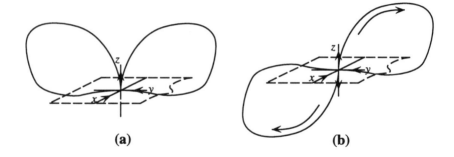

Figure 16.9. Two types of homoclinic orbit in systems with symmetry.

(see figure 16.10(a)). Note that the direction of the flow has been reversed with respect to figure 16.6 and figure 16.9. At the homoclinic parameter value $\mu = 0$, the left half of Π_0 will flow towards Π_1 while the right half will flow towards Π_1'. The line S_0 determined by the intersection of Π_0 and the local stable manifold flows towards the fixed point. Consider a thin rectangular strip on Π_0 parallel and very close to S_0, such as H_1 in figure 16.10. The local flow will contract this strip to the shape of a triangular sector on Π_1 and later the global flow will transport and stretch this triangle back to Π_0. If the strip is 'tangent' to S_0, the tip of the triangle will land back on S_0 at $\mu = 0$. A rectangular strip on the right half of Π_0 will suffer a similar fate because of the symmetry, passing through Π_1' and back again to Π_0. When $\mu \neq 0$ the two triangles will return to Π_0 *shifted* a distance proportional to μ, i.e., the tips will not land on S_0 but rather at some point nearby. If one tip lands at (a, b), the other will land at $(-a, -b)$ because of the symmetry. Hence, for values of μ at one side of zero up to a maximal value μ_0 where the present approximation breaks down, we will have the situation depicted in figure 16.10, i.e., that 'horizontal' bars on Π_0 are mapped by the flow roughly onto themselves as 'vertical' bars. The picture shown in figure 16.10 was made assuming that the global map has positive determinant for both branches. The general result still holds, with the proper modifications if one (or both) return map(s) would have negative determinant.

The reader may guess here the presence of something like a horseshoe. Indeed, when the contraction and expansion rates of the eigenvalues of the fixed point at the origin are in the proper ratio, the images of Π_0 compress and stretch sufficiently in order to guarantee that each of them will cross *both* rectangular strips as displayed in the figure. It is sufficient that $|\lambda_3/\lambda_1| < 1$ where λ_1 is the positive eigenvalue and λ_3 is the negative eigenvalue with largest modulus, i.e., it is *not* the eigenvalue associated with the incoming part of the homoclinic orbit, but the remaining one instead. The final result is:

Theorem (Sparrow) [spar82]: There exists an open interval $(0, \mu_0)$ where the approximate return map constructed by way of the local and global partition has an invariant Cantor set topologically equivalent to the full shift on two symbols.

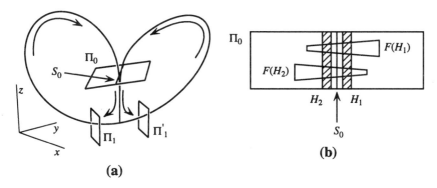

Figure 16.10. The symmetric homoclinic connection of the Lorenz system. (a) Sketch of the flow and the model. (b) Approximate return map on Π_0.

Exercise 16.7: Check that the stretching and bending of the strips in the Lorenz system is different from the one in the horseshoe map of chapter 10.

Exercise 16.8: Challenge! Compare the horseshoe template (recall chapter 12) with the Lorenz template. Figure 16.10 may help you to compute this template.

Other computer-assisted techniques to establish the existence of a horseshoe-like dynamics in the Lorenz system can be found in [misc95].

While the non-symmetric homoclinic loop only gave a periodic orbit before or after homoclinicity, the symmetric case is far richer, since immediately after homoclinicity we obtain a horseshoe-like complex structure having an infinite number of periodic orbits! In contrast with the model of chapters 9 and 10, where a sequence of bifurcations near a manifold tangency prepared the way to the formation of the horseshoe, the chaotic structure appears here all of a sudden for μ values arbitrarily close to zero. This sudden appearance is called a *homoclinic explosion*.

The reader should not believe, however, that non-symmetric homoclinic orbits are completely featureless. An interesting phenomenon resembling a period-doubling bifurcation may occur for a suitable choice of parameter values, namely that a double-loop homoclinic orbit (an orbit that makes two passes through Π_0 and Π_1 before returning towards the fixed point) may branch from (or die at) a homoclinic curve [yana87]. This mechanism allows for having more than one periodic orbit close (in parameter space) to the homoclinic orbit. As a consequence, more bifurcations may occur.

16.4.3 Homoclinic orbit with complex eigenvalues

We turn now to the second possible alternative of 3-d flows, namely that the saddle fixed point has two complex conjugated eigenvalues with negative real part. The original (and thorough) treatment of this problem goes back to Šilnikov [siln65]. Consequently, homoclinic orbits to focus saddle-points are also called the *Šilnikov phenomenon* or *Šilnikov orbits*.

The approach we will follow is analogous to the treatment of homoclinic orbits earlier in this chapter. Some (minor and major) differences arise due to the fact that the local flow is spiralling around the fixed point because of its focus character. We will consider as incoming 'plane' Π_0 the surface of a small cylinder totally contained within the region where the linear approximation of the flow is valid. The cylinder has its axis parallel to the unstable eigenvector (the z axis) and its base tangent to the stable manifold (the xy plane). The outgoing plane Π_1 is the disc $z = z_1$ building the upper lid of the cylinder. See figure 16.11.

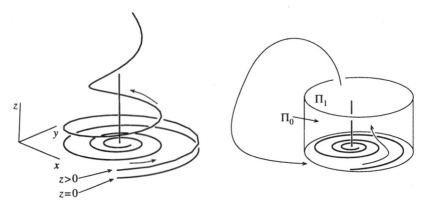

Figure 16.11. Local and global approximation of the flow near a focus saddle-point with a homoclinic connection.

The local flow reads

$$\begin{pmatrix} \dot{x} \\ \dot{y} \\ \dot{z} \end{pmatrix} = \begin{pmatrix} \alpha & -\beta & 0 \\ \beta & \alpha & 0 \\ 0 & 0 & \lambda \end{pmatrix} \begin{pmatrix} x \\ y \\ z \end{pmatrix} \tag{16.14}$$

where $\lambda > 0 > \alpha$. A point p entering Π_0 at (x_0, y_0, z) will leave Π_1 at $q = (x_1, y_1, z_1)$. Integrating the local flow equation (16.14) for a general initial condition p, we obtain

$$\begin{pmatrix} x \\ y \\ z \end{pmatrix}(t) = \begin{pmatrix} r_0\, e^{\alpha t} \cos(\beta t + \theta) \\ r_0\, e^{\alpha t} \sin(\beta t + \theta) \\ e^{\lambda t} z \end{pmatrix} \tag{16.15}$$

where we performed the substitution $(x, y) \rightarrow (r_0 \cos \theta, r_0 \sin \theta)$ on Π_0.

The time of flight is, as usual, $F = (1/\lambda) \ln(z_1/z)$. Recalling that F depends on the height of the initial point, we will write $F(z)$ in what follows to bear this dependence in mind. Defining $\delta = |\alpha/\lambda|$ we can write the coordinates (x, y) of the image of p on Π_1 under the action of the local map P_0:

$$x = r_0(z/z_1)^\delta \cos(\beta F(z) + \theta) \tag{16.16}$$

$$y = r_0(z/z_1)^\delta \sin(\beta F(z) + \theta). \tag{16.17}$$

Let us restrict ourselves to a small section on Π_0 where $|\theta| < \theta_0 \ll 1$. Points from this section map according to the above equations on Π_1. For $\mu = 0$, the global reinjection will map a centred disc on Π_1 so that its centre re-enters the cylinder at $z = 0$. We may define $\theta = 0$ as the direction where the homoclinic orbit enters Π_0. The global reinjection is modelled by an affine map P_1:

$$\begin{pmatrix} z \\ r_0 \sin \theta \end{pmatrix} = \begin{pmatrix} a \cos \alpha_1 & a \sin \alpha_1 \\ b \cos \alpha_2 & b \cos \alpha_2 \end{pmatrix} \begin{pmatrix} x \\ y \end{pmatrix} + \begin{pmatrix} \mu \\ 0 \end{pmatrix} \tag{16.18}$$

where we approximated the surface of the cylinder with a plane. We have $a, b \neq 0$ and $\alpha_1 + \alpha_2 \neq \pi/2$ since P_1 has to be non-singular.

Exercise 16.9: Why was it possible to have a zero in the second component of the constant term in P_1?

Now comes the usual thing. First, we put together P_0 and P_1 in a return map $P = P1 \circ P_0$. The map reads

$$\begin{pmatrix} z \\ r_0 \sin \theta \end{pmatrix} = \begin{pmatrix} a \cos \alpha_1 & a \sin \alpha_1 \\ b \cos \alpha_2 & b \cos \alpha_2 \end{pmatrix}$$
$$\times \begin{pmatrix} r_0(z/z_1)^\delta \cos(\beta F(z) + \theta) \\ r_0(z/z_1)^\delta \sin(\beta F(z) + \theta) \end{pmatrix} + \begin{pmatrix} \mu \\ 0 \end{pmatrix}. \tag{16.19}$$

Then, we solve equation (16.19) for $\sin \theta$ and $\cos \theta$ and insert the result in the z equation to obtain, after renaming some constant factors,

$$\mu = z - \frac{Az^\delta(\cos(\beta F(z) - \alpha_1) + Bz^\delta \sin(\alpha_1 - \alpha_2))}{\sqrt{1 + 2Bz^\delta \sin(\beta F(z) - \alpha_2) + B^2 z^{2\delta}}}. \tag{16.20}$$

The location of the periodic orbits is obtained as the intersection of constant μ lines with the right hand side of equation (16.20). Our first observation is that, concentrating on a given value of z where the right hand side of equation (16.20) has a local maximum, we note that for μ values higher than the maximum there is no periodic orbit. Decreasing μ we have one periodic orbit at the intersection. This orbit splits in two immediately after the intersection, i.e., we are in the presence of a saddle-node bifurcation on the return map.

16.4.3.1 Parameter versus period plots in the Šilnikov phenomenon

Writing z as a function of F we can obtain the periodic orbits of the return map in the form of a parameter versus period plot. After possibly another rescaling, we obtain that the periodic orbits occur for (μ, F) values such that

$$\mu = e^{-\lambda F} - \frac{A\,e^{\alpha F}(\cos(\beta F - \alpha_1) + B\,e^{\alpha F}\sin(\alpha_1 - \alpha_2))}{\sqrt{1 + 2B\,e^{\alpha F}\sin(\beta F - \alpha_2) + B^2\,e^{2\alpha F}}} \equiv h(F). \quad (16.21)$$

As noted previously, $F \to \infty$ when $z \to 0$. Hence, with the possible exception of small F values (where z might be too large to consider our approximated return map as valid), all curve intersections $\mu = h(F)$ will give valid periodic orbits. Also note that the right hand side of equation (16.21) goes to zero for large F. Hence, for any constant $\mu \neq 0$ we will at most have finitely many intersections and hence finitely many orbits.

In contrast to what happened in the real case, we may have infinitely many periodic orbits for $\mu = 0$, because of the circular functions in $h(F)$. If $h(F)$ oscillates infinitely many times for large F, it will necessarily cross the $\mu = 0$ line infinitely many times (the oscillations cannot occur only on one side of the line except for small F, because the non-oscillating term $e^{-\lambda F}$ in $h(F)$ goes to zero). The scenario is that for $\mu \neq 0$ we have finitely many periodic orbits of higher and higher period while for $\mu = 0$ we have infinitely many periodic orbits or arbitrarily high period and a homoclinic orbit (of 'infinite' period).

Is there a situation where we will *not* have infinitely many orbits at all? The answer is positive: if $h(F)$ is a monotonic function for large F, it will tend to zero from one side, i.e., except for a finite number of intersections for medium or small F, $h(F)$ will intersect $\mu = 0$ just at '$F = \infty$', i.e., at the homoclinic orbit. In this case, we will have no periodic orbits of large period for $\mu < 0$, the homoclinic orbit for $\mu = 0$ and a periodic orbit 'expelled' by the homoclinic connection.

In order to assess whether $h(F)$ tends monotonically or oscillatorily towards zero for large F, it is enough to estimate the derivative of h:

$$\frac{dh}{dF} \simeq -\lambda\,e^{-\lambda F} + A\,e^{\alpha F}(\beta\sin(\beta F - \alpha_2) - \alpha\cos(\beta F - \alpha_1)) + \dots. \quad (16.22)$$

Recalling that $\alpha < 0$, we see that if $\lambda > |\alpha|$ the first term goes to zero faster than the second term, and the derivative oscillates because of the circular functions. Conversely, if $\lambda < |\alpha|$ the oscillatory term is much smaller than the non-oscillatory one and dh/dF remains negative for all values of F larger than a fixed value F_0. Hence, we can summarize the observations stating that [wigg90] for $\delta < 1$ we have an oscillating function and infinitely many periodic orbits, while for $\delta > 1$ we have finitely many periodic orbits. The general shape of the curves for $\delta > 1$ and $\delta < 1$ is shown in figure 16.12.

Exercise 16.10: Show that there is an infinite number of periodic orbits for

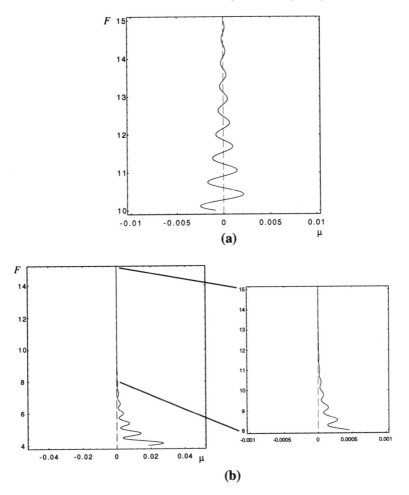

Figure 16.12. Parameter versus period plots for the Šilnikov phenomenon. (a) $\delta < 1$ (chaotic case). (b) $\delta > 1$ (non-chaotic case).

$z \to 0$ when $\delta < 1$, while there is only a finite number for $\delta > 1$, by considering the derivative of the right hand side of equation (16.20) with respect to z as in subsection 16.4.1 and the frequency of the angular functions.

Exercise 16.11: Check the stability of the orbits (taking the Jacobian of the return map of equation (16.18)), and show that the parameter versus period curve is built by a sequence of saddle-node and period-doubling bifurcations.

16.4.3.2   The Šilnikov's scenario and horseshoes

The striking feature about the Šilnikov phenomenon is that it has more than just infinitely many periodic orbits. Let us try now a geometric approach, similar to the one applied in the real, symmetric case.

Let us estimate how the small 'square' $|\theta| < \theta_0$ and $0 < \epsilon < z < z_0 + \epsilon$ on Π_0 maps onto itself at $\mu = 0$. A vertical line (constant θ) on this square will map onto Π_1 as a logarithmic spiral with one of its ends ϵ-close to the origin, as is depicted in figure 16.13.

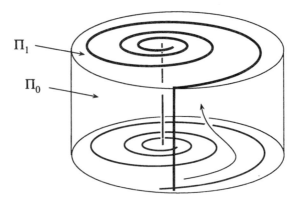

Figure 16.13. How a vertical line of Π_0 maps on Π_1.

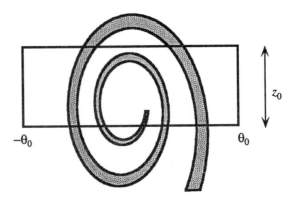

Figure 16.14. Geometric model of the return map at a saddle-focus homoclinic orbit.

The whole squarelike region will be mapped onto a thin and long spiral strip in a similar manner. After the global excursion, the spiral strip will be mapped back onto Π_0, having its centrum at ($z = 0, \theta = 0$) for $\mu = 0$. If the ratio of contraction and expansion rates δ is small enough, the spiral will be higher than

z_0 and the situation shown in figure 16.14 will occur. The reader may guess from the figure that a horseshoe-like structure appears. The value of δ is crucial, since, in order to get a map that resembles a horseshoe, we need that part of the spiral strip lands *above* the upper edge of the square. The critical condition to have horseshoes is $\delta < 1$ [siln65, wigg90]. Since the spiral may have many turns that land 'correctly', the structure that is formed is in fact more complex than the standard horseshoe of chapter 10. In fact, there are infinitely many horseshoes which persist for small enough values of μ around zero [glen84].

16.5 The quest for homoclinic connections

This chapter has so far dealt with the consequences of having a homoclinic orbit in different set-ups. For the theorist that is studying a natural system, the problems appear usually in a different way, namely that one has a model of a system that experimentally has shown (definite or strong) indications of being chaotic. Moreover, one might suspect the presence of a homoclinic orbit due for example to the verification of a sequence of bifurcations of periodic orbits such as those discussed in this chapter. In such a case, the first task to complete at this point is to show that the homoclinic orbit exists. In a sense, the problem is *complementary* to the one we have been analysing. When studying a particular problem we have to be aware that:

• If we guess the homoclinic orbits by their effects we do not know whether the same effects can be produced by other dynamical situations. Hence, guessing the existence of the homoclinic orbits is not the same as showing their presence.

• The unfolding parameters controlling the reinjection of the homoclinic orbit, in the above presentation, are taken *ad hoc* and bear no relation with the physical parameters of the model. Moreover, it might very well be that the model does not contain the homoclinic structure but it is close to a model that does contain a homoclinic orbit, i.e., the parameter that controls the reinjection is not a function of the parameters present in our model. In turn, the model presenting the homoclinic orbit could be unreasonable because of physical considerations.

Unfortunately, there are no infallible ways of asserting the existence of homoclinic orbits. One obvious method is to search directly for them with a suitable numerical algorithm. The second method has been already illustrated in chapter 8 with the Lorenz problem: the sudden change of behaviour of the unstable manifolds of the saddle, figure 8.11, can only be reconciled with the smooth character of the vector field by the presence of a homoclinic orbit.

There are yet other situations where the existence of homoclinic orbits or heteroclinic cycles can be addressed analytically. Most important among them are co-dimension-two local bifurcations. We refer the reader to [guck86, wigg90] for the discussion of this topic.

We discuss in what follows two situations in which the search for a mechanism which contains homoclinic orbits and heteroclinic cycles can be pursued with analytical methods, as both cases involve symmetries.

16.5.1 Heteroclinic cycles in symmetric systems

In symmetric systems, it is possible to state algebraic conditions for the existence of homoclinic loops and heteroclinic cycles. Suppose that the problem has the following structure (or its generalization involving n fixed points):

- The system, $\dot{x} = f(x)$ is equivariant with respect to the group Γ, i.e., for $\gamma \in \Gamma$ $\gamma f(x) = f(\gamma x)$.
- There are two inequivalent subgroups Σ and Ψ with fixed point subspaces S_Σ and S_Ψ with $\dim(S_\Sigma) = \dim(S_\Psi) = 2$ (recall from chapter 15 that the fixed points subspace with respect to a subgroup H is the set of points $\{x : hx = x , \forall h \in H\}$). $S_\Sigma \neq S_\Psi$ and $S_\Sigma \cap S_\Psi \neq \emptyset$.
- There are two non-zero fixed points of the flow, namely a and b, that belong to $S_\Sigma \cap S_\Psi$.
- When the restriction of the flow to the invariant subspace S_Σ is considered a is a sink (stable node) and b is a saddle.
- When the restriction of the flow to the invariant subspace S_Ψ is considered, a is a saddle and b is a sink.

The situation is illustrated in figure 16.15.

The key to the construction is to consider the restriction of the differential equation to each of the fixed point subspaces S_Σ and S_Ψ. The restricted problem is a bi-dimensional one and we can use the full machinery developed in chapter 6. If the region of phase space under consideration is inside a trapping region the intersection of the trapping region with S_Σ (or S_Ψ) produces a positively invariant region for the restricted flow. Using the Poincaré–Bendixson theorem the ω-limit of any point in the region is either a fixed point, a periodic orbit or a heteroclinic cycle. Thus, if we can show that for the particular system we are considering there are no other fixed points except zero (which is unstable in all directions) and that there are no periodic orbits, we can guarantee that there is a heteroclinic connection from a to b in S_Σ and a heteroclinic connection from b to a in S_Ψ, and, moreover, that these connections are structurally stable within the class of systems with symmetry group Γ.

The structural stability, in turn, assures us that if we can find such a heteroclinic cycle in an approximated system (usually a normal form) we can be sure that the heteroclinic cycle is not an artifact of the normal form but persists under small enough perturbations, i.e., it persists for our problem.

The discussion and application of the procedure just outlined can be found in [melb89, guck88, fiel80, kirk94] and references therein.

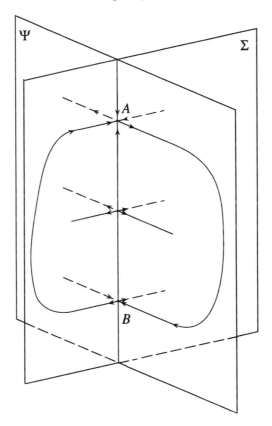

Figure 16.15. Stable heteroclinic cycle in a symmetric system.

16.5.2   Homoclinic orbits in the laser

We undertake the discussion of the laser with injected signal (LIS) for the last time.

We observed on the bifurcation diagram of figure 14.1 that at the point $(\eta = -\kappa, \kappa = \nu/\sin 2\alpha)$ the curves of saddle-node and Hopf bifurcations of the periodic orbit at $r = 0$ coincide for the averaged model of the laser. Both bifurcations occur at $\psi = \pi$. We recall that such a point in parameter space is called a point of *co-dimension two*. It is not enough with just one bifurcation parameter to describe the flow completely in the surroundings of the bifurcation point. This fact is clear from figure 14.1. Using only one parameter would correspond to drawing a 1-d curve on the 2-d parameter space of the diagram. We see that there is no 1-d curve that can completely describe both the co-dimension-two point *and* its surroundings.

The periodic orbit at $r = 0$ is a fundamental characteristic of our system.

The reader may remember that it arose from the (broken) $\mathbb{O}(2)$ symmetry of the unperturbed problem. The fact that we have a local bifurcation will not alter the general shape of the flow except in the local part (this is what in fact is meant by the expression *local* bifurcation) so we can safely assume that in the surroundings of the bifurcation point the flow still 'follows' the periodic orbit at $r = 0$.

We observe that some of the flow types depicted in figure 14.2 are 'suspicious' when considered in terms of the three-dimensional model. We should question, in principle, that the saddle-node bifurcations of fixed points break the invariant circle with the non-local part of the unstable manifold of the saddle being part of the stable manifold of the node. Is this connection an artifact of the averaging technique? Or, alternatively, is the connection structurally stable?

We have already some experience with this sort of connection since in our discussion of the torus break-up (chapter 8) we have seen that such connections can be broken with sufficiently large perturbations.

One way to investigate what really happens in the three-dimensional model of the LIS is to decompose the flow into local and global parts, the local part represented by the dynamics near the saddle-node bifurcation (that we take close to the co-dimension-two bifurcation) and the global part represented by an affine transformation as usual.

In particular we want to investigate the presence of homoclinic orbits in this context.

16.5.2.1 Local and global maps

The local flow near the co-dimension-two bifurcation is represented by

$$\begin{aligned}
\dot{r} &= (\mu' + az)r \\
\dot{z} &= v' + br^2 - z^2 \\
\dot{\zeta} &= \omega
\end{aligned} \qquad (16.23)$$

as already discussed in chapter 14 (see equation (14.13)). Note that we have performed the substitution $\psi = \pi + z$ in order to set the origin of coordinates at the co-dimension-two point.

The new bifurcation parameters are

$$\mu' = \left(-\frac{v}{\kappa} + \sin(2\alpha)\right)$$

$$v' = 2\left(1 + \frac{\eta}{\kappa}\right)$$

which depend on the old bifurcation parameters $\eta, v = \mu(1 + gA^2), \kappa = \epsilon/(A^2 \sin\alpha)$ and on the detuning θ via $\alpha = \tan^{-1}(1/\theta)$. The constants a, b

depend only on θ in the following way:

$$a = \cos(2\alpha)$$
$$b = 1 - \frac{\cos(2\alpha)}{2\sin^2(\alpha)}.$$

For the case of very large detuning, $\theta > \sqrt{3}$, b can be further reset to -1 by rescaling r.

Note that ν' describes the saddle-node bifurcation, while μ' is involved in the Hopf bifurcation. The co-dimension-two point is at $\mu' = \nu' = 0$. We will restrict the remaining analysis to $\mu' < 0$. The behaviour for positive μ can be recovered by a symmetry transformation of equation (16.23).

The stable manifold of the saddle-node fixed point (SN) at the bifurcation can be approximated (locally) by

$$W^s(\text{SN}) = \left\{ (r, z) \mid z = \frac{b}{2\mu'} r^2 + O(r^4) \right\}. \tag{16.24}$$

A sketch of the manifold organization can be seen in figure 16.16.

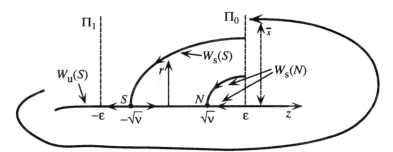

Figure 16.16. (r, z) projection of the manifold organization of the laser system for $\nu' > 0$.

For $\nu' > 0$ the saddle-node fixed point breaks into saddle and node, the saddle located roughly at $(r, z) = (0, -\sqrt{\nu'})$ and the node at $(r, z) = (0, \sqrt{\nu'})$. Their respective stable manifolds move with $\nu' > 0$ to

$$W^s(S) = \left\{ (r, z) \mid z + \sqrt{\nu'} = qr^2 + O(r^4) \right\} \tag{16.25}$$

with

$$q = \frac{b}{2(\mu' - \sqrt{\nu'}(1 + a))}$$

for the saddle and to

$$W^s(N) = \left\{ (r, z) \mid z - \sqrt{\nu'} = q'r^2 + O(r^4) \right\} \tag{16.26}$$

with

$$q' = \frac{b}{2(\mu' + \sqrt{v'}(1 + a))}$$

for the node (we are considering $b, \mu' < 0$).

We observe that the stable manifolds are roughly parabolic with a coefficient that strongly depends on μ' for $\mu' \approx 0$. The intersection of $W^s(S)$ with a control plane taken at $z = \epsilon$ is given by

$$r = \left(\frac{\sqrt{v'} + \epsilon}{q} \right)^{1/2}.$$

In order to discuss the possible existence of homoclinic orbits we have to consider only where the unstable manifold of the saddle is mapped by the global map.

The unstable manifold of the saddle will be mapped approximately from $(-\epsilon, 0)$ into $(\epsilon, \bar{x})$ with $\bar{x} \approx x_{v'} + \gamma\mu'$. Here $x_{v'}$ depends on v' and $x_{v'}$ is a small number since the complete problem is a perturbation of the averaged problem where $x_{v'} = 0$, hence the smallness of $\bar{x}$ is controlled by the size of the perturbation, i.e., κ.

There will be a homoclinic orbit if the condition

$$\bar{x} = x_{v'} + \gamma\mu' \approx \left(\frac{\sqrt{v'} + \epsilon}{q} \right)^{1/2} \tag{16.27}$$

is satisfied.

For small $x_{v'}$ there is always a solution of equation (16.27) for

$$\mu' \approx \frac{bx_{v'}^2}{2(\sqrt{v'} + \epsilon)} + (1 + a)\sqrt{v'}.$$

Note that this result is independent of γ to the lowest order. We disregard the other possible solution with large $|\mu'|$ since it is out of the scope of our discussion.

16.5.2.2   *Homoclinic orbits and the Šilnikov phenomenon*

We have shown the existence of homoclinic orbits in a geometrical realization of the LIS model. The model can be regarded as an approximation to the real system made under highly general considerations (that the bifurcations are *local* and hence the flow near the periodic orbit at $r = 0$ will 'persist' outside the local region for parameter values sufficiently close to the bifurcation). Even for rough estimations of the constants $\bar{x}, a$, the system will have homoclinic orbits.

Since the homoclinic fixed point is of focus type, we will encounter the Šilnikov phenomenon. Indeed, the ratio of the eigenvalues of the saddle focus at ($z = -\sqrt{v'}, r = 0$) can be computed from the local flow, to obtain $\lambda_1 = 2\sqrt{v'} > 0$ and $\lambda_2 = \mu' - a\sqrt{v'} < 0$. The condition for the chaotic Šilnikov phenomenon is then $|\mu' - a\sqrt{v'}| < 2\sqrt{v'}$, or $\mu' + (2 - a)\sqrt{v'} > 0$. We refer to [zimm95] for more details.

16.6 Summary

Homoclinic orbits in 2-d flows are quite featureless, unlike their counterparts in 2-d maps. They just 'emit' or 'absorb' a periodic orbit which in its turn can be either a sink or a source. However, their study gave us the basis to pursue the analysis of homoclinic orbits in 3-d flows with real eigenvalues, where we saw a little more structure.

Concerning symmetric systems, we learned while studying the Lorenz system that symmetry drastically changes the features of homoclinic connections with the appearance of chaotic sets for adequate values of the contraction and expansion rates of the flow.

We may wonder what is it that one actually sees in numerical observations on the Lorenz system. An invariant chaotic set is formed right after the homoclinic connection at $\rho \sim 13.926\ldots$. However, this set coexists with the attractor of the system, namely the two symmetry-related focus fixed points. In numerical experiments one typically sees the orbits that approach the attractor. The two fixed points become unstable after colliding with an unstable periodic orbit in a Hopf bifurcation at $\rho = 470/19$. After this point, i.e., when there no longer exist attractive fixed points, the effects of the chaotic set are easily observed in the famous 'butterfly' (see figure 8.10).

Finally, we saw that when the fixed point has a pair of complex conjugated eigenvalues there is a possibility of having a strange set associated with the homoclinic orbit if the local contraction and expansion rates are appropriated. In this case the situation has a closer resemblance to the horseshoe formation described in chapter 9, i.e., a series of bifurcations generates more and more orbits as $\mu \to 0$ and eventually a horseshoe appears.

We have finally discussed how the presence of symmetries in the system can help us to detect the existence of homoclinic orbits. In this context we discussed both a case of perfect symmetries and the case of the LIS where symmetries are slightly broken.

A final remark for the curious reader is pertinent. In chapter 7 we learned that Ladis's theorem [ladi73] allows us to map the flow around an arbitrary hyperbolic fixed point to the flow around a fixed point with real eigenvalues by a continuous coordinate and time transformation. Moreover, the real eigenvalues can be chosen to be 1 and -1. Can we invoke Ladis's theorem instead of Hartman–Grobman's and use as linearized form close to the fixed point the topological normal form? We may guess that there is something strange in the proposal since we have seen that homoclinic orbits to focus saddle points are quite different from homoclinic orbits to real-valued saddles. In order to find out why this is not possible we encourage the reader to rewrite the case of a saddle in two dimensions in terms of Ladis's coordinates $(X, Y) = (x^{1/\gamma}, y^{-1/\lambda})$ for $x, y \geq 0$, and to realize that the global map is not longer an affine transformation but a nonlinear map. What really entitle us to use an affine transformation for the reinjection combined with a linearized expression for the local flow is that the Hartman–Grobman change of coordinates is a near-identity one. Hence, by adopting the linear expression for the local flow we only have to modify the nonlinear terms of the *actual* reinjection which are not used in the discussion. By contrast, if we adopt the topological (Ladis) form of the saddle we would have to change from a linear reinjection to a nonlinear one containing most of the interesting information.

References

[abar93]   Abarbanel H D I, Brown R, Sidorowich J J and Tsimring L Sh 1993 The analysis of observed chaotic data in physical systems *Rev. Mod. Phys.* **65** 1331

[abra95]   Abraham N B, Allen U A, Peterson E, Vicens A, Vilaseca R, Espinosa V and Lippi G L 1995 Structural similarities and differences among attractors and their intensity maps in the laser–Lorenz model *Opt. Commun.* **117** 367

[ande69]   Anderson O H and Kunkel W B 1969 Tubular pinch and tearing instability *Phys. Fluids* **12** 2099

[arec84]   Arecchi F T, Lippi G L, Puccioni G P and Tredicce J R 1984 Deterministic chaos in laser with injected signal *Opt. Commun.* **51** 308

[arno68]   Arnold V I and Avez A 1989 *Ergodic Problems of Classical Mechanics* (Redwood City, CA: Addison-Wesley) (Original edition 1968, publisher W A Benjamin)

[arno73]   Arnold V I 1989 *Ordinary Differential Equations* (Cambridge, MA: MIT Press) (6th printing from original edition of 1973)

[arno83a] Arnold V I 1983 Remarks on the perturbation theory for problems of Mathiew type *Usp. Mat. Nauk.* **38** 215 (Original in *Russian Math. Surveys*)

[arno83b] Arnold V I 1983 *Ordinary Differential Equations* (Cambridge, MA: MIT Press) (1st printing from original edition of 1973)

[arno88a] Arnold V I and Il'yashenko Yu S 1988 Dynamical systems I V I Arnold ed *Ordinary Differential Equations* (Berlin: Springer)

[arno88b] Arnold V I 1988 *Geometrical Methods in the Theory of Ordinary Differential Equations* (Berlin: Springer)

[arno89]   Arnold V I 1989 *Mathematical Methods of Classical Mechanics. 2nd edition.* (New York: Springer) (1st edition 1978)

[aron82]   Aronson D G, Chory M A, Hall G R and McGehee R P 1982 Bifurcation from an invariant circle for two-parameter families of maps of the plane: a computer-assisted study *Commun. Math. Phys.* **83** 303

[arti47]   Artin E 1947 Theory of braids *Ann. Math.* **48** 101

[bald69]   Baldwin G C 1969 *An Introduction to Nonlinear Optics* (New York: Plenum)

[bald87]   Baldwin S 1987 Generalizations of a theorem of Sarkovskii on orbits of continuous real-valued functions *Discrete Math.* **67** 111

[barn04]   Barnes H T and Coker E G 1904 The flow of water through pipes. experiments on stream-line motion and the measurement of critical velocity *Proc. R. Soc.* **74** 341

[barn88]   Barnsley M 1988 *Fractals Everywhere* (San Diego, CA: Academic)

335

[beie83]   Beiersdorfer P, Wersinger J-M and Treve Y 1983 Topology of the invariant manifolds of period-doubling attractors for some forced nonlinear oscillators *Phys. Lett.* **96A** 269

[bena01]   Bénard H 1901 Les tourbillons cellulaires dans une nappe liquide transportant de la chaleur par convection en régime permanent *Ann. Chem. Physique* **23** 62

[birm83]   Birman J and Williams R 1983 Knotted periodic orbits in dynamical systems I: Lorenz equations *Topology* **22** 47

[bleh90]   Bleher S, Greboggi C and Ott E 1990 Bifurcation to chaotic scattering *Physica* **46D** 87

[bloc80]   Block L, Guggenheimer J, Misiurewicz M and Young L S 1980 Periodic points and topological entropy of one dimensional maps *Lecture Notes Math.* **819** 18

[blum89]   Bluman G W and Kumei S 1989 *Symmetries and Differential Equations* (New York: Springer)

[bowe70]   Bowen R 1970 Markov partitions for Axiom A diffeomorphisms *Am. J. Math.* **92** 725

[bowe76]   Bowen R and Franks J 1976 The periodic points of maps of the disk and the interval *Topology* **15** 337

[boyl84]   Boyland P 1984 Braid types and a topological method of proving positive topological entropy *Technical report, Department of Mathematics, Boston University, Preprint*

[broo86]   Broomhead D S and King G P 1986 Extracting qualitative dynamics from experimental data *Physica* **20D** 217

[buss78]   Busse F H 1978 Non-linear properties of thermal convection *Rep. Prog. Phys.* **41** 1929

[carr81]   Carr J 1981 *Applications of Centre Manifold Theory* (Berlin: Springer)

[cecc93]   Cecchi G A, Gonzalez D L, Magnasco M O, Mindlin G B, Piro O and Santillan J 1993 Periodically kicked hard oscillators *Chaos* **3** 51

[chan84]   Chang K W and Howes F A 1984   *Nonlinear Singular Perturbation Phenomena: Theory and Application* (Berlin: Springer)

[chen87]   Chen C, Györgyi G and Schmidt G 1987 Universal scaling in dissipative systems *Phys. Rev.* **35A** 2660

[chom87]   Chomsky N 1987 *The Chomsky Reader* J Peck ed (New York: Pantheon) Chomsky cites Humboldt on p 149

[chos88]   Chossat P and Golubitsky M 1988 Symmetry-increasing bifurcation of chaotic attractors *Physica* **32D** 423

[chow82]   Shui-Nee Chow and Hale J K 1982 *Methods of Bifurcation Theory* (Berlin: Springer)

[cohe77]   Cohen-Tannoudji C, Dieu B and Laloë F 1977 *Quantum Mechanics* (New York: Wiley)

[coll86]   Collet P and Eckman J-P 1986 *Iterated Maps of the Interval as Dynamical Systems* (Basel:Birkhäuser)

[cour53]   Courant R and Hilbert D 1953 *Methods of Mathematical Physics* (New York: Interscience)

[curr78]   Curry J H and Yorke J A 1978 A transition from hopf bifurcation to chaos: computer experiments with maps on $\mathbb{R}^2$ *Lecture Notes Math.* **668** 48

[dang91]   Dangelmayr G and Knobloch E 1991 Hopf bifurcation with broken circular symmetry *Nonlinearity* **4** 399

[dang92]   D'Angelo E J, Izaguirre E, Mindlin G B, Huyet G, Gil L and Tredicce J R 1992 Spatiotemporal dynamics of lasers in the presence of an imperfect $O(2)$ symmetry *Phys. Rev. Lett.* **68** 3702

[darw88]   Darwin C 1988 *The Voyage of the Beagle* (New York: Nal Penguin) (Original edition 1836)

[duff18]   Duffing G 1918 *Erzwungene Schwingungen bei Veränderlicher Eigenfrequenz und ihre Technische Bedeutung* (Braunschweig: Vieweg-Verlag)

[dwye90]   Dwyer G, Levin S A and Buttel L 1990 A simulation model of the population dynamics and the evolution of myxomatosis *Ecol. Monographs* **60** 423

[esch89]   Eschenazi E, Solari H G and Gilmore R 1989 Basins of attraction in driven dynamical systems *Phys. Rev.* **39A** 2609

[feig78]   Feigenbaum M J 1978 Quantitative universality for a class of nonlinear transformations *J. Stat. Phys.* **19** 25

[feig79]   Feigenbaum M J 1979 The universal metric properties of nonlinear transformations *J. Stat. Phys.* **21** 669

[fend92]   Fendrik A J and Sanchez J M 1992 Decay of quasibounded classical Hamiltonian systems populated by scattering experiments *Technical report, Department of Physics, Buenos Aires University, Preprint*

[fiel80]   Field M 1980 Equivariant dynamical systems *Trans. Am. Math. Soc.* **259** 185

[gamb89]   van Strien S, Gambaudo J M and Tresser C 1989 The periodic orbit structure of orientation preserving diffeomorphisms on D^2 with topological entropy zero *Ann. Inst. Henri Poincaré Phys. Théor.* **49** 335

[gant59]   Gantmacher F R 1959 *The Theory of Matrices* (New York: Chelsea)

[gilm74]   Gilmore R 1974 *Lie Groups, Lie Algebras, and Some of Their Applications* (New York: Wiley)

[glan71]   Glansdorff P and Prigogine I 1971 *Thermodynamic Theory of Structure, Stability and Fluctuations* (New York: Wiley)

[glas89]   von Glasersfeld E 1989 Cognition, construction of knowledge and teaching *Synthese* **80** 121

[glei87]   Gleick J 1987 *Chaos: Making a New Science* (New York: Penguin)

[glen84]   Glendinning P and Sparrow C 1984 Local and global behavior near homoclinic orbits *J. Stat. Phys.* **35** 645

[gold80]   Goldstein H 1980 *Classical Mechanics* (Reading, MA: Addison-Wesley)

[golu85]   Golubitsky M and Stewart I 1985 Hopf bifurcation in the presence of symmetry *Arch. Ration. Mech. Anal.* **87** 107

[golu88]   Golubitsky M, Stewart I and Schaeffer D G 1988 *Singularities and Groups in Bifurcation Theory* (New York: Springer)

[gonz89]   Gonzalez D L, Magnasco M O, Mindlin G B, Larrondo H A and Romanelli L 1989 A universal departure from the classical period doubling spectrum *Physica* **39D** 111

[gras84]   Grassberger P and Procaccia I 1984 Dimensions and entropies of strange attractors from a fluctuating dynamics approach *Physica* **13D** 34

[greb83]   Grebogi C and Ott E 1983 Crisis, sudden changes in chaotic attractors and transient chaos *Physica* **7D** 181

[gree90]   Green C, Mindlin G B, D'Angelo E J, Solari H G and Tredicce J R 1990
            Spontaneous symmetry breaking in a laser: the experimental side *Phys.
            Rev. Lett.* **65** 3124

[guck86]   Guckenheimer J and Holmes P J 1986 *Nonlinear Oscillators, Dynamical
            Systems and Bifurcations of Vector Fields* (New York: Springer) (1st
            printing 1983)

[guck88]   Guckenheimer J and Holmes P 1988 Structurally stable heteroclinic cycles
            Math. Proc. Cambridge Phil. Soc. **103** 189

[hake75]   Haken H 1975 Analogy between higher instabilities in fluids and lasers *Phys.
            Lett.* **53A** 77

[hake83]   Haken H 1983 *Advanced Synergetics* (Berlin: Springer)

[hale69]   Hale J K 1969 *Ordinary Differential Equations* (New York: Wiley)

[hale88]   Hale J K 1988 *Asymptotic Behavior of Dissipative Systems, Mathematical
            Surveys and Monographs* (Providence, RI: American Mathematical Society)
            p 25

[hall94]   Hall T 1994 The creation of horseshoes *Nonlinearity* **7** 861

[hame62]   Hamermesh M 1962 *Groups Theory and its Application to Physical Problems*
            (Reading, MA: Addison-Wesley)

[hao 82]   Bai-Lin Hao and Shu-Yu Zhang 1982 Hierarchy of chaotic bands *J. Stat.
            Phys.* **28** 769

[hend71]   Henderson J M and Quandt R E 1971 *Microeconomics Theory* (New York:
            McGraw-Hill)

[heno76]   Hénon M 1976 A two-dimensional mapping with a strange attractor *Commun.
            Math. Phys.* **50** 69

[hill77]   Hill G W 1877 On the part of motion of the lunar perigee which is function
            of the mean motions of the sun and moon *Acta Math.* **8** 1

[hirs78]   Hirsh M W and Smale S 1978 *Differential Equations, Dynamical Systems,
            and Linear Algebra* (Orlando, FL: Academic)

[holm79]   Holmes P J 1979 A non-linear oscillator with a strange attractor *Phil. Trans.
            R. Soc.* **292A** 419

[holm84]   Holmes P and Whitley D 1984 Bifurcatioons of one- and two-dimensional
            maps *Phil. Trans. R. Soc.* **311A** 43

[holm85]   Holmes P and Williams R F 1985 Knotted periodic orbits in suspensions of
            Smale's horseshoe: torus knots and bifurcation sequences *Arch. Rational
            Mech. Anal.* **90** 115

[ioos80]   Ioos G and Joseph D D 1980 *Elementary Stability and Bifurcation Theory*
            (Berlin: Springer)

[kato80]   Katok A 1980   Liapunov exponents, entropy and periodic orbits for
            diffeomorphisms *Publ. Math. IHES* **51** 137

[kauf91]   Kauffman L H 1991 *Knots and Physics* (Singapore: World Scientific)

[kenn92]   Kennel M B, Brown R and Abarbanel H D I 1992 Determining embedding
            dimension for phase-space reconstruction using a geometrical construction
            Phys. Rev. **45A** 3403

[kirk94]   Kirk V and Silber M 1994   A competition between heteroclinic cycles
            Nonlinearity **7** 1605

[kitt76]   Kittel C 1976 *Solid State Physics* (New York: Wiley)

[kolm57]   Kolmogorov A N and Fomin S V 1957 1954 *Elements of the Theory of Functions and Functional Analysis Volume I, Metric and Normed Spaces* (Rochester, NY: Graylock) Translated from the first Russian edition

[kosi93]   Kosinski A A 1993 *Differential Manifolds* (San Diego: Academic)

[kris70a]   Krishnamurti R 1970 On the transition to turbulent convection. Part 1. The transition from two- to three-dimensional flow *J. Fluid Mech.* **42** 295

[kris70b]   Krishnamurti R 1970 On the transition to turbulent convection. Part 2. The transition to time-dependent flow *J. Fluid Mech.* **42** 309

[krue93]   Kruel Th-M, Eiswirth M and Schneider F W 1993 Computation of Liapunov spectra: effect of interactive noise and application to a chemical system *Physica* **63D** 117

[ladi73]   Ladis N N 1973 Topological equivalence of linear flows *Differential Equations* **9** 938

[land69]   Landau L and Lifchitz E 1969 *Physique Théorique, tome I. Mécanique* (Moscow: Mir)

[lath89]   Lathrop D P and Kostelich E J 1989 Analyzing periodic saddles in experimental strange attractors *Measures of Complexity and Chaos, NATO ASI Series* vol B 208, p 147, N B Abraham, A M Albano, A Passamante and P E Rapp ed (New York: Plenum)

[lefr94]   Lefranc M, Glorieux P, Papoff F, Molesti F and Arimondo E 1994 Combining topological analysis and symbolic dynamics to describe a strange attractor and its crises *Phys. Rev. Lett.* **73** 1364

[li-y75]   Li T-Y and Yorke J 1975 Period three implies chaos *Amer. Math. Monthly* **82** 985

[libc82]   Libchaber A and Maurer J 1982 A Rayleigh-Bénard experiment: helium in a small box *Nonlinear Phenomena and Phase Transitions and Instabilities* T Riste ed (New York: Plenum)

[lope94]   Lopez-Ruiz R, Mindlin G B, Perez Garcia C and Tredicce J R 1994 Nonlinear interaction of transversal modes in a CO_2 laser *Phys. Rev.* **49A** 4916

[lore63]   Lorenz E N 1963 Deterministic non-periodic flows *J. Atmos. Sci.* **20** 130

[lore80]   Lorenz E N 1980 Noisy periodicity and reverse bifurcation *Ann. NY Acad. Sci.* **357** 282

[magn79]   Magnus W and Winkler S 1979 *Hill's Equation* (New York: Dover)

[mann91]   Manneville P 1991 *Structures Dissipatives, Chaos et Turbulence* (Gif-sur-Yvette: Commissariat a l'Energie Atomic)

[mars76]   Marsden J E and McCracken M 1976 *The Hopf Bifurcation and Its Applications* (New York: Springer)

[may 75]   May R and Leonard W 1975 Nonlinear aspects of competition between three species *SIAM J. Appl. Math.* **29** 243

[may 76]   May R M 1976 Simple mathematical models with very complicated dynamics *Nature* **261** 459

[mcdo85]   McDonald S W, Greboggi C, Ott E and Yorke J A 1985 Fractal basin boundaries *Physica* **17D** 125

[melb89]   Melbourne I, Chossat P and Golubitsky M 1989 Heteroclinic cycles involving periodic solutions in mode interactions with $O(2)$ symmetry *Proc. R. Soc. Edinburgh* **113A** 315

[metr73]   Metropolis N, Stein M L and Stein P R 1973 On finite limit sets for transformations on the unit interval *J. Combinatorial Theory* **15A** 25

[mill26]   Mill J 1826 *Elements of Political Economy* (London: Baldwin, Cradock and Joy)

[mind91]   Mindlin G B, Solari H G, Natiello M, Gilmore R and Hou X 1991 Topological analysis of chaotic time series data from the Belousov-Zhabotinskii reaction *J. Nonlinear Sci.* **1** 147

[mind93]   Mindlin G B, Lopez-Ruiz R, Solari H G and Gilmore R 1993 Horseshoe implications *Phys. Rev.* **48A** 4297

[misc95]   Mischaikow K and Mrozek M 1995 Chaos in the Lorenz equations: a computer-assisted proof *Bull. Am. Math. Soc.* **32** 66

[mise71]   von Mises R and Friedrichs K O 1971 *Fluid Dynamics* (Berlin: Springer)

[molo90]   Moloney J V and Newell A C 1990 Nonlinear optics *Physica* **44D** 1

[most63]   Mostov G D, Sampson J H and Meyer J-P 1963 *Fundamental Structures of Algebra* (New York: McGraw-Hill)

[murr89]   Murray J D 1989 *Mathematical Biology* Biomathematics Texts, vol 19 (Heidelberg: Springer)

[nard88]   Narducci L M and Abraham N B 1988 *Laser Physics and Laser Instabilities* (Singapore: World Scientific)

[nati94]   Natiello M A and Solari H G 1994 Remarks on braid theory and the characterisation of periodic orbits *J. Knot Theory Ramifications* **3** 511

[nite71]   Nitecki Z 1971 *Differential Dynamics* (Cambridge, MA: MIT Press)

[pack80]   Packard N, Crutchfield J and Shaw R 1980 Geometry from a time series *Phys. Rev. Lett.* **45** 712

[papo91]   Papoff F, Fioretti A, Arimondo E, Mindlin G B, Solari H G and Gilmore R 1991 Structure of chaos in the laser with saturable absorber *Phys. Rev. Letters* **68** 1128

[pran34]   Prandtl L and Tietjens O G 1934 *Applied Hydro- and Aeromechanics* (New York: McGraw-Hill)

[purc65]   Purcell E M 1965 *Berkeley Physics Course V 2, Electricity and Magnetism* (New York: McGraw-Hill)

[rayl16]   Lord Rayleigh (Strutt J W) 1916 On convection currents in a horizontal layer of fluid, when the higher temperature is on the under side *Phil. Mag.* **32** 529

[reyn83]   Reynolds O 1883 An experimental investigation of the circumstances which determine whether the motion of water shall be direct or sinuous, and of the law of resistance in parallel channels *Phil. Trans. R. Soc.* 935

[rich87]   Richetti P, de Keeper P, Roux J C and Swinney H L 1987 A crisis in the Belousov–Zhabotinskii reaction: experiment and simulation *J. Stat. Phys.* **48** 977

[risk89]   Risken H 1989 *The Fokker–Planck equation, Methods of Solution and Application (Springer Series in Synergetics, V 18), 2nd edition* (Heidelberg: Springer)

[rod 73]   Rod D L Pathology of invariant sets in the monkey saddle 1973 *J. Diff. Equations* **14** 129

[roma75]   Roman P 1975 *Some Modern Mathematics for Physicists and Other Outsiders* (New York: Pergamon)

[salt62]   Saltzman B 1962 Finite amplitude free convection as an initial value problem I *J. Atmos. Sci.* **19** 329

[sand85] Sanders J A and Verhulst F 1985 *Averaging Methods in Nonlinear Dynamics* (Berlin: Springer)

[sark64] Šarkovskii A N 1964 Coexistence of cycles of a continuous map of a line into itself *Ukr. Mat. Z.* **16** 61

[sart66] Sartre Jean-Paul 1966 *Being and Nothingness* (New York: Washington Square) (Original edition of 1956)

[saue91] Sauer T, Yorke J A and Casdagli M 1991 Embedology *J. Stat. Phys.* **65** 579

[schw88] Schwartz I 1988 Infinite primary saddle-node bifurcation in periodically forced systems *Phys. Letters* **126A** 411

[sieg86] Siegman E 1986 *Lasers* (Mill Valley: University Science)

[siln65] Šilnikov L P 1965 A case for the existence of a denumerable set of periodic motions *Sov. Math. Dokl.* **6** 163

[siln66] Šilnikov L P 1966 On the generation of a periodic motion from a trajectory which leaves and re-enters a saddle-saddle state of equilibrium *Soviet Math. Dokl.* **7** 155

[smal67] Smale S 1967 Differentiable dynamical systems *Bull. Am. Math. Soc.* **73** 747

[smal76] Smale S 1976 On the differential equations of species in competition *J. Math. Biol.* **3** 5

[sola87] Solari H G, Eschenazi E, Gilmore R and Tredicce J R 1987 Influence of coexisting attractors on the dynamics of a laser system *Opt. Commun.* **64** 49

[sola88] Solari H G and Gilmore R 1988 Relative rotation rates for driven dynamical systems *Phys. Rev.* **37A** 3096

[sola90] Solari H G and Gilmore R 1990 Dynamics in the transverse section of a CO_2 laser *J. Opt. Soc. Am.* **7B** 828

[sola94] Solari H G and Oppo G-L 1994 Laser with injected signal: the perturbation of an invariant circle *Opt. Commun.* **111** 173

[sosi73] Šošitaĭšvili A N 1973 Bifurcations of topological type of a vector field near a singular point *Functional Anal. Applications* **6** 97

[spar82] Sparrow C 1982 *The Lorenz Equations: Bifurcations, Chaos, and Strange Attractors* (New York: Springer)

[take81] Takens F 1981 Detecting strange attractors in turbulence *Lecture Notes Math.* D A Rand and L-S Young ed, vol 898, p 366 (Berlin: Springer)

[tang91] Tang D Y, Li M Y and Weiss C O 1991 Field dynamics of a single-mode laser *Phys. Rev.* **44A** 7597

[thei92] Theiler J, Eubank S, Longtin A, Galdrikian B and Farmer J D 1992 Testing for nonlinearity in time series: the method of surrogate data *Physica* **58D** 77

[thur88] Thurston W P 1988 On the geometry and dynamics of diffeomorphisms of surfaces *Bull. Am. Math. Soc.* **19** 417

[tred86] Tredicce J R, Arecchi F T, Puccioni G P, Poggi A and Gadomski W 1986 Dynamic behavior and onset of low dimensional chaos in a modulated homogeneously broadened single mode laser: experiments and theory *Phys. Rev.* **34A** 2073

[uezu82] Uezu T and Aizawa Y 1982 Topological character of a periodic solution in three-dimensional ordinary differential equation system *Prog. Theor. Phys.* **68** 1907

[weis95]   Weiss C O, Vilaseca R and Tang D Y 1995   Models, predictions, and experimental measurements of far-infrared nh3-laser dynamics and comparisons with the Lorenz–Haken model *Appl. Phy.* **61B** 223

[whit36]   Whitney H 1936 Differentiable manifolds *Ann. Math.* **37** 654

[wigg90]   Wiggins S 1990 *Introduction to Applied Nonlinear Dynamical Systems and Chaos* (New York: Springer)

[yana87]   Yanagida E 1987 Branching of double pulse solutions from single pulse solutions in nerve axon equations *J. Diff. Equations* **66** 243

[zeng84]   Zeng W-Z, Hao B-L, Wang G-R and Chen S-G 1984 Scaling property of period-n-tupling sequences in one-dimensional mappings *Commun. Theor. Phys* **3** 283

[zimm95]   Zimmermann M G, Natiello M A and Solari H G 1995 Šil'nikov-saddle-node interaction near a codimension-2 bifurcation: laser with injected signal *Technical report* (Uppsala, Sweden: Uppsala University)

Index